Natural Gas in Asia

Natural Gas in Asia

The Challenges of Growth in China, India, Japan and Korea

Second Edition

Edited by
JONATHAN STERN

Contributors
MICHAEL BRADSHAW
ANDY FLOWER
DAVID FRIDLEY
SUNJOY JOSHI
NAJEEB JUNG
DAVID LEDESMA
AKIRA MIYAMOTO
KEUN-WOOK PAIK
JONATHAN STERN

Published by the Oxford University Press
for the Oxford Institute for Energy Studies
2008

OXFORD
UNIVERSITY PRESS
Great Clarendon Street, Oxford OX2 6DP
Oxford University Press is a department of the University of Oxford.
It furthers the University's objective of excellence in research, scholarship
and education by publishing worldwide in
Oxford New York
Auckland Cape Town Dar es Salaam Hong Kong Karachi
Kuala Lumpur Madrid Melbourne Mexico City Nairobi
New Delhi Shanghai Taipei Toronto
with offices in
Argentina Austria Brazil Chile Czech Republic France Greece
Guatemala Hungary Italy Japan Poland Portugal Singapore
South Korea Switzerland Thailand Turkey Ukraine Vietnam
Oxford is a registered trade mark of Oxford University Press
in the UK and in certain other countries
Published in the United States
by Oxford University Press Inc., New York

First edition published 2002
Second edition published 2008

British Library Cataloguing in Publication Data
Data available
Library of Congress Cataloguing in Publication Data
Data available
Cover designed by Clare Hofmann
Typeset by Philip Armstrong, Sheffield
Printed by The Alden Group, Oxford
ISBN 978-0-19-954141-6
1 3 5 7 9 10 8 6 4 2

CONTENTS

Tables, Maps and Figures vi
Acknowledgements xiii
Units and Conversion xiv
Notes on Contributors xv

1 Introduction 1
Jonathan Stern

2 Natural Gas in China 7
David Fridley

3 Natural Gas in India 66
Sunjoy Joshi and Najeeb Jung

4 Natural Gas in Japan 116
Akira Miyamoto

5 Natural Gas in Korea 174
Keun-Wook Paik

6 Russian and Central Asian Gas Supply for Asia 220
Jonathan Stern and Michael Bradshaw

7 South East Asian and Australian Supply to Asia 279
David Ledesma

8 Natural Gas from the Middle East 330
Andy Flower

9 Conclusions 379
Jonathan Stern

Appendix 1: Cost Increases for LNG Projects 402
David Ledesma

Appendix 2: LNG Pricing in Asia – Japan Crude Cocktail (JCC) and 'S'-Curves 405
Andy Flower

Index 410

TABLES, MAPS AND FIGURES

Chapter 1

Tables

1.1 Gas Demand and Share of Energy Demand in Asian Markets, 2000 and 2006 2

Chapter 2

Tables

2.1 Primary Energy Production and Consumption (million tonnes standard coal equivalent) 9
2.2 Natural Gas Resources 10
2.3 Natural Gas Reserves and Production, 2006 11
2.4 Proven Natural Gas Reserves by Basin (1999) 12
2.5 Natural Gas Production and Distribution by Province 15
2.6 Natural Gas Production and Growth 16
2.7 Shanghai Gas Supply and Consumption, 2005 24
2.8 Planned Pricing of West–East Pipeline Gas 28
2.9 Proven Reserves and Production Capacity in the Tarim Basin (2000) 30
2.10 Natural Gas Consumption by Sector 32
2.11 Structure of National Gas Consumption, 2005 33
2.12 Natural Gas Power Generation Capacity, 2006 34
2.13 Supply and Demand Projections to 2020 43
2.14 Summary of Pre- and Post-2010 Supply Proposals 53
2.15 Structure of Gas Demand to 2020 55
2.16 Scenarios of China's Energy Mix to 2020 59

Map

2.1 Pipelines and LNG Infrastructure 25

Figure

2.1 Structure of Gas Demand to 2020 56

Chapter 3

Tables

3.1	Gas Production and Imports in India	69
3.2	Production of Natural Gas by Region 2005–06	70
3.3	Gas Demand Projections Eleventh Plan 2007–12	72
3.4	Projections of Domestic Natural Gas Production	96
3.5	Gas Prices in July 2007: A mixed bag	97
3.6	Pipeline Capacity (2006)	101
A3.1.1	Demand Projections for Natural Gas in India by Different Agencies	108
A3.3.3	Dollar equivalence of Consumer Prices for APM Gas	110

Maps

3.1	New Discoveries	79
3.2	Transnational Pipelines	86
3.3	Pipeline Infrastructure – Existing and Future	102

Figures

3.1	Primary Energy Consumption in 2006	66
3.2	Sectoral Shares of Gas Consumption 2006	70
3.3	Installed Generating Capacity March 2007	71
3.4	Gas Reserves	80
3.5	The Declining Share of APM Gas	96

Chapter 4

Tables

4.1	Primary Energy Supply	119
4.2	Supply/Demand Balance for Natural Gas	120
4.3	Changes in Power Generation of 10 Incumbent Electric Power Utilities by Energy Source	122
4.4	Fuel Consumption by Electric Utilities	123
4.5	Installed Generating Capacity of Power Plants. Excluding private power generation	123
4.6	LNG Imports by Region 2000–06 (excluding supplies for power generation).	124
4.7	Gas Utility Sales by Sector and Region 2002–06	126
4.8	LNG Imports by Source 2000–06	128
4.9	Short-term and Spot LNG Trades in the Pacific Basin 2000–06	128
4.10	New LNG Terminals	131

4.11 Secondary LNG Terminals supplied by Coastal Shipping 135
4.12 PPS Entrants in the Electric Power Market 2001–06 137
4.13 Gas Market Penetration by New Entrants 139
4.14 Potential Energy Demand by Sector which could Switch to LNG 144
4.15 Potential Gas Demand by Region 145
4.16 LNG Consumption in the Power Sector in 2020, at different load factors for nuclear stations 153
4.17 Natural Gas Demand Outlook. 157
A4.1.1 New Japanese Gas Pipelines 161
A4.2.1 Long-term Energy Outlooks 162
A4.3.1 Trends in Peak Power Demand for the 10 Incumbent Electric Utilities 164
A4.3.2 Projected Growth in Peak Power Demand over the next 10 Years 164
A4.3.3 Nuclear Power Plant Utilisation in Major Countries 2000–05 165
A4.3.4 Nuclear Power Plant Utilisation by Japanese Electric Utilities 2000–05 165

Map
4.1 LNG Terminals and Gas Pipelines 132

Figures
4.1 City Gas Sales by Sector 2001–07 127
4.2 LNG Prices for Spot Transactions in Asia Pacific Market 129
4.3 Import Prices for LNG and Oil 130
4.4 Opening of the Electricity Market to Competition 136
4.5 Opening of the City Gas Market to Competition 138
4.6 Electricity Prices for Large Customers 2002–07 147
4.7 Contracted LNG volumes 2005–20 149
4.8 Price Competitiveness of Natural Gas against Oil Products 150

Chapter 5

Tables
5.1 Primary Energy Consumption by Source 175
5.2 Energy Demand Projections by Source 176
5.3 Korea's Long-term Natural Gas Demand Projection 7th Plan 178
5.4 Korea's Long-term Natural Gas Demand Projection 8th Plan 178

5.5 Gas Demand Expansion in Korea: 1987–2005 180
5.6 City Gas Companies as at the end of 2005 182
5.7 Trunk Pipeline Construction Plan 184
5.8 Korea's Gas Industry Infrastructure: As of Dec 2006 184
5.9 Gas Industry Investment Plan 185
5.10 Seasonal Demand Fluctuations 185
5.11 Kogas Long-term LNG Import Contracts 187
5.12 Korea's LNG Import Contract Structure 188
5.13 Projected Gas Supply Shortage 190
5.14 Gas Price Comparison 191
5.15 Price Advantage during the Buyer's Market period 192
5.16 Options of KNOC and Kogas Merge 198
5.17 LNG Import Plans of Power Generators 200
5.18 Projection of LNG Demand by Sector 201
5.19 Korea's LNG Demand and Supply Gap 210

Map
5.1 Pipelines and LNG Regasification Plants 181

Chapter 6

Tables
6.1 Natural Gas Reserves and Resources in Siberian and the Far Eastern Federal Districts (at January 1, 2005) 226
6.2 Reserves in Eastern and Western Siberian Gas Fields Available for North East Asian Markets 227
6.3 Projected Gas Production in Eastern Siberia and the Far East 228
6.4 The Current Status of the Sakhalin Projects (as of 1 May 2007) 229
6.5 Sakhalin 2: Contractual Commitments for LNG Exports (June 2007) 236
6.6 Rusia Petroleum Shareholdings 1997–2007 240
6.7 Gas Demand in Eastern Siberia and the Far East, 2004 246
6.8 Central Asian Gas Reserve Estimates 260

Maps
6.1 Sakhalin Gas Fields, Pipelines and the Sakhalin 2 LNG Project 231
6.2 The Eastern Gas Programme Plans for Pipeline Developments in Eastern Siberia and the Russian Far East 252

Chapter 7

Tables

7.1 South East Asian and Australian Gas Reserves and Economic Data 279
7.2 Indonesia–Arun LNG Contractual Position (as at September 2007) 287
7.3 Indonesia–Bontang LNG Contractual Position (as at September 2007) 288
7.4 Indonesia–Tangguh LNG Contractual Position (as at September 2007) 290
7.5 Malaysia LNG Contractual Position (as at September 2007) 304
7.6 North West Shelf LNG Contractual Position (as at September 2007) 311
7.7 Darwin LNG Contractual Position (as at September 2007) 311
7.8 Pluto LNG Contractual Position (as at September 2007) 312
7.9 Gorgon LNG Contractual Position (as at September 2007) 313
7.10 Browse LNG Contractual Position (as at September 2007) 314
7.11 Brunei LNG Contractual Position (as at September 2007) 318

Maps

7.1 Asian Gas Pipeline Network 282
7.2 ASEAN Powergrid 283
7.3 LNG and Domestic Gas Pipelines in Western Australia 309

Figures

7.1 LNG Supply to Asia 284
7.2 LNG Supply to Asian Markets 1980–2006 285
7.3 Estimated Gas Price Comparison, Domestic vs. Export 292
7.4 Indonesian LNG Production, Historical & Projected (as at September 2007) 297
7.5 Australian LNG Production, Historical & Projected (as at September 2007) 316
7.6 Regional Energy Policy Considerations 323
7.7 Regional LNG Production, Historical & Projected (as at September 2007) 326

Chapter 8

Tables

8.1 Proven Gas Reserves in the Middle East 331
8.2 Middle East Proven Gas and Oil Reserves 333
8.3 LNG Export Capacity in the Middle East 338
8.4 Qatargas Liquefaction Trains 346
8.5 RasGas Liquefaction Trains 346
8.6 Volumes Contracted on a Medium and Long-term Basis from Qatargas 347
8.7 Volumes Contracted on a Long-term Basis from RasGas 348
8.8 Qatar LNG Sales Profile 349
8.9 Expected Gas Production from Qatar's North Field 354
8.10 Long-term Sales Contracts for Omani LNG Production 356
8.11 Contracted Volumes from Yemen LNG 359
8.12 Development Phases of Iran's South Pars Field 365
8.13 The Share of Middle Eastern LNG in Asian Markets 371
8.14 Shares of Middle Eastern LNG Output 375

Map

8.1 LNG Pipelines and Infrastructure, Existing, under Construction and Proposed 336

Figures

8.1 Natural Gas Production in the Middle East 1990–2006 334
8.2 Uses of Gas Produced in Iran, Saudi Arabia, Qatar, UAE, Oman and Yemen 335
8.3 Middle Eastern LNG Exports 1990–2006 339
8.4 Middle East Share of Global LNG Exports 1990–2006 339
8.5 LNG Exports from Qatar 1997–2006 348
8.6 Middle Eastern LNG Exports 2000–2020 from Projects in Operation and Under Construction in 2007 372
8.7 Possible Middle Eastern LNG Exports 2006–2020 374

Chapter 9

Table

9.1 Gas Demand Scenarios for the Major Asian Markets, 2020 399

Figure

9.1 Global Spot and Short-term LNG Trade, 1997–2006 382

Appendices

Figures

A.1.1: Industry Cost Escalation 402
A.1.2: Liquefaction Plant Capital Costs 403
A.2.1: Basic Japanese LNG Price Formulae 406
A.2.2: 'S'-Curve Price Formula 407

ACKNOWLEDGEMENTS

My grateful thanks to everybody at the Oxford Institute for Energy Studies, in particular Chris Allsopp for his encouragement and intellectual support, and Kate Teasdale for her organisational support in seeing the book through to a speedy conclusion.

I would like to pay particular tribute to my co-authors not just for the effort they have put into their chapters, and the impressive self discipline they all displayed in meeting agreed deadlines throughout the preparation of this book, but also their overall contribution. Ian Wybrew-Bond, my co-author for the first edition of this book, was with us again both in spirit and at the meeting of authors to discuss early drafts. Much of my introduction and conclusions draw heavily on ideas and suggestions which arose in discussions and in correspondence. I count myself very fortunate to have managed to assemble the collective expertise of these authors, each of whom brings a particular insight born of personal experience of living, working and studying in these countries over a long period of time.

Anita Gardiner's maps and charts have again contributed to understanding the significance of regional geography, and the conclusions that are drawn from the research. Grateful thanks also to Judy Mabro, Bob Sutcliffe and Philip Armstrong for their editorial and typesetting skills.

Jonathan Stern Oxford, March 2008

UNITS AND CONVERSIONS

This book uses two gas units: billion cubic metres (Bcm) and million tonnes (Mt) of LNG. An acceptable statistical generalisation used in this book is that 1Mt of LNG = 1.35 Bcm of natural gas. Conversions for other units used are defined in the relevant chapters.

Prices in this book are generally quoted in US dollars per million British thermal units ($/MMBtu). Conversions for gas priced in local currency are given in the chapters.

Crude oil prices are quoted in dollars per barrel ($/bbl); an approximate conversion between gas and oil prices is that $1/MMBtu = $5.8/bbl

NOTES ON CONTRIBUTORS

Michael Bradshaw is Professor of Human Geography at the University of Leicester. He is also Honorary Senior Research Fellow at the Centre for Russian and East European Studies at the University of Birmingham and an Associate Fellow at Chatham House.

Andy Flower is an independent consultant specialising in LNG and gas. He has more than thirty years experience of working in gas and LNG industries worldwide, of which many were spent with BP.

David Fridley, former researcher of Chinese Energy at the East–West Center and Business Manager for Caltex China, is currently managing the China Energy Group at the Lawrence Berkeley National Laboratory.

Sunjoy Joshi is a civil servant with the Government of India. He has been closely associated with the changes occurring in the oil and gas sector since 2001. He is at present on a sabbatical as Senior Fellow with the Observer Research Foundation in New Delhi.

Najeeb Jung is a former civil servant with the Government of India. He has worked on various energy issues and projects with the Asian Development Bank and is currently a Senior Research Advisor at the Oxford Institute for Energy Studies.

David Ledesma is an independent energy and strategy consultant specialising in LNG and gas. In his twenty-five years of experience in the energy and utility sector in Shell and other companies, David worked on the development of complex integrated energy projects, negotiations at government level, and in the management of joint ventures.

Akira Miyamoto is a General Manager of Osaka Gas Co. Ltd. He was formerly a Visiting Fellow of the Royal Institute of International Affairs (UK) and a Senior Researcher at the Institute of Energy Economics, Japan.

Keun-Wook Paik is a Senior Research Fellow at the Oxford Institute for Energy Studies, an Associate Fellow of Chatham House and an Honorary Fellow of the University of Aberdeen.

Jonathan Stern is Director of Gas Research at the Oxford Institute for Energy Studies; Honorary Professor at the Centre for Energy, Petroleum & Mineral Law & Policy, University of Dundee and Visiting Professor at Imperial College's Centre for Environmental Policy.

INTRODUCTION

Jonathan Stern

The Purpose of the Book

The purpose of this book is to investigate whether natural gas can and will become a much more substantial energy source in four of the major countries in Asia: China, India, Japan and Korea. The position of each country is different: in China, gas is a minor fuel accounting for less than 3 percent of energy demand; in India it is established only in relatively few locations; Korea and Japan have well-established industries but growth prospects are uncertain particularly in Japan where there is no national pipeline network. China and India are countries with massive populations and huge potential energy markets, where gas has a relatively low share of primary energy demand. In these countries, high rates of economic growth, increasing demand for power generation and severe urban air quality problems could add up to substantial additional requirements for the cleanest of the fossil fuels over the next decade. Gas resources within these countries range from negligible in Japan and Korea, to modest (in comparison with potential demand) in China and India, but huge potential resources of gas exist – which could be delivered in liquefied and piped form – from within and outside the region. Despite this potential, gas trade on the Asian Continent has been largely restricted to liquefied natural gas (LNG) with very little pipeline trade.

It can fairly be asked why we have chosen these four countries when gas demand in other Asian countries – Bangladesh (15.2 Bcm), Indonesia (39.6 Bcm) Malaysia (40.3 Bcm), Pakistan (30.2 Bcm), Taiwan (11.9 Bcm) and Thailand (30.6 Bcm) – is very significant and rapidly growing.[1] There is no wholly satisfactory explanation other than the judgement that Japan is the largest market in Asia and, along with the other three countries promises to dominate the future gas trade of the Continent, which is a large part of the focus of this book. Of the countries listed above, Indonesia and Malaysia feature as exporters in the South East Asian chapter; while Pakistan and Bangladesh are included in the India chapter as potential transit (and in the case of Bangladesh supplier) countries.

The principal research question of the first edition of this book was whether in the markets under investigation, natural gas could become a much more significant part of each country's energy balance by 2020, and the developments that would be necessary to bring that about. We defined 'significant' as reaching the average OECD level, around 20 percent, of primary energy demand. As Table 1.1 shows, all four countries were some way short of this level in 2000 and, six years later, despite the fact that Chinese gas demand had more than doubled and Indian demand had risen by nearly 50 percent, only Korea had made appreciable progress in increasing the share of gas in primary energy demand. This book will look not just at how much progress has been made in the region over the past six years, but how the perception of future prospects has changed over the same period.

Table 1.1: Gas Demand and Share of Energy Demand in Asian Markets, 2000 and 2006

	Gas Demand (Bcm)*		*Share of Gas in Primary Energy Demand percent*	
	2000	*2006*	*2000*	*2006*
Japan	76.2	84.6	13.3	14.6
Korea	21.0	34.2	9.9	13.6
China	24.5	55.6	2.7	2.9
India	26.0	39.7	7.5	8.5
	147.7	214.1		

Source: BP Statistical Review of World Energy, June 2002 and 2007.

* Gas demand figures are not consistent between different annual editions of the source but this makes little difference to the shares of primary energy demand.

The Changed Context of the late 2000s

The first edition of this book was written in the wake of the 1997 Asian economic crisis with crude oil prices in the range of $20–30/bbl and a consensus that they would remain at this level for the foreseeable future. At that time there were few signs that international gas, especially LNG, trade was developing into a global market. Another dimension of the subject which has changed significantly is the importance of security issues on domestic and international energy agendas. Following the terrorist attacks of 11 September 2001, security of supply issues moved to the centre of the policy and particularly the energy agenda. But in Asia, those concerns have been overtaken by more traditional supply adequacy and import dependence issues.

Of the four markets under consideration, Japan and Korea have always imported the vast majority of their oil and gas needs and lived with import dependence as a fact of life. China and India have substantial and growing oil import dependence with security concerns focused not only on the consequences of that dependence, but also whether access to sufficient resources can be created and sustained to meet both rapidly increasing demand and the aspirations of regions and end-users in these very large countries.

It was these issues, rather than creating liberalised and competitive utility markets, which loomed large in Asian countries in the late 2000s. In most of Asia, the principal competition issue has been one of access to acreage, both domestic and international, with Chinese (and to a lesser extent Indian) companies competing with Multinational Oil Companies and National Oil Companies for global opportunities. Another development of note has been the discovery of new domestic gas resources in China and India in excess of what was expected in the early 2000s.

But many countries that could provide long-term gas supplies, or through which such supplies will need to transit, are beset by long-running political conflicts. The conflicts that will have an ongoing impact include all of those affecting the stability of the Middle East region; historical and territorial problems between India and Pakistan and between India and Bangladesh affect pipeline prospects from the east. In East Asia, long-running bilateral political and territorial problems between Russia and Japan have not been resolved. Post-Soviet Russia's relations with China make it possible to envisage interdependence based on large-scale pipeline supplies which would also allow pipeline gas to reach South Korea. But should that prove impossible, South Korean pipeline gas from Russia will depend on the resolution of political difficulties on the Korean Peninsula.

As far as the overall policy agenda since the first edition was published in the early 2000s is concerned, perhaps the greatest change has been the increase in the importance of local, regional and global environmental issues on national and international policy agendas of most OECD countries. The impact of environmental issues on energy – and particular fuel – choices will vary by country with some focusing on carbon reduction targets and others on local air quality problems.

The Four Markets

The four countries selected as the main focus of this book represent substantial current and potential gas markets, but with very different starting points and approaches to future development. In Japan, a number of private power and gas utilities worked with government to develop external supplies. Each utility had its own monopoly market with its own local distribution network. The construction of a national transmission system has been regularly discussed but topographical and cost constraints have prevented it from being realised.

The Korean market is dominated by the state-owned Korean Gas Company (Kogas). Unlike the other countries in this book, Korea has a national pipeline network. In the late 1990s, the government decided to privatise Kogas and liberalise the gas market, but progress towards both these goals has been slow. The South Korean gas market is highly seasonal, leading Kogas to become very active in short-term LNG purchases.

China and India have gas markets with limited pipeline networks confined to relatively small regions of the country and shares of the population, but both have plans which promise to change this situation radically. China's huge population, rapidly increasing demand for energy and urban air quality, are well known problems. The decision to develop domestic gas reserves located in the west of the country led to the construction of the 4000 km West–East pipeline to the main demand areas in the east. Since the early 2000s, substantial additional domestic gas has been discovered, a range of pipeline gas imports have been under consideration, one LNG terminal is operating, two more will be in operation by 2009, and many more have been proposed. The market therefore seems likely to grow based on both discoveries of indigenous gas and substantial imports.

India presents a unique challenge to any potential gas supplier and market participant. The Indian central and regional governments have been slow to face up to the structural difficulties that exist in the utility sector. Power and fertiliser subsidies have placed many national and state companies in a parlous financial position. It has proved difficult to push through unpopular measures to create an attractive investment climate. Since the early 2000s, very substantial additional gas resources have been located offshore the east coast, transforming the country's near-term production outlook. There have been negotiations on pipeline imports from Iran, Turkmenistan, Bangladesh and Myanmar. Two LNG terminals are operating and another two are under construction. Key questions of social policy and the pace of economic development will continue to determine how far and how fast market reforms will enable commercially viable gas and power markets to develop on a national basis, and whether this can be achieved if it requires substantial imports at high prices.

Pricing is a major issue throughout the region. Japan's gas prices to final customers are the highest in the OECD, while China's are also very high for a country with a relatively low GDP/capita, leading to questions as to how quickly markets can grow, particularly in the power sector. In India, the main problem is how to phase out subsidies and raise prices to levels that will be commercially attractive for suppliers. The relationship of gas prices to the prices of other fuels will be crucial given the link between oil and international LNG prices, and domestic coal and electricity prices in China and India. The future of gas-fired power generation looms large in all four countries but the questions are different:

- For China and India, the main issue will be the extent to which gas can displace new coal plant as an economically viable source of power, even if the coal plant is fitted with sulphur and NOx controls;
- For Japan and Korea a major issue will be whether those countries will press ahead with nuclear power to the extent the government plans suggest, and the load factor at which nuclear stations will operate. Should nuclear power generation prove less viable than current government targets suggest, gas-fired power generation may be the beneficiary.

Aims of, and Lessons for, the Second Edition

The second edition of this book provides an opportunity to learn some lessons from the period 2002–07. As noted above, the energy context of the first edition has changed dramatically over the past five years, and these changes are not only continuing but may be accelerating. Although the detail of project developments and policies is probably condemned to become rapidly outdated as events unfold, our aims in this edition are to:

- analyse the developments that have taken place in the 2000s;
- provide a snapshot of the situation towards the end of 2007;
- identify shorter- and longer-term trends, specifically to look at what has changed over the past five years since the publication of the first edition, and then to look forward to 2020 to try to identify longer-term trends.

One important lesson which may be drawn from rereading the first edition is that, even given the best efforts of the authors, projecting development over five – let alone twelve – years may be too ambitious a task.

Structure of the Book

The book is structured in eight chapters:

- four deal with the major markets: China, India, Japan and Korea;
- three deal with supplying countries and regions: Russia, Central Asia, the Middle East, South East Asia and Australia.

These chapters:

- weigh the balance between the policies and market forces required to bring about a substantial increase in gas demand in each of the four markets;
- assess the availability of supply to support increased market penetration and the geopolitical obstacles that have to be overcome for different supply options to be realised.

The final chapter draws together the conclusions from the previous seven and addresses the aims and research questions noted above. Because of the pace of regional and global energy developments, readers should bear in mind that much of the research was carried out in the early part of 2007 and completed in September of that year.

Notes

1 2006 demand figures from *BP Statistical Review of World Energy, June 2007.*

CHAPTER 2

NATURAL GAS IN CHINA

David Fridley

Introduction

China's modern natural gas industry marked its birth with the establishment of the People's Republic in 1949. Although historical records exist noting the use of natural gas as early as 200 BC in making salt from brine, it never supplied more than small-scale local use over the centuries. In 1949, output totalled a mere 7 million cubic metres (Mcm).

Early industrial policy in the 1950s did not favour the fuel; priority in the early days of rapid industrialisation focused on development of the plentiful reserves of easily accessible coal. Although the following four decades witnessed several major shifts in energy policy, natural gas remained a secondary consideration behind coal, oil, and hydropower expansion. By 2000, natural gas production had expanded to nearly 28 billion cubic metres (Bcm), over 4000 times the amount in 1949.[1]

China entered the twenty-first century touting natural gas as the 'fuel of the century', stressing the environmental benefits of its use compared to coal and promoting it as a solution to intense urban pollution and committing to expanding its use to account for 10 percent of the national energy mix in 2020, up from 2 percent in 2000. In the previous edition of this chapter, written in 2001, these goals were considered in the context of resource, institutional, political, geopolitical, economic, infrastructural, and social issues to assess China's capacity to achieve them. At that time, the conclusion was hesitant: although plans were already laid for development of LNG and a basic national pipeline network, the market appeared immature, the consumer base fragmented, and pricing structure too high. Looking out to 2020, it also appeared difficult for China to increase consumption commensurate with a 10 percent target.

In the intervening six years since the first edition was written, much has changed, new challenges have arisen, and the prospects for expansion of the natural gas to 2020 have become clearer. The West–East Pipeline, considered a 'boondoggle' by some for its small scale and high prices, has been completed and natural gas is flowing from Xinjiang to

Shanghai. A second West–East Pipeline is being planned. Production has continued to rise sharply, with output reaching nearly 60 Bcm in 2006, more than double the 2000 level, and the reserve base has expanded with the discovery of several very large gas fields in Sichuan and Xinjiang. The first LNG terminal went into operation in 2006, with two more under construction. The oversupply potential seen in the early 2000s has disappeared, replaced by increasing market tightness, expected to continue well into the next decade. What were originally considered overly high city gate prices in 2001 now look reasonable, in international terms, with the doubling and tripling of natural gas prices in many countries around the world since then.

Pricing challenges continue, however. The structure of natural gas prices continues to be controlled centrally without recognition of supply and demand changes or value of use. This has held back expansion of the gas-based power generation and led to prioritisation of supply to higher-paying residential and commercial consumers. Government studies call for fundamental reform, but pricing policy 'incrementalism' remains the rule. Laws and regulations guiding the participation of foreign firms in China's domestic gas market development also remain fragmented and incomplete.

Internationally, China's push to acquire foreign energy resources to support economic growth that has exceeded all expectations in 2000 has led to renewed discussions with its Central Asia and Russian neighbours to secure large-scale pipeline imports. Pricing here, too, remains an issue, but China's own rising wealth and economic influence provide greater latitude to accept higher gas prices than was the case six years ago.

This chapter examines these changes in detail and returns to an assessment of China's 2020 target for natural gas use in context of the dramatic changes in the country's energy economy in the last half-decade. It does not attempt to formulate detailed energy forecasts but outlines the likely development paths of natural gas and other energy types in light of the renewed emphasis on reducing the energy efficiency of the economy and the increasing recognition that China, too, will need to respond to global warming.

China's Natural Gas Industry

Natural Gas in China's Energy Balance

Natural gas remains a minor fuel in China's overall energy mix

(Table 2.1). At 3.2 percent, natural gas accounts for only a small part of China's primary energy production, and, at 2.8 percent, an even smaller proportion of primary energy consumption. Although consumption of natural gas grew by 88 percent between 2000 and 2005, the

Table 2.1: Primary Energy Production and Consumption (million tonnes standard coal equivalent)

	Coal	*Oil*	*Natural Gas*	*Electricity**	*Total*	*Natural Gas Percentage*
Production						
1990	771	198	20	51	1,040	2.0
1991	777	201	21	50	1,050	2.0
1992	797	203	21	53	1,074	2.0
1993	821	207	22	62	1,113	2.0
1994	886	209	23	73	1,191	2.0
1995	972	214	24	82	1,292	1.8
1996	998	225	27	82	1,331	2.0
1997	981	230	30	85	1,325	2.3
1998	893	230	31	90	1,244	2.5
1999	914	229	34	88	1,265	2.6
2000	928	233	36	97	1,294	2.8
2001	987	234	40	119	1,381	2.9
2002	1,039	239	43	126	1,448	3.0
2003	1,230	242	47	132	1,651	2.8
2004	1,423	251	55	163	1,893	2.9
2005	1,575	259	66	182	2,081	3.2
Consumption						
1990	754	164	21	51	989	2.1
1991	789	177	21	50	1,037	2.0
1992	815	191	21	53	1,079	1.9
1993	854	211	22	62	1,149	1.9
1994	914	214	23	73	1,224	1.9
1995	983	230	24	82	1,319	1.8
1996	1,034	249	25	82	1,390	1.8
1997	995	281	26	85	1,386	1.8
1998	925	283	29	90	1,327	2.2
1999	929	301	29	88	1,347	2.1
2000	943	321	33	97	1,393	2.3
2001	964	326	36	119	1,446	2.5
2002	1,011	354	39	126	1,531	2.5
2003	1,209	388	45	132	1,774	2.5
2004	1,383	453	53	163	2,052	2.6
2005	1,548	465	62	182	2,257	2.8

* includes hydropower and nuclear power.
Note: Natural gas consumption does not include volumes sold to Hong Kong.

Source: China National Bureau of Statistics, *China Energy Statistical Yearbook 2006*

volumetric rise in consumption was equivalent to only 5 percent of the increase in coal consumption over this period and just 20 percent of the rise in oil consumption. Since 2000, however, the nascent form of a national market has emerged as the first long-distance gas pipeline from Western to Eastern China was completed, and the first LNG import terminal came into operation. Originally an industrial fuel and fertiliser feedstock for the most part, natural gas now has strongly penetrated the residential sector and provides cleaner transport fuels in a number of major cities.

Natural Gas Resources and Reserves

In 2003, China launched the third national resource survey to reassess the state of oil, gas, and coal resources in the country. Although the survey has not been formally completed, preliminary results indicate an expansion of China's resource base to 52.7 trillion cubic metres (Tcm), up from the previous estimate of 38 Tcm in the 1994 national survey and 33 Tcm in the first national survey of 1987. Unlike the previous surveys, however, unconventional gas resources, including coal-bed methane (CBM), have been included in these figures.

The bulk of these reserves are found in the central and western parts of China, primarily in Sichuan in central China, the Ordos Basin of Shaanxi, Inner Mongolia, and Ningxia, and Qinghai and Xinjiang in the west (Table 2.2).

Table 2.2: Natural Gas Resources

	Resources		*Undiscovered*	
	Tcm	*Percent*	*Tcm*	*Percent*
Onshore	38.8	74	14.5	83
East	4.4	8	1.8	10
Central & West	31.3	59	11.9	69
South	3.2	6	0.8	4
Offshore	13.8	26	2.9	17
Total	52.7	100	17.4	100

Source: Chen Mingshuang, Future Oil & Gas Resources of China, Sino-US Oil and Gas Industry Forum, Sept 2006.

The resource survey has also increased the volume of China's proven reserves, estimated at 2.85 Tcm, up from 2.72 Tcm in 2000. Over this same period, international estimates (BP) of China's proven reserves have risen from 1.27 Tcm to 2.45 Tcm (Table 2.3).

Table 2.3: Natural Gas Reserves and Production, 2006

	Reserves Tcm	*Production Bcm*	*R/P Ratio*
China (BP)	2.45	58.6	42
Malaysia	2.48	60.2	41
Australia	2.61	38.9	67
Indonesia	2.63	74.0	36
China (2006 Survey)	2.85	58.6	49
Turkmenistan	2.86	62.2	46
Norway	2.89	87.6	33
Kazakhstan	3.00	23.9	126

Source: *BP Statistical Review of World Energy 2007.*

Accelerated exploration over the last decade has resulted in notable successes in identifying new gas reserves. Among the most notable of these is China's largest natural gasfield at Sulige, in the Inner Mongolia area of the Ordos Basin. Containing 535 Bcm of proven reserves, the field now produces 1 million cubic metres per day. In 2006, the China National Petrochemical Corporation (Sinopec) announced the discovery of the Puguang field in Sichuan, with proven reserves of 356 Bcm, and in mid-2007, the China National Petroleum and Natural Gas Corporation (CNPC) announced a further discovery at the Longgang field in western Sichuan, estimated to hold up to 700 Bcm of reserves. The field remains under assessment but is likely to help ease Sichuan's growing gas deficit.

About 67 percent of China's proven reserves are classified as non-associated, or gasfield, gas. China's oldest gas-producing fields in northeast and northern China are for the most part the centres of crude oil production, and the gas is used extensively within the oilfields themselves. In the southwest, the Sichuan basin was historically China's largest non-associated gas-producing area, but exploration in the last fifteen years has uncovered a number of other significant gasfield basins. Qiongdongnan and Yingehai off the south China coast, the East China Sea basin, and the Ordos, Qaidam and Tarim basins contain almost exclusively non-associated gas (Table 2.4).

The third national survey also estimates that the remaining undiscovered gas resources total 17.4 Tcm, of which 2.7 Tcm are to be found in Sichuan, and 4.1 Tcm in the Ordos Basin. Offshore undiscovered resources are estimated at 2.9 Tcm. Between 2004 and 2020, reserve growth is estimated to average 184 Bcm per year, to reach 5.6 Tcm by 2020.

Table 2.4: Proven Natural Gas Reserves by Basin (1999)

	Total Proved Reserves (Bcm)	*Non-Associated (percent)*
Songliao	290.56	15
Bohai Gulf	566.05	35
Turpan-Hami	69.78	45
Tarim	215.02	87
Jungar	163.67	17
Qaidam	141.97	95
Sichuan	570.06	100
Ordos	323.45	95
Bohai	54.09	-
East China Sea	31.49	100
Qiongdongnan	88.50	100
Yinggehai	160.66	100
Other Basins	45.19	70
National Total	2,720.47	67

Source: Qian Kai, Ma Xinhua, et al. 1999. *China's Natural Gas Resources*. Beijing: Petroleum Industry Press.

Coal-Bed Methane Resources

In addition to conventional natural gas resources, China possesses extensive resources of coal-bed methane (CBM), concentrated in China's vast coal basins of northern and northwest China, particularly Qinshui and Ordos in Shanxi and Shaanxi, and in Xinjiang. Resource estimates now stand at about 36 Tcm.

China began exploration and development of CBM resources in the late 1980s, but only in January 2001 authorised the first CBM development scheme in the Qinshui Basin of Shanxi province. The CBM field in Qinshui contains proven reserves of 40.2 Bcm and is expected to eventually reach a production level of 1 Bcm per year. This field is being developed by the China United Coalbed Methane Corporation (CUCBM), established by the government in March 1996 as the sole state company for exploration, production, pipeline construction, sales and foreign cooperation.[2] CUCBM's monopoly was ended in late 2007 as the government moved to open up the sector to further foreign investment in conjunction with other Chinese companies as designated by the State Council.

To date, CUCBM has signed 27 cooperative contracts with foreign companies, including Chevron, Phillips, Greka, Lowell, Orion, Far

East Energy, and Hong Kong Towngas. In total, ¥1.7 billion (US$213 million) has been invested by foreign companies, 70 percent of the total investment in CBM in China. Foreign companies, however, have been hampered by restrictive relinquishment requirements and have urged the government to adopt 'roll up' overall development plans that can be reevaluated each year in order to provide the incentive to develop appropriate resource development plans. To encourage CBM projects, the government has put into place several favourable policies for the industry, including a reduced 5 percent VAT rate for CBM projects with foreign partners, exemption from import and other duties on materials and equipment for CBM exploration and development, exemption from royalties for CBM projects producing less than 1 Bcm per year (the royalty rate is the same as conventional natural gas), and free market pricing of the gas with no state controls, in contrast to conventional natural gas. This means that CBM transported outside a local producing region to areas where conventional natural gas is available would be subject to the economics of conventional natural gas prices.

Despite the abundance of resources and favourable development policies, CBM development in China has proceeded more slowly than similar schemes in the United States, Canada or Australia. Originally projected to reach 3–4 Bcm in 2005, production is expected to grow to just 200 million cubic metres in 2007. Nonetheless, CUCBM maintains an unchanged forecast of 10 Bcm production in 2010 and 20 Bcm in 2015. (In comparison, US CBM production in 2005 reached 51 Bcm.) Compared to the fairly simple geology of Western US CBM basins, Chinese basins are much more complex and fractured, yields are lower, and fewer detailed geological studies have been undertaken. Existing US CBM well and field modelling tools to optimise development schemes have not been found completely appropriate for Chinese use. Data sharing has also been an issue, particularly where CUCBM tracts overlap with those of the National Coal Corporation. Technically, CUCBM requires assistance in dealing with water treatment problems and other environmental issues. CUCBM also lacks the financial resources to pay for extensive exploration, while potential production is constrained by the general lack of infrastructure, particularly pipelines. As the infrastructure for conventional natural gas expands, this latter problem may be alleviated. The key areas of current CBM exploration and development are fairly proximate to the West–East Pipeline. In areas that lack such infrastructure, CBM could become an important local source for both residential and power plant use.

International interest in CBM jumped in 2006 and 2007 as China signed a series of Clean Development Mechanism (CDM) projects for the development of CBM. Because methane releases from coal mines are more potent as a greenhouse gas than carbon dioxide, every tonne of captured methane attracts 20 tonnes of CO_2 credits in the CDM scheme. According to the World Bank, China obtained 63 percent of the total UN-certified carbon credits in 2006, amounting to US$3 billion. In late 2007, a new CBM-fired 120 MW power plant at Jincheng, Shanxi, built with the assistance of $150 million in carbon credits, will begin operation, the largest of its kind in Asia.

Production and Trade

Historically, the output of natural gas has been closely linked to the development of oilfields. Until the recent focus on development of gas stratum fields, most natural gas has been produced in association with crude oil. Major oilfields such as Daqing and Shengli have used the gas for oilfield fuel and to heat their heavy crude streams for long-distance transport, which has left little available commercially. As crude production in these mature eastern fields peaked in the late 1990s, natural gas output from the oilfields stagnated as well. At the beginning of China's economic reforms in 1980, the oil-producing regions of northeast, east, and northern China accounted for over half of China's natural gas production (Table 2.5). Today the three regions account for only 15 percent.

The Sichuan basin is China's largest gas-stratum producing area and has the most developed transmission infrastructure. Developed primarily to supply a number of key fertiliser manufacturers in Sichuan and neighbouring provinces, the pipeline system also supplies gas to urban residents in Chongqing and Chengdu, where residential gas usage is the highest in the country. A major World Bank-supported project in the 1990s assisted in the revamp of exploration, production, and transmission facilities, focusing as well on the environmental and safety concerns of pipeline leakages, leading to a rebound in Sichuan natural gas production. After stagnating in the early 1990s at about 7 Bcm per year, production rose rapidly in the early 2000s, reaching 14.2 Bcm in 2005. Historically, the southwest region of China produced 40–50 percent of China's natural gas, and it is home to the new Puguang and Longgang gasfields.

Following Sichuan, Xinjiang is now China's second largest producer. By the time the West–East Pipeline from Xinjiang to Shanghai was completed in late 2004, Xinjiang's output had hit 5.7 Bcm; in 2005 alone, production jumped 86 percent to 10.7 Bcm as gas began to flow to eastern China. Shaanxi and Qinghai are the other major producing

Table 2.5: Natural Gas Production and Distribution by Province

	1980	*1985*	*1990*	*1995*	*2000*	*2005*
Production (Bcm)						
Northeast	5.27	4.07	4.39	4.89	3.98	4.16
Heilongjiang	3.39	2.50	2.25	2.59	2.30	2.44
Jilin	0.09	0.08	0.10	0.18	0.21	0.54
Liaoning	1.79	1.49	2.04	2.11	1.47	1.17
North	0.80	0.84	0.72	1.15	1.99	1.89
Tianjin	0.80	0.57	0.37	0.76	0.91	0.88
Hebei		0.21	0.29	0.35	0.51	0.69
Shanxi		0.07	0.06	0.05	0.11	0.32
Inner Mongolia					0.46	
East	1.44	1.18	1.48	1.30	0.97	1.59
Shanghai					0.26	0.60
Jiangsu	0.02	0.04	0.04	0.02	0.02	0.06
Shandong	1.42	1.14	1.44	1.29	0.69	0.93
Central-South	0.05	0.66	1.47	1.32	5.05	6.51
Henan	0.03	0.63	1.39	1.14	1.50	1.76
Hubei	0.02	0.03	0.08	0.08	0.09	0.11
Guangdong				0.10	3.46	4.48
Hainan						0.17
Southwest	6.33	5.66	6.65	8.03	9.14	14.63
Chongqing					0.19	0.33
Sichuan	6.28	5.64	6.62	7.66	8.87	14.23
Guizhou	0.05	0.03	0.03	0.18	0.07	0.05
Yunnan				0.18	0.01	0.02
West	0.38	0.60	0.60	1.41	6.07	21.04
Shaanxi	0.01	0.01	0.01	0.02	2.11	8.06
Gansu	0.02	0.03	0.03	0.11	0.02	0.08
Qinghai		0.00	0.05	0.06	0.39	2.23
Ningxia		0.01	0.01	0.06	0.02	0.00
Xinjiang	0.35	0.55	0.50	1.15	3.54	10.67
Total	14.27	12.93	15.30	17.95	27.20	49.32
Distribution (percent)						
Northeast	37	31	29	27	15	8
North	6	7	5	6	7	4
East	10	9	10	7	4	3
Central-South	0	5	10	7	19	13
Southwest	44	44	43	45	34	30
West	3	5	4	8	22	43

Source: *China Energy Statistical Yearbook*, various years.

Note: Age totals may not add to 100 because of differing reporting bases between provincial and national totals.

provinces in the west, and the region now is the centre of China's gas production, accounting for 43 percent of national output in 2005.

The natural gas sector has experienced five main periods of development (Table 2.6). The first stage of 'socialist reconstruction' witnessed a modest expansion of output from only 7 Mcm in 1949 to 290 Mcm in 1959, the year that the Daqing oilfield was discovered. The development of the Daqing field – and the subsequent discovery of other major gas-bearing oilfields such as Shengli and Huabei over the next decade – opened a second period of development characterised by a surge of production; in 1960 alone, output jumped 3.5 times over the 1959 level. By the peak in 1979, production had increased by an average 23 percent a year and had reached a historical high of 14.5 Bcm. The extensive growth during this period was underpinned by government control of exploration, production, consumption and pricing. Output was allocated to industrial users, primarily the oil and gasfield bureaus for self-use and to fertiliser manufacturers; residential and power plant use were excluded.

Table 2.6: Natural Gas Production and Growth

Period	*Year*	*Production (Bcm)*	*Average Annual Growth over Period (percent)*
	1949	0.01	
1949–59	1959	0.29	45
1959–79	1979	14.51	23
1979–92	1992	15.79	1
1992–2000	2000	27.70	7
2000–2006	2006	58.56	16

Source: *China Statistical Yearbook*

The initiation of economic reforms in 1979 ushered in a third period of development characterised by consolidation in the industry and initial operational and policy reforms. Recognising the damage to oilfield integrity as a result of extensive use of water flooding, the government cut back on oil production, and natural gas output fell as well. From a peak of 14.5 Bcm in 1979, output retreated to 11.9 Bcm in 1982. The Ministry of Petroleum maintained control of exploration, production and allocations, but initial pricing reform was introduced, encompassing a two-tiered system of 'quota' and 'above-quota' prices, designed to encourage production beyond the quota set by the state. In 1988, the government merged the Ministry of Petroleum into a new Ministry of Energy, and separately formed a new corporate state-owned and state-run entity named the China National Petroleum and Natural Gas Corporation (CNPC) to manage upstream oil and gas operations, while downstream refining and marketing operations remained the responsibility of the China National Petrochemical Corporation, or Sinopec. Responsibility for both oil and gas production in the offshore areas was devolved to the newly established China National Offshore Oil Corporation (CNOOC).

This period presented a growing set of challenges to CNPC, setting the stage for many of the reforms and new policies that came into effect in the early 1990s. Of greatest concern were the growing financial losses at CNPC owing to sharp increases in exploration and drilling costs without compensating increases in wellhead prices, a decline in the government budget for exploration and production, and the shift to reliance on domestic bank loans to cover operating costs. By 1992, CNPC's total losses had mounted to nearly ¥4.9 billion (US$890 million at 1992 exchange rate), up nearly ¥2 billion from 1990, and worsening. At the same time, output of oil and natural gas was stagnating; natural gas production in 1992 was only 8.8 percent higher than it had been in 1979.

By 1992 as well, it was becoming apparent that China was facing a growing oil shortfall and would soon become a net oil importer, in sharp contrast to its position as Asia's largest oil exporter in 1986. The financial losses at CNPC in the face of low oil and gas prices, and growing concern about domestic energy self-sufficiency and energy security led to a reevaluation of domestic energy policy, which resulted in a sharp rise in oil and gas prices, loosening of trade restrictions, increased investment in exploration and production, further opening to foreign company participation, and reconsideration of gas consumption restrictions, marking the fourth period of natural gas development. The turn-around was fairly quick: CNPC returned to the black by 1994, the number of operating gasfields jumped by 14 percent and proven reserves doubled between 1992 and 1996 alone.[3]

Although significant progress was being made to increase gas supply, the government put little focus on the foundations needed to increase consumption. This was reflected in the ongoing negotiations between CNOOC and ARCO about the development of the Ya 13-1 gasfield in the Qiongdongnan basin of the South China Sea, south of Hainan Island, which was discovered in 1983. This discovery presented a major challenge to China's traditional views towards natural gas development. Hoping to find oil, but faced with a major new supply of gas, China was unprepared to deliver the gas to consumers in China's nearest major market of Guangdong, citing concerns of cost for new infrastructure and payment in foreign currency. After years of discussion and negotiation with ARCO, the field's operator, CNOOC agreed to support the laying of an underwater pipeline to Hong Kong and to supply 85 percent of the gas to power plants in the former British colony, starting in 1996. Numerous market studies by domestic and foreign consultants had determined that the booming economy of Guangdong had huge potential demand, and more importantly, the income needed to pay for gas; but the large scale of the infrastructure needed to bring it to homes and factories was considered too daunting at the time. Part of the gas, however, did end up in Guangdong in the form of electric power generated in Hong Kong and delivered over the border.

The fourth phase also saw 'supply chasing demand'. Natural gas was being promoted as a clean alternative to coal and restrictions on power plant usage were lifted. Urban residents were encouraged to switch to natural gas to reduce the use of coal and its subsequent pollution in cities. However, supply-side measures outpaced the development of consumer markets. One of the first major pipeline projects to supply a large urban market was the 868 km 3 Bcm pipeline from the Changqing gasfield in Shaanxi to Beijing (Shaanjing-1 pipeline). The pipeline was completed in 1997 and was intended to spur a rapid shift to residential, power, and industrial gas use. Not having first contracted users for the entire pipeline volume of up to 3 Bcm, and faced with higher than expected costs in renovating the city's gas distribution system, the pipeline remained heavily underutilised for the first few years of operation, and it was reported at times that the pipeline gas was being reinjected into the Huabei oilfield south of Beijing. By 2000, when consumption from the pipeline grew to an annual 1 Bcm, operators found that they could not adequately manage the peak periods in Beijing when demand jumped above pipeline capacity owing to a lack of storage.

This supply-side orientation was also apparent in the initiation of the 12 Bcm West–East Pipeline project to bring western gas to the east coast. The project was approved in 2000 and construction began in 2001. At the outset, planners encountered numerous challenges in finding suitable markets for the gas owing to the relatively high price (over $5.00/MMBtu at city gates) set by the state to ensure adequate return to CNPC for its construction. In particular, disputes arose with power generators, who were squeezed by the regulated price of electricity and claimed that a gas price above $3.80/MMBtu would make them lose money. In addition, few cities along the pipeline route had adequate reticulation to deliver the volume of gas available to residential and commercial users.

By the time the pipeline was completed in December 2004, however, the market situation had changed completely. The construction was accelerated to meet rapidly rising demand in the Yangtze River Delta region, including from power producers in the Shanghai area who complained of inadequate supply from the small Pinghu gasfield offshore Shanghai. The emergence of another power supply crisis after 2003 and the adoption of new and more flexible power tariffs had given a boost to demand in the generation sector.

This transition from oversupply to ascendant demand characterises the fifth and current period of China's gas development, where demand chases supply. Since 2000, consumption of natural gas has risen on average 14 percent per year, with residential use soaring at a 20 percent annual rate of growth, and power sector use by 16 percent per year. Two fundamentals underlie this shift: the response to the electricity crisis of 2003–2006, and the doubling of the size of the national economy since 2000.

The two issues are closely related: unexpectedly high rates of economic growth – led by heavy industrial investment unleashed with a relaxation of credit availability in 2001 – overwhelmed the power system's supply capacity. The power sector, having reduced investment in the late 1990s as economic growth moderated, found itself facing up to 35 GW of capacity shortfall in 2004 alone amid double-digit growth in electricity demand. In response, some cities and provinces reformed their electricity tariff structures, adopting peak, shoulder and valley pricing structures that effectively made natural gas-based peak-power generators economical despite initial concerns about relatively high city gate gas prices.

The boom in economic growth led primarily by industry resulted in a scramble for all forms of energy. Coal demand rose by 10 percent per year from 2000 to 2005, while oil went up 8 percent a year; but electricity – hydro and nuclear power – soared 13 percent a year. From an average annual capacity addition of 30 GW per year in the late 1990s, a boom in construction of new power plants added 60 GW of new capacity in 2005, 102 GW in 2006, with 90 GW additional slated for 2007. Nearly all of these were coal-fired, but China added 15.6 GW of gas-fired generation during this period, with 19 GW additional planned.

Even with the sharp rise in domestic production of natural gas, demand now exceeds what the country can produce. Anticipating growing markets for gas particularly along the eastern coastal area, China began planning in the late 1990s for additional supplies in the form of LNG.

LNG and Trade

In 2006 the first shipment of liquefied natural gas (LNG) arrived at the Dapeng LNG terminal in Guangdong province from Australia. The successful launch of this project comes twenty years after the discovery of the large Ya 13-1 natural gasfield on the southern coast of Hainan raised the promise of bringing natural gas to the province. The route to the delivery of the first LNG cargo, however, reflects many of the policy constraints that have made natural gas a latecomer among China's energy mix.

Guangdong Dapeng LNG

Ever since ARCO's discovery of the large Ya 13-1 gasfield off the southern end of Hainan island in the 1980s, China has investigated the options for supplying Guangdong with natural gas. However, infrastructure, market, and political conditions were not sufficiently mature to permit direct delivery of natural gas during the 1980s or early 1990s, and one of the biggest challenges in the negotiations to develop the field was the settling of the foreign exchange costs that would be incurred by the sale of offshore gas to Guangdong. At that time, joint-venture projects were required to balance foreign exchange expenditures and revenues, and this regulation would have required Guangdong to expend foreign exchange to compensate ARCO's sales in the province. The issue was almost completely avoided by redirecting the gas to Hong Kong, where payment would be made in a convertible currency.

Guangdong, as one of China's most open and outward-oriented provinces, did not lack income to pay for natural gas; home to China's earliest Special Economic Zones, the province became China's main export engine and earner of foreign exchange. By 1994, when the agreement to build a pipeline to Hong Kong had been reached, Guangdong accounted for one-third of foreign investment into China, and the foreign-invested enterprises in the province accounted for nearly 60 percent of total exports from such enterprises nationally. The magnitude of the payments required for the natural gas project, however, exceeded the local and national government's willingness to modify the foreign exchange balancing requirements for the project.

Nonetheless, Guangdong rapidly became China's largest importer of gas in the form of LPG under a policy relaxing LPG import restrictions and freeing domestic prices. As imports grew from 120,000 tonnes in 1990 to 3.9 million tonnes in 2005, Guangdong now accounts for two-thirds of national imports and 30 percent of national consumption. Already in 2005, 32 million people in Guangdong were using LPG, compared to just 1.6 million using town gas, and 785,000 with access to natural gas.[4] Residential gas use alone accounted for 54 percent of the total consumption of all gases in all sectors (on a heat-equivalent basis). Industry end-use accounted for another 39 percent in the form of coke oven gas, town gas, LPG, refinery gas, and natural gas; only 1 percent was used for power generation.

The foundation for natural gas use in Guangdong is sparse, and the absence of an infrastructure on which to introduce natural gas has shaped the nature of the first major gas project in the province. In 1995, CNOOC commissioned a preliminary study of an LNG terminal in Guangdong, submitting a full feasibility study to the State Development Planning Commission[5] in 1998. Later that year, China's State Council formally approved the importation of LNG in the southern and eastern parts of the country, boosting the prospects for approval of the Guangdong project subject to standard review. The SDPC gave final approval in late 1999, at which point the bidding and selection process for a foreign partner began. In March 2001, the project executive office selected BP as exclusive negotiating partner to form a joint-venture LNG company with a 30 percent share valued at about US$180 million. One of the first tasks of the partnership was to review tenders for LNG supplies. CNOOC placed Australia LNG Ltd, BP Gas in Indonesia and Qatar's Ras Laffan Co. on their shortlist of suppliers, and the final award went to Australia in 2002.

The Chinese partners in the venture are China National Offshore Oil Corporation, Shenzhen Gas, Guangdong Electric Power, Guangzhou Gas Company, Shenzhen Energy, Hong Kong Electric, Hong Kong & China Gas, Dongguan Fuel and Foshan Gas. These partners hold 70 percent of the project; CNOOC is the lead partner with 33 percent share.

Total investment in the project was ¥29.1 billion (about US$3.5 billion), the first phase of the project encompassed a 3.7 million tonne receiving terminal and 385 km trunkline for US$866 million, four power plants of a total 4.56 GW capacity for $1.6 billion, city gas grids supplying residential users in Foshan, Guangzhou, Shenzhen, Dongguan and Hong Kong for $635 million, and $400 million for the construction of China's first LNG tankers, delivery of the first to be at the end of 2007. At start-up, LNG consumption is expected to be split 2.3 Bcm for power generation, 1.2 Bcm for the urban grid and industrial use, and 1.0 Bcm sent to Hong Kong, or about a 55 percent reliance on power generation. Foreign participation was sought for the terminal and trunk line project only; Chinese companies built the power plants, of which the Huizhou plant was financed with the assistance of a World Bank loan. The local gas companies were responsible for the expansion of the distribution grid in their areas, as is the practice elsewhere in China.

The terminal was formally dedicated by the Premiers of China and Australia in June 2006, after which an application for Phase I expansion and a feasibility study for Phase II were submitted for review and approval. The Phase I expansion would add an additional 2 million tonnes capacity to the existing terminal, and Phase II would involve a new receiving terminal of an initial 3 million tonne capacity built at Zhuhai. The trunk line with the second station would connect with the existing one from Dapeng, 'ringing' the Pearl River Basin with gas. The Zhuhai project, however, is in competition with Sinopec's proposed 2 million tonne terminal near Macau. In October 2007, neither had received final approval.

The Guangdong LNG project was a pilot and a learning ground for China's initial entry into the LNG business. Similar to the approach taken with offshore oil exploration and development, nuclear power, petroleum retailing, and the automobile industry, pilot efforts have aimed at maximising Chinese participation and opportunities to learn managerial, financial, negotiation, marketing and other operational practices, accelerating the transfer of desired technical skills and technologies, expanding the degree of 'domesticisation' (the proportion of material inputs provided from Chinese firms), minimising the national or regional impact of project failure, and maintaining a separation of the foreign partner from the domestic end-users. Projects achieving these goals along with operational profitability have often laid the groundwork for more such projects, with or without foreign participation. Indeed, China's second LNG terminal in Fujian is proceeding without foreign ownership.

Fujian LNG

The Fujian LNG terminal is located in Xiuyu Port of Putian city, Fujian. The first phase includes a 2.6 million tonne LNG receiving terminal and a 315 km trunk line, three gas-fired power plants, and branch lines to gas users in Fuzhou, Putian, Quanzhou, Xiamen, and Zhangzhou. Ground was broken in April 2005, with completion scheduled for December 2007, on an initial investment of ¥5.5 billion (US$0.69 billion). The 25-year LNG supply contract for the terminal was signed in September 2006 with Tangguh LNG of Indonesia, in which BP, CNOOC's partner in the Guangdong Dapeng LNG project, is a part owner. The Fujian project is a joint investment between CNOOC (60 percent) and the Fujian Investment and Development

Company (40 percent). The planned second phase of the project, scheduled for completion in 2012, would bring the total capacity of the terminal up to 5 million tonnes, with the additional gas slated for four new power plants.

Unlike the case in Guangdong, where millions of residents were already accustomed to using LPG or town gas and are now switching to natural gas, Fujian's gas market is fairly undeveloped. CNOOC found instant demand for its LNG imports into Guangdong, but Fujian presented it with the need to develop new markets and new users. In 2005, total LPG consumption in Fujian reached 1 million tonnes – 15 percent of that in Guangdong – over half of which was consumed in the residential sector. The strong focus on power plant use – which is expected to take up to 75 percent of the supply – has provided a stable base for both phases of the project. Fujian, like Guangdong, has few provincial coal resources and is the end of a long coal supply chain from Northern China. The high transport cost of coal to Fujian provides natural gas with economic competitiveness in the power generation sector.

Two additional LNG terminals have been approved by the National Development and Reform Commission (NDRC) for construction in Shanghai and Zhejiang Province. The Shanghai terminal, located offshore in the Yangshan Deepwater Port, is planned for two phases of 6 million tonnes per year capacity, with the first phase of 3 million tonnes to be completed by the end of 2008 at a cost of ¥7 billion (US$890 million). CNOOC is taking a minority share in this project (45 percent) with its partner Shenergy (55 percent), a major owner of the Shanghai gas grid. A contract for initial LNG supply was signed between Shenergy and Petronas of Malaysia in late 2006, for deliveries of 1.1 million tonnes starting in 2009, rising to 3 million tonnes by 2012.

Although Shanghai is already supplied with natural gas from the offshore Pinghu gasfield (0.6 Bcm in 2005) and from the West–East Pipeline (1.27 Bcm in 2005), the city intends to use the increase in natural gas supply to completely replace the use of low-heat-value coal gas and coke oven gas. Currently, coal and coke oven gas provides over 1.8 Bcm of natural gas equivalent for power generation, and 1.6 Bcm of natural gas equivalent energy to other sectors, mainly industry and residential uses (Table 2.7). As gas supply grows – including a boost in the supply from Pinghu to 0.8 Bcm per year starting in 2007 – these uses will be converted to natural gas, and additional natural gas power generation will be added. Shanghai's first natural gas-fired power plant of 1.2 GW went online in late 2006, and 3.6 GW additional capacity

Table 2.7: Shanghai Gas Supply and Consumption, 2005

	Supply	*Final Consumption*	*Of which: Residential*	*Of which: Industry*	*Power Generation*
(Bcm)					
Coal Gas	11.60	2.95	1.12	1.18	8.37
Coke Over Gas	2.81	2.16	0.03	2.13	0.17
Natural Gas	1.87	1.20	0.29	0.70	0.14
LPG	0.79	0.79	0.25	0.40	-
(Bcm natural gas equivalent)					
Coal Gas	2.51	0.64	0.24	0.25	1.81
Coke Over Gas	1.25	0.96	0.01	0.95	0.07
Natural Gas	1.87	1.20	0.29	0.70	0.14
LPG	1.02	1.02	0.33	0.52	-
Total	6.66	3.82	0.87	2.43	2.03

Source: China National Bureau of Statistics, China Energy Statistical Yearbook 2006. Beijing: Statistical Press, 2006

is planned by 2010, when Shanghai's total consumption is expected to reach 6–7.5 Bcm.

CNOOC's fourth terminal is located in Ningbo, Zhejiang province, south of Shanghai. Here, CNOOC has a 51 percent equity stake with partners Zhejiang Energy (29 percent) and Ningbo Power (20 percent). The first phase of the project, estimated to cost ¥4.99 billion (US$640 million) involves a 3 million tonne per year terminal, trunk line connecting with the existing Zhejiang gas grid, and a 1.4 GW power plant on the terminal site. Completion was originally planned for the end of 2008, but because CNOOC has not been able to secure a supply contract, they announced a delay in completion in early 2007, pushing the date out to mid-2009.

Zhejiang's existing gas market is largely LPG-based, with far less coal gas and coke oven gas consumption (0.33 Bcm natural gas equivalent) compared to Shanghai. LPG consumption is 75 percent residential based, and gas use in the power sector has accounted for less than 2 percent of the total. About half of the initial LNG supply will be used for power generation.

The completion of the Fujian, Shanghai, and Zhejiang LNG projects will provide CNOOC with the foundation of its long-term plan to connect and integrate the gas markets of China's southeast coast, eventually connecting Shanghai in the north to Hainan Island in the south (Map 2.1).

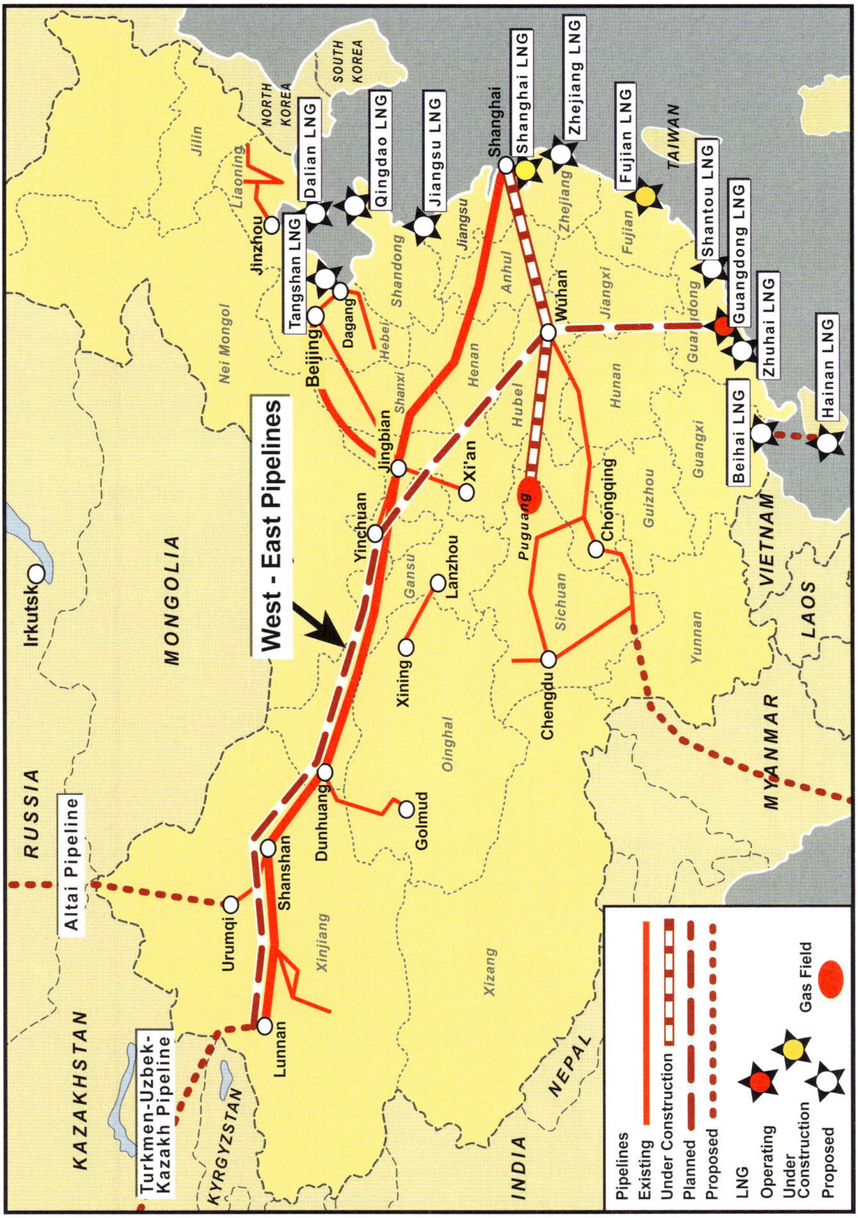

Map 2.1: Pipelines and LNG Infrastructure

Transportation and Storage

China's natural gas transport infrastructure developed primarily on a regional basis, serving mainly the producing oil and gasfields and local industries. In 2005, total pipeline length reached 22,451 km, with a total transport capacity of 37.4 Bcm per year, up from 11,630 km and 14.1 Bcm per year in 1999. Until the completion of the 12 Bcm West–East Pipeline in 2004, the majority of transport capacity had been built in Sichuan, where the industry had developed a large ring pipeline system around the province for the distribution of up to 11 Bcm of natural gas production. This pipeline system includes branch lines to major fertiliser producers, other industrial customers, and to residential consumers in Chongqing, Chengdu and other cities.

The late 1990s and early 2000s marked a wave of construction of long distance pipelines for supply to urban centres, with a major focus on supply to Beijing as part of its goal to displace coal and reduce air pollution. In 1997, the first Shaanxi-Beijing pipeline of 3.6 Bcm capacity went into operation, followed in 2005 by the second larger line of 12 Bcm capacity. Extensions to the pipeline have been built to supply Tianjin, Shandong and Hebei provinces. In total, the two pipeline projects will supply Beijing with up to 6 Bcm of natural gas by the time of the Olympics in 2008, with about 60 percent used in residential heating, cooling, and cooking applications, substantially replacing coal use in the city.

In southwest China, the first major pipeline out of Sichuan was the 780 km 3 Bcm line from Chongqing (now an independent province-level city) to Wuhan, with spurs to Changsha and other cities in Hubei province. Originally designed as a joint venture between PetroChina and Enron in 1999, the pipeline was built by PetroChina alone after Enron's bankruptcy led to its withdrawal from the project in 2001. Wuhan began to receive natural gas in 2004.

In Qinghai, a 1000 km pipeline of 2 Bcm capacity from the Liangbei field in the Qaidam basin was completed in late 2001, supplying natural gas to Xining, capital of Qinghai, and the industrial city of Lanzhou, capital of Gansu. In late 2003 a 1.1 Bcm capacity pipeline from the Changqing gasfield in the Ordos Basin was completed to Huhhot, capital of the Inner Mongolia Autonomous Region. Because of the high demand for gas along the pipeline route through the industrial city of Baotou, Huhhot has found itself experiencing problems of supply, pressure and peak management, and has decided to build a dedicated line from Wushen Banner to supply the city. In 2005, the original 1 Bcm pipeline from Jingbian to Xi'an was increased in capacity to 2.5 Bcm.

In 2005, PetroChina completed the construction of a 780 km linkage line connecting the second Shaanxi-Beijing pipeline and the West–East Pipeline, linking the provinces of Hebei, Shandong, and Jiangsu. This linkage line will serve to coordinate supply from the two major pipelines and supply additional customers along its route.

West–East Pipeline

The second major project designed to significantly advance the development of the natural gas market was the West–East Pipeline, a key national project approved in 2000 and completed in October 2004. This immense project under the direction of PetroChina links the gasfields of Western China to demand centres in the Yangtze Delta region of Shanghai, Jiangsu and Zhejiang over 4000 km away. The trunk pipeline of 1118 mm diameter was initially designed to carry 12 Bcm annually, of which 1.2 Bcm was contracted at the terminus point in Shanghai.

The West–East Pipeline project was intended to boost China's domestic pipeline manufacturing and construction industry, which heretofore had never attempted a project of this magnitude. A key challenge, however, was the limited domestic capacity to produce pipe larger than 36 inches in diameter. Although some pipe was imported, the project allowed BaoSteel in Shanghai, China's largest steel producer, to upgrade its steel pipe production capabilities, but the diameter limitation is a likely reason for the limited capacity of the pipeline. Internationally, the economics of long-distance pipeline transmission, such as from Russia to Western Europe, indicates the use of large-diameter pipelines of 30 Bcm capacity or more; proposals for even shorter gas lines from Iran to India have considered construction of 48-inch lines carrying 20 Bcm annually. With increases in compression, the capacity of the West–East Pipeline has risen to 17 Bcm, but it is insufficient to serve as a conduit for large volumes of Russian or Central Asian gas to China. The high transmission costs of the pipeline also hampered initial market development in the Lower Yangtze region, and PetroChina was eventually forced to reduce prices to spur market uptake.

The prices originally proposed are shown in Table 2.8. At the time the pricing schedule was issued in 2001, the prices appeared considerably out of step with the development of new markets and the average income level of Chinese households, and indeed, the NRDC was forced to lower prices by ¥0.02/m3 (US$0.07/MMBtu) in 2003, capping prices for industrial users such as chemical companies at ¥1.2–1.3/m3 (US$4.40–$4.75/MMBtu) and to power producers at ¥1.1–1.2/m3 (US$4.40–$4.40/MMBtu). City gate prices for residential distribution were, on average, unchanged.

Table 2.8: Planned Pricing of West–East Pipeline Gas

	Well-head	*P&T**	*City Gate Prices*						*Residential*
			Henan	*Anhui*	*Jiangsu*	*Shanghai*	*Hangzhou*	*Average*	
¥/m^3	0.45	0.877	1.16	1.24	1.31	1.35	1.37	1.327	2.00
$/MMBtu	1.94	3.78	5.00	5.34	5.64	5.81	5.90	5.72	8.61

* processing and transportation. The transport portion is the pipeline average. The wellhead price and processing costs have now been combined into an 'ex-factory' price.

Source: *China Oil, Gas & Petrochemicals*, Beijing, China OGP, October 2001.

Originally planned to run directly from Xinjiang though Lanzhou, Xi'an and Luoyang to Shanghai, the development of the large Changbei natural gasfield in the Ordos Basin of northern Shaanxi and the discovery of the new Sulige field in nearby Inner Mongolia, prompted a revised plan that divided the pipeline project into two stages. The first stage involved the completion of a pipeline link from Jingbian in Shaanxi to Shanghai, with natural gas supplied from the Ordos Basin gasfields, while the second phase linked the new gasfields in the Tarim Basin in Xinjiang to Jingbian, thus completing the transcontinental link. The State Development Planning Commission (now the NDRC) estimated the total cost of the project at $17.8 billion, including $3.5 billion for development of the natural gas reserves, $6 billion to build the pipeline, and $8.4 billion to construct the supply systems from city gate to end users (Map 2.1).

The scope and technical challenge of the pipeline project – a first for China – led the government to propose foreign partnership in it, against PetroChina's wishes. Under Premier Zhu Rongji, a strong proponent of such international cooperation, China issued new guidelines to promote interest in participation. These guidelines included the lifting of the restrictions on foreign ownership limits in the project and the form that the ownership could take, such as joint venture or association. It also opened up construction of urban gas grids in cities to be supplied by the pipeline, overturning one element of the 'Guidelines for Foreign Invested Industries', which forbid foreign investment in constructing and operating urban water supply and sewage, gas and thermal energy supply networks, and required Chinese controlling interest in any gas pipeline projects. In addition, the new guidelines exempted imported equipment from the value-added tax and Chinese customs duties and relaxed land acquisition rules. Without providing any details, the guidelines opened the door partway to investment beyond the pipeline by allowing foreign invested ventures specialised in 'transportation, energy, and other facilities' to expand their business scope 'upon approval'. This last element was an acknowledgment of the interest of international companies in integrated operations, such as gas production and marketing, or pipeline operation and local distribution.

PetroChina, following Premier Zhu's lead, issued a tender offer in early 2001 for participation in the pipeline project which included options to participate in upstream exploration in six gas blocks in the Tarim Basin, including the Kela-2 gasfield reserved for the winning bidder, but only on a minority basis. Nonetheless, the tender brought bids in March 2001 from nineteen foreign companies, including BP, Royal Dutch/Shell, ExxonMobil, TotalFinaElf, Gaz de France, Itochu, Marubeni, Mitsubishi, Mitsui, Nissho Iwai, Gazprom, Petronas, Sumitomo, Hong Kong & China Gas, Transgas, China Light and Power, Energomachexport Russia, Houston Inspection International, and United Technologies. A number of companies later withdrew their bids, and the remaining formed bidding consortia for final consideration of the PetroChina shortlist of participants. The shortlisted companies announced in mid-2001 included an ExxonMobil-led group with China Light and Power; a BP-led group with Petronas, Mitsubishi, Itochu and Nissho Iwai; and Shell, who submitted its bid alone at the time. According to PetroChina, the choice of companies was based on 'commercial experience and financial strength' in the face of the immense cost and complexity of the pipeline. Perhaps with a nod to other elements of consideration, PetroChina later invited Gazprom (in consortium with Stroitransgaz), with whom possible pipeline projects from Russia had been discussed, and Hong Kong & China Gas, already selected to participate in the Guangdong LNG project, to enter into negotiations as well.

In late 2001, BP withdrew their participation in the project, citing the advanced stage of project design and their inability to contribute their expertise in an appropriate way. Shell, in turn, teamed up with Gazprom, and in late 2001 signed an interim agreement with PetroChina to take a 45 percent stake in the project, including upstream gas development, the pipeline, and downstream gas marketing. In early 2002, Shell and ExxonMobil notified PetroChina of their intention to negotiate ExxonMobil's possible participation in the project. By 2003, however, the foreign partnerships were moribund. As negotiations continued with the companies, PetroChina continued to build the pipeline, even accelerating the schedule from a 2006 completion to 2004. With the retirement of Premier Zhu in 2003 and the ascension of the new Premier Wen Jiabao, the pressure from the government to include foreign participation collapsed.

The motivation for and justification of the West–East pipeline project extended further than energy policy alone. Given the concentration of natural gas resources in the western part of China, it became the flagship project to promote the development of the region, which has increasingly fallen behind the national average in income and development. The growing disparity between the economically advanced eastern provinces and the stagnant interior provinces has created growing concern among the leadership, who recall the role of the rural and dispossessed in their own revolution. The interior provinces have failed to attract the level of foreign investment that has provided such impetus for economic growth in the east, and the east has received the lion share of national investment in infrastructure projects. The actual impact of the pipeline project on local development, however, is unclear. As a highly capital-intensive project demanding either technically skilled employees or labourers for the period of construction, the completed pipeline serves mainly to stimulate the exploration and development of gas resources in Xinjiang.

By the time of the completion of the pipeline, PetroChina had developed nearly 450 Bcm of proven reserves to support an ultimate production level of around 14 Bcm (Table 2.9).

Although the pace of pipeline construction has accelerated significantly, gas storage capacity necessary to regulate the flow of gas to consumers and to even out daily and seasonal variations lagged behind.

Table 2.9: Proven Reserves and Production Capacity in the Tarim Basin (2000). Bcm

Gasfield	*Reserves*	*Production Capacity*	*Production for Pipeline Project*
Kela-2	250.61	10.00	10.00
Hetianhe	61.69	2.00	
Yaha Condensate Field	35.78	1.20	1.20
Yingmai-li Condensate Field	29.57	1.05	1.05
Yangtake Condensate Field	24.91	1.00	0.99
Jilake	12.71	0.49	0.49
Yudong Condensate Field	7.33	0.33	0.33
Tazhong-6 Condensate Field	8.53	0.25	
Jinan-4 Condensate Field	2.58	0.10	
Ti'ergen	1.60	0.07	
Hongqi	0.95	0.03	
Total	436.25	16.52	14.05

Source: China National Petroleum Corporation, West-East Natural Gas Transportation Pipeline Project, August 2000

In 2001, CNPC built its first underground storage facility, located in the Dagang oilfield in Tianjin, to regulate supply to Beijing and Tianjin. Total storage capacity is 1.6 Bcm. The lack of storage for the Beijing-Tianjin network had worsened the performance of the Shaanxi-Beijing pipeline, as maximum daily demand in winter exceeded the 3 Bcm capacity of the pipeline, while summer demand remained less than 1 Bcm. Additional storage facilities have been built in the abundant salt structures in the lower Yangtze valley for management of supply from the West–East Pipeline.

Natural Gas Consumption

Natural gas remains primarily a fuel of the industrial sector. At the time that economic reforms started in the early 1980s, industry accounted for over 90 percent of total consumption, and nearly half of that amount was used in the oil and gas sector itself. As the value of commercial gas has grown, natural gas increasingly penetrated other sectors, including residential, power generation, and transportation. Today, the share of industrial consumption has declined to 63 percent, but over half of industrial use is in the production of chemicals, including chemical fertilisers, as both fuel and feedstock. Fertiliser plants remain priority recipients of natural gas, and their supply is guaranteed through both allocation and pricing controls. Since 2000, gas has become increasingly important in other industrial subsectors such as non-metal mineral processing (e.g. magnesite, graphite, fluorite, talc, asbestos, gypsum, cement limestone), and use in the iron and steel industry has grown strongly as well (Table 2.10).

The high percentage devoted to the industrial sector reflects the government's historical priorities for energy allocation, and its current share roughly parallels the industrial share of other energy forms. Not until the late 1980s did the government start promoting alternatives to coal in the residential sector, and natural gas consumption began to grow steadily, though to less of an extent than LPG or town gas owing to a lack of distribution infrastructure and limited supply. Since 1995, transportation and power generation consumption have seen steady growth, albeit from a very low base. In cities such as Beijing, nearly all public buses now run on compressed natural gas (CNG). In total, China now has about 30,000 CNG-fuel transit buses and 70,000 CNG-fuel taxis.[6]

The large iron and steel industry and the plentiful supply of coal laid the foundation for China's development of a large coking and coal gas industry. Together with LPG, first introduced in significant

Table 2.10: Natural Gas Consumption by Sector. Bcm

	2000	*2001*	*2002*	*2003*	*2004*	*2005*	*2005 Structure*	*A.A.I. 2000-2005*
Total Production	27.20	30.33	32.66	35.02	41.46	49.32		13
Total Consumption	24.50	27.43	29.18	33.91	39.67	46.35	100	14
Power Generation	1.52	1.30	1.11	1.32	1.90	3.15	7	16
Heating Supply	1.43	1.64	1.70	1.40	2.04	2.32	5	10
Industry Total	17.07	18.69	19.79	23.90	25.26	29.18	63	11
Chemical Production*	4.12	3.70	4.20	6.52	5.39	6.39	14	9
Chemical Feedstock	4.92	5.85	6.00	6.68	7.63	8.77	19	12
Petroleum & Gas Extraction	5.12	5.94	6.09	6.37	5.16	5.34	12	1
Non-Metal Mineral Processing	0.25	0.28	0.35	0.39	1.96	2.60	6	60
Machinery & Electronics Manufacturing	0.73	0.84	1.01	1.12	1.25	1.69	4	18
Petroleum Refining	1.02	1.17	1.15	1.60	1.52	1.48	3	8
Ferrous Metallurgy	0.17	0.17	0.23	0.33	0.76	1.07	2	44
Other Industrial Use	0.75	0.75	0.76	0.90	1.59	1.85	4	20
Commerce, Construction, Agriculture	0.43	0.57	0.68	0.76	1.06	1.23	3	24
Transportation	0.10	0.12	0.16	0.18	0.50	0.58	1	42
Residential	3.23	4.41	5.12	5.69	6.72	7.94	17	20
Other (including losses)	0.73	0.69	0.63	0.67	2.19	1.95	4	22

* Non-feedstock energy use

Source: *China Energy Statistical Yearbook 2006*

volumes in the late 1980s, these other gases total twice the volume of natural gas consumed, calculated on a heat equivalent basis (Table 2.11). Although use of coal or coke oven gas in residential and power generation applications – a total of 8.6 Bcm natural gas equivalent in 2005 – is targeted for substitution by natural gas, production of these gases has nearly doubled since the late 1990s, primarily for industrial use. Taken in total, households are now receiving nearly 30 percent of all gas in China, compared to 17 percent of natural gas alone, and LPG has become the most common cooking and water heater fuel in most coastal cities. Many cities that are already reticulated for home gas use, however, have developed on the basis of town and coke gas, both of which contain much lower heat content than natural gas, and many cooking and heating appliances are adapted for this low-value gas use. In Guangzhou, Shenzhen, Shanghai, Wuhan and other cities where natural gas is increasingly replacing both town gas and LPG, city governments have generally provided assistance to residents in

Table 2.11: Structure of National Gas Consumption, 2005

	Total Supply	*Final Consumption*	*Of which: Residential*	*Of which: Industry*	*Power Generation*
Coal Gas (Bcm)	90.6	69.7	10.1	59.0	13.1
Coke Oven Gas (Bcm)	47.9	42.4	4.4	37.4	3.6
Natural Gas (Bcm)	49.3	39.8	7.9	29.2	3.2
LPG (MMt)	20.5	20.4	13.3	5.3	0.0
(Bcm natural gas equivalent)					
Coal Gas	19.6	15.1	2.2	12.8	2.8
Coke Oven Gas	21.3	18.9	2.0	16.7	1.6
Natural Gas	49.3	39.8	7.9	29.2	3.2
LPG	26.4	26.3	17.1	6.8	0.0
Total	116.7	100.1	29.2	65.4	7.6

Source: *China Energy Statistical Yearbook 2006*

conversion of their existing gas appliances to compensate for the differing heat value of natural gas.

Although consumption of natural gas has doubled between 2000 and 2006, the domestic market remains in a formative stage. Allocation continues to play a major role in directing annual flows of natural gas and will probably continue as such because of the shortfalls in supply. Power plants contract directly with the gas producers – Petrochina, Sinopec, and CNOOC – as do provincial and city gas companies who supply end-users in the residential and commercial sectors. Fertiliser producers, long a loss-leader for gas suppliers, are allocated annual volumes based on planned production levels. In all cases, prices are controlled, and mechanisms such as forward purchasing, price hedging and other market tools have yet to be developed. Lacking a network of both long-distance transport and regional grids, producers have few choices on the location and volume of sales. A key target sector has been the power generation sector, but expansion of sales to this sector has been constrained by competition from coal power generation, which forms the foundation of the electricity pricing structure.

Power Generation

As of 2006, China reached an installed capacity of 622 GW of power generation capacity, of which 15.6 GW, or 2.5 percent, was natural gas based (Table 2.12).The majority of capacity has been built in the Yangtze River delta region around Shanghai, where the promise of natural gas from the West–East Pipeline spurred the construction of 9 GW of capacity. In 2006, however, it was reported that only 3 GW of

Table 2.12: Natural Gas Power Generation Capacity, 2006. MW

Region	*Installed capacity*	*Under construction*	*Planned capacity*	*Total*
Bohai coast	814	1,750	-	2,564
Yangtze river delta	9,707	600	1,170	11,477
Southeast coast	3,735	1,500	8,250	13,485
Northwest	1,053	280	1,000	2,333
Other regions	300	-	4,910	5,210
Total	15,609	4,130	15,330	35,069

Source: Interfax-China, 4 April 2007

this capacity was in use because of the shortage of natural gas supply.[7] By 2010, total gas-fired capacity is expected to rise to at least 30 GW, with significant increases in the southeast coastal region of Guangdong and Fujian, home to China's first two LNG terminals.

Although China has planned to install up to 70 GW of natural gas-fired power plants by 2020, significant challenges remain for the sector. These include a lack of competitiveness and supply contract and reliability issues. According to a recent study by the State Grid Corporation, natural gas is not competitive with coal when the price of natural gas is 2.5 times that of coal on a Btu basis (accounting for different heat rates in power plants).[8] In 2004, when coal delivered to a power plant in north China averaged ¥230 per tonne ($1.30/MMBtu), power producers were being offered gas at $3.80 per MMBtu, or a 2.9:1 ratio between gas and coal. Since then, however, the price of coal has strengthened significantly as demand has boomed and transportation capacity bottlenecks have emerged, narrowing the competitive advantage over gas; in south China, at the end of the long coal supply chain from the north, the gas price ratio to delivered coal has fallen to as low as 1.2: 1. At the same time, reforms in electricity pricing, including the adoption of peak and off-peak pricing systems in certain areas and the linkage of electricity prices to the coal price (allowing an adjustment in the electricity price if the coal price rises more than 5 percent over a six-month period, passing 70 percent of the cost change to the consumer) have helped increase the competitiveness of gas-fired generation, particularly south of the Yangtze River. In mid-2007, fob steam coal prices at China's major port of Qinhuangdao in Hebei province for 5500-kcal Shanxi coal reached ¥470/tonne, more than double the price in 2004. If such high coal prices persist, this implies that natural gas may be competitive even in northern China at as much as $6–6.50/MMBtu.

Pricing

Prices in China historically were used for accounting functions, divorced from their role of providing signals for resource allocation. As was the case with coal, oil and electricity, natural gas has been priced to keep industrial production costs – particularly that of fertiliser – low, placing the burden of subsidy on the national government. However, as production costs rose rapidly across the board in the late 1980s and early 1990s, leading to growing financial losses for energy producers, the pricing system underwent general reform, in most cases allowing prices to rise to internationally comparable levels. In the case of natural gas, China has retained pricing controls but varies prices by sector. In general, the residential sector pays the highest prices, followed by large industry, power generators and fertiliser manufacturers. The pricing structure incorporates elements of national industrial policy, subsidising the production of fertilisers for farmers, for example, and increasing the competitiveness of chemical producers, but it has also resulted in opposite incentives for the natural gas producers. The residential market is their preferred market, while power producers and chemical plants complain of a lack of natural gas supply.

Natural gas prices are composed of two main elements: a combined 'ex-factory' price (well-head price plus processing costs), and transport costs, and both elements remain under control of the NDRC. As production and transportation costs have risen, however, Sinopec, Petrochina, and CNOOC have all lobbied the government to raise prices and to move towards a more flexible market-based pricing system. In late 2005, the NDRC agreed to move in this direction, but only after a period of transitional price reform. Concerned about the oligopolistic power of the main natural gas producers but mindful of their need for higher revenue to support expansion of exploration and development, NDRC adjusted natural gas prices higher, raised 'ex-factory' prices by an average 10 percent (¥0.05–¥0.15/cu m; $.0.18–$0.55/MMBtu) and announced that prices would be adjusted on an annual basis starting in 2006 with an 8 percent per year maximum adjustment (no adjustment took place in 2006 in consideration of the 'ability of consumers to shoulder higher prices'). For fertiliser manufacturers, this has raised the ex-factory price basis of their gas supply to $2.55/MMBtu; for industrial users to $3.25/MMBtu, and for residential users, prices of natural gas have now reached about $7.50/MMBtu (¥2.05/cu m) in Beijing and as high as $9.50/MMBtu (¥2.6/cu m) in the southern island province of Hainan. Ultimately, the intention of the NDRC is to develop a pricing mechanism for natural gas – and other energy forms – that reflects both scarcity and the resultant pollution and encourages conservation, although an initial proposal in 2007 called for a replacement of the current cost-plus pricing basis by linking the price of natural gas to that of its substitutes, using a weighted average of 40 percent crude price, 20 percent coal prices, and 40 percent LNG prices over a five-year period to set natural gas prices.

As China moves to become an importer of LNG, it will become more difficult to maintain domestic prices substantially below those of imports. After CNOOC negotiated an LNG supply contract with Australia Northwest Shelf containing a pricing formula with weaker links to crude oil prices than others in the region, reported at $2.20/MMBtu fob or about $3.00/MMBtu cif, expectations that future projects might receive similar treatment led to a boom in new LNG proposals by CNOOC, PetroChina, and Sinopec.[9] By the time that the Fujian supply contract from Tangguh was finally settled in 2006, the LNG market had changed considerably, with prices rising in tandem with global oil prices. Originally set at $2.40/MMBtu fob with a cap at $4.30/MMBtu, the final pricing formula from Tangguh agreed in early 2007 raised the cap to $6.15/MMBtu. Reportedly, the Shanghai LNG terminal will be paying around $7/MMBtu. These higher prices for imports increase pressure for reform in the domestic natural gas pricing system.

The rise in LNG prices has been a factor in the cancellation or deferral of a number of LNG projects in China as the companies balked at paying higher prices. In turn, this has led to difficulties in securing supply contracts for new projects, although resistance to the higher prices appears to be fading, and CNOOC began buying spot cargoes of LNG for its Dapeng facility at over $8.00/MMBtu in 2007. Nonetheless, China's Ministry of Commerce issued new rules in mid-2007 that imposed new licensing requirements for LNG imports (none were required heretofore) and central review and approval of all future import licences, including the price level. Resistance to the higher level of LNG prices was broken in late 2007 as PetroChina signed a deal with Woodside Petroleum to buy as much as 3 million tonnes of LNG a year for fifteen to twenty years 'at market prices'. At the same time, PetroChina agreed with Royal Dutch Shell to import LNG from the Gorgon venture in Australia. The destination of the LNG was not specified but will most likely go to one of PetroChina's three proposed terminals at Rudong, Dalian, or Tangshan (discussed below).

Natural Gas Industry: Policies and Development Plans

Industry Structure and Governing Policies

Natural gas production is almost entirely in the hands of China's three major oil companies: PetroChina, Sinopec and CNOOC. All three are outgrowths of the earlier Ministry of Petroleum, which in the 1970s had near total control over the upstream production and downstream processing and marketing of oil and natural gas. The first round of major reform in the industry came in 1981, when the government created two monopolies in the upstream and downstream sectors. The China National Petrochemical Corporation, or Sinopec, was established to manage oil refining, petrochemicals, and marketing, while crude and natural gas production remained under the control of the Ministry of Petroleum, which shortly afterwards established the China National Petroleum and Natural Gas Corporation, or CNPC, to operate its producing assets.[10] Further, in 1982, the State Council established the China National Offshore Oil Corporation (CNOOC), the first firm fully open to foreign cooperation in China's offshore regions.

While the focus of Sinopec and CNPC remained on the domestic oil and gas market, few attempts were made to alter their monopoly positions, but as external import dependency grew rapidly in the 1990s, the lack of domestic competition and international competitiveness became major factors underlying the second major reform of 1998. At this time, the upstream/downstream duopolies were abolished, and CNPC and Sinopec were given monopolistic control of integrated upstream and downstream operations based on a geographical division, with Sinopec dominating the east and south, and CNPC dominating the northeast and west. In general, however, Sinopec maintained a larger share of refining capacity, while CNPC remained the dominant producer of crude oil and natural gas. At the time of the reorganisation, Sinopec's proven natural gas reserves totalled only 25 Bcm compared to CNPC's 688 Bcm,[11] but in 2000, Sinopec's reserves total jumped as the Sinopec Group Company acquired the China National Star Petroleum Corporation (CNSPC), renaming it Sinopec National Star Petroleum. CNSPC was established by the government in January 1997 to provide competition to CNPC, awarding it acreage in the Sichuan, west Tarim, Ordos Basin and the offshore East China Sea basins. This acquisition expanded Sinopec's gas reserve base by 19 percent, placing it ahead of CNOOC.[12] In 2001, Sinopec Group Company in turn sold the company to Sinopec.

The intent of the 1998 reorganisation was to separate government administration of the sector from operations, allowing the two companies to decide on investment and production plans based on financial considerations instead of government fiat. Indeed, 1998 and 1999 witnessed for the first time the deliberate shut-down of some high-cost producing oil wells as prices plunged internationally. At the same time Sinopec and CNPC were loosened from direct government control. The administrative responsibilities for the sector devolved from the Ministry of Energy, which was abolished, to a petrochemical bureau within the State Economic and Trade Commission (SETC), a powerful national commission in charge of day-to-day management of the economy and of state assets, which was headed until 2001 by Sheng Huaren, former chairman of Sinopec. In 2000, this bureau was consolidated with numerous other industrial bureaus into the Department of Sectoral Planning, further reducing the status of the government managers vis-à-vis the operating companies. During the government reforms of 2003, when the responsibilities of the SETC were transferred to the reorganised National Development and Reform Commission (NDRC), control over stated-owned assets, including the national oil companies, was transferred to the State-owned Assets Supervision and Administration Commission (SASAC) under the State Council. This commission has responsibility for the management of the state's shareholding interests in 178 state-owned enterprises (SOE) groups.

The national oil companies are 'pillar' enterprises within the ranks of China's SOEs and produce about one-fifth of the net profits of all SOEs. They are also the vehicles for China's international competition in the oil and gas sectors. A major step in their internationalisation came with the restructuring of Sinopec, CNPC, and CNOOC into joint stock companies in 1999 (with PetroChina as CNPC's listed arm), with a subsequent offering of a portion of their stock to the international market in 2000 and early 2001. Though encountering some glitches – such as a sudden withdrawal of CNOOC's first offering, and controversy over human rights and CNPC's investment in Sudan – the companies attracted strong international support from major oil companies; ExxonMobil, BP and Shell all bought signficant portions of the public shares. While China publicly welcomed the benefits that participation of the international majors would bring the investments apparently did not provide the benefits of 'strategic partnerships' expected by some of the international companies; in 2004 and 2005, Shell, BP and Exxon all sold their shares in Sinopec and PetroChina.

Prior to China's accession to the World Trade Organisation (WTO) in late 2001, foreign companies faced restrictive policies on participation in Chinese energy markets. Since first opening the offshore areas of China to foreign exploration and development, the government permitted investment in onshore oil and gas basins, petrochemicals, and to a limited extent, pipeline construction. Investments in refining, direct marketing, trade, wholesaling or retailing of gas or oil products were not permitted, except in limited and exceptional cases that serve to provide domestic companies with opportunities to learn new technical, management, or operational practices. Examples of such exceptions included a limited concentration of foreign-branded gasoline stations in the Special Economic Zones and other coastal areas, joint investment in the West Pacific (WEPAC) refinery (with Total) in Dalian, and a joint investment in the expansion of the Fuj refinery (with ExxonMobil and Saudi Aramco, but including petrochemicals).

China's accession to the WTO promised a scheduled relaxation of some restrictive policies, particularly those related to market access, extending national treatment to foreign firms and allowing foreign participation in direct product trade and retailing, though with quantitative restrictions remaining in place until the end of 2004 (retailing) and 2006 (wholesaling). Nonetheless, the establishment of significant barriers to entry, such as capital requirements, storage minimums, port facilities, and other facilities has resulted in less foreign participation than originally expected.

Beyond market access, China lacks a comprehensive law governing participation in the upstream and downstream oil and gas industries. A new Oil and Gas Law has been under deliberation for several years, but the relationship between it and the Energy Law, proposed and drafted in 2007, is unclear. China's current laws and regulations on oil and natural gas were developed largely during the 1980s and 1990s as the economy moved from reliance on central planning to the market, but the governing regulations remained scattered among a range of different ministries, including the NDRC, the Ministry of Land and Resources, Ministry of Finance, and the State Environmental Protection Administration. Lacking a central Ministry of Energy with defined authority over the energy sector, China will probably find it difficult to implement a comprehensive Energy Law or Oil and Gas Law owing to the historical difficulty of achieving inter-ministry coordination on regulation development.

In 2005, China established the Energy Leading Group and the State Energy Office under the State Council, largely in response to the energy crisis of 2003–04.[13] The Energy Leading Group provides a forum for inter-ministerial consensus building and information exchange, but it lacks the authority to issue policies and regulations or for enforcement; its work is largely that of developing guiding principles for policy development. Although the goal of the Energy Law is to provide 'transparency' in the development and implementation of China's energy-related policies and regulations, the absence of a central body with responsibility for both regulation development and implementation, makes it uncertain how effective the new energy and oil and gas law would be in providing clarity and assurance to foreign investors in the oil and gas sectors and reducing the level of perceived policy risk in these sectors.

Energy Security

Energy security has become an increasingly common term in China's national discourse. In the fifteen years since it grew from being a net oil exporter to a net oil importer, China has also moved into the ranks of natural gas importers in 2006 and net coal importers in 2007. As reliance on imported energy has grown, concerns of adequacy of supply, transport security, affordability and energy diversity have received greater focus.

China's responses to these concerns have reflected its own domestic considerations and international worldview, and have shaped its strategies for acquiring oil and gas imports. Domestically, economic growth must be sustained by all means because it is required to absorb the annual growth in the workforce – and to avoid problems stemming from higher unemployment – and the rising wealth from growth has become the foundation of the Communist Party's popular legitimacy. Moreover, growth has brought China to the verge of great power status, erasing the legacy, often bitterly expressed even today, of more than a century of great power domination and national humiliation. Underpinning growth is an increasing supply of energy, and the imperatives of growth have sent China on an international quest to secure it. Physical ownership, in many cases, has been a preferred method of acquisition, including equity stakes and oil production agreements in over two dozen countries and equity stakes in natural gasfields in Australia, Indonesia, Kazakhstan and other locations. Generally, China has not accepted conventional western advice that their energy security is best served by reliance on the international market. Perceiving increased instability in the Middle East subsequent to the US invasion of Iraq and having been rebuffed by the US government in their attempted purchase of Unocal in 2005, the Chinese leadership remains distrustful of a market seen as controlled by and for Western interests. Indeed, one analysis by a Sinopec official in 2006 published in the *Shanghai Securities News* claimed that the USA was manipulating oil prices in order to 'attack China's economy'.[14]

Similarly, China's growing concern over the proportion of its imports that must transit the Straits of Malacca and its inability to project its own sea power into the region has led to proposals for overland transit options. In 2004, China proposed an oil transit pipeline route crossing peninsular Thailand to avoid the Malacca chokepoint, as well as an oil pipeline from the coast of Myanmar north to Chongqing in southwest China. Although mooted as saving on transit times to northeast Asia, the projects were designed to decrease the threat of a Malacca blockade. In this vein, China and Myanmar in 2007 agreed to a project to build a pipeline to bring Myanmar natural gas north to China, and China has proposed an 'Energy Silk Route' gas pipeline network that would connect Iran in the west to China in the east.

Security interests are also served through diversification – not only of import sources, but of energy sources. This issue extends to the domestic situation as well. At over 70 percent, China's reliance on coal to fuel the economy has been seen as a major challenge to reducing pollution, especially sulphur dioxide, and the environmental impacts of coal mining and production, and to cleaning up the air in urban areas. The shift to maximising the use of domestic natural gas and the initiation of a planned large-scale LNG import programme in the 1990s both reflect the push to reduce the role of coal in the economy.

China looks to Central Asia as a potential long-term source of both oil and gas and has been developing closer ties to the region through investment and political exchange as it increasingly regards the region as its 'back yard'. In 2001, the leaders of China, Russia, Kazakhstan, Kyrgyzstan, Tajikistan, and Uzbekistan signed the Declaration of the Shanghai Cooperation Organisation in Shanghai. This Declaration formalised and strengthened cooperative ties among the nations, expanding from the original 'Shanghai Five' grouping (minus Uzbekistan) organised after border demilitarisation pacts were signed in 1996. The Shanghai Cooperation Organisation (SCO) targeted mutual security issues, primarily terrorism and separatism, and in 2004 SCO established the Regional Anti-Terrorism Structure (RATS). Joint military exercises have been held since 2003, and in late 2007, SCO signed an agreement with the Collective Security Treaty Organisation (CSTO), comprised of the Central Asian SCO member countries and Russia, along with Armenia and Belarus, in order to expand cooperation on security and crime.

Since the establishment of the SCO, Mongolia, Pakistan, India and Iran have joined as observers, with Mongolia, Pakistan and Iran applying for full membership. The United States applied for observer status in 2005, but the application was rejected. In that same year, the SCO called for a timetable of withdrawal of US troops from Central Asia, followed shortly by Uzbekistan's dispelling of the USA from one of its air bases. Turkmenistan remains neither a member nor an observer, following the policy put forth by former President Niyazov. Although his successor, President Berdymukhamedov attended the 2007 summit, and Turkmenistan appears to be overturning some of Niyazov's neutrality policies, in all likelihood it will remain outside the SCO structure until the organisation can reach a consensus on how to respond to current requests for expansion.

At the Russian summit in 2005, the SCO announced that it would give priority to joint energy projects, including oil and gas exploration and development. On this foundation, President Putin of Russia in 2006 called for the establishment of an 'Energy Club' as the framework for cooperation and economic development. Although the 'Club' was not formally established in 2007, the SCO did agree to create a unified energy market that would 'bring energy resources from member countries rich in oil and natural gas to those that need such resources to promote their development.'[15] Iran in turn pledged to organise a meeting of the SCO oil and gas ministers. Russia also sees an 'energy club' as a conduit for participation by non-members such as Turkmenistan.

Although China strongly supports the role of the SCO in structuring energy relationships in Central Asia, its interests are not likely to be fully shared by Russia. Indeed, in the case of Turkmen natural gas, or Kazakh petroleum, Russia would be competing directly with China to get these resources flowing west instead of east. China can use the SCO structure to assist in solving the multi-country transit issues for pipeline natural gas. Russia, though, to date has preferred to deal with such issues on a bilateral basis, and may see the SCO energy club as a way to offset China's growing influence.

China's 2020 Plan

China's long-term plan for natural gas development was laid out in the *National Energy Policy and Strategy* report from the State Council's Development Research Centre in 2004.[16] This report strongly promoted the use of natural gas as a clean alternative to coal, primarily in the power sector and in residential use. Although it avoided quantitative targets, it reiterated an earlier goal set for the 10th Five-Year Plan (2001–2005) of raising natural gas to 10 percent of the energy mix in 2020, based on expectations of national economic growth. These growth expectations included a quadrupling of GDP between 2000 and 2020 with only a doubling of energy consumption, essentially, replicating China's success in the period from 1980 to 2000. In 2000, China's energy consumption reached 1.39 billion tonnes of coal equivalent (tce); a doubling of this by 2020 would raise demand to 2.78 billion tce, of which 278 million tce would thus be targeted as natural gas. Using China's conversion conventions, this is equivalent to 209 Bcm of natural gas.

Since 2000, however, China's explosive economic and energy consumption growth has made this proportionate penetration of the future consumption mix highly unlikely. Not only did energy growth exceed the rate of economic growth through 2005, but also total energy consumption in 2006 had already reached nearly 2.4 billion tce, only 14 percent less than the original 2020 goal.

Nonetheless, as seen in Table 2.13, recent demand forecasts, particularly those from China, remain centred around the 200 Bcm figure, in part because of the government's hope to maintain import dependency for natural gas at no more than 50 percent in 2020. CNPC, the largest producer of natural gas in China, expects production will reach a maximum of 120 Bcm in 2020.

Table 2.13: Supply and Demand Projections to 2020. Billion cubic metres

Source	*Date of Forecast*	*Supply*	*Demand*	*Gap*
US Energy Information Administration	2006	98	133	(35)
Institute for Energy Economics, Japan	2005	138	180	(42)
IEA	2005	-	106	
China Engineering Institute	2006	80	200	(120)
China National Development and Reform Commission (NDRC)	2005	150	250	(100)
CNPC	2004	120	160-210	(40-90)
CNOOC	2006	120	200	(80)

In all cases, China's supply gap in 2020 will range from 35 Bcm – assuming low rates of growth in demand – to as high as 120 Bcm – assuming a slowdown in the rate of production growth. As noted earlier, it is likely that China will have a total of 11–12 Bcm of supply available from imports by 2010; increasing this figure to 80–100 Bcm will require dramatic increases in either LNG terminal construction, breakthroughs in international pipeline projects, or, most likely, both.

Supply Developments

Other LNG Projects

After an initial flush of over ten proposed LNG projects in the early 2000s, the rush to LNG has faded as higher LNG prices, rising terminal construction costs, duplicative investment proposals, and the prospects of higher domestic gas availability from new finds in Sichuan led to postponement or cancellation of some projects. The leading projects in the near-term – having received approval for a full feasibility study – are the Rudong LNG terminal in Jiangsu, Qingdao LNG terminal in Shandong, and the Zhuhai terminal in Guangdong.

All three oil companies – CNPC, Sinopec, and CNOOC – competed for the opportunity to build an LNG terminal in the booming province of Jiangsu, north of Shanghai. Sinopec proposed a project at Lianyungang, CNPC at Rudong, and CNOOC at Yancheng (Binhai). In early 2007, CNPC's feasibility study was approved by the NDRC, and the company plans a ¥7.4 billion (US$950 million) investment in a 3.5 million tonne terminal, holding 55 percent ownership, with Pacific Oil holding 35 percent and Jiangsu Guoxin Investment holding the remaining 10 percent. The terminal would feed a 2.4 GW power plant and connect to CNPC's West–East Pipeline, a strong point of the project in the eyes of the government. Supply has not been determined, although CNPC finalised a Memorandum of Understanding (MOU) with Iran on the development of a block in Iran's South Pars natural gasfield. CNPC also has approvals for feasibility studies for terminals in Tangshan, near Tianjin, and in Dalian, Liaoning province, and a proposal for a terminal in Guangxi province.

Sinopec is a relative latecomer to the LNG terminal business. Its leading project is the proposed Qingdao LNG terminal in Shandong province, approved by the government for a feasibility study in 2004 and slated for an initial 3 million tonne capacity, along with an associated 1.5 GW power station to be built by Huaneng. In early 2007, however, Sinopec delayed the project as negotiations with Iran to buy 10 million tonnes a year of LNG and for Sinopec's involvement with the Yadaveran oilfield ended on disagreements over LNG prices, and focus shifted to the development of the large Puguang gasfield in Sichuan. Sinopec, however, has continued discussions with Iran over a refinery upgrading project to increase Iranian gasoline production, and has held talks with Russia over potential exports from both the Sakhalin 1 and 2 projects.

Sinopec is also competing with CNOOC in Guangdong with a proposed 2–3 million tonne LNG terminal to be built near Macao, with whom Sinopec has already signed a gas supply contract. In 2006, the NDRC announced that they would approve only one LNG project per province, and two in Guangdong. This leaves only one of the two Guangdong proposals viable in the short term, either Sinopec's Macao project or the CNOOC Zhuhai project. This restriction also puts Sinopec's Tianjin project up against CNPC's Tangshan, CNPC's Dalian project against CNOOC's Yingkou terminal, and has effectively ended Sinopec and CNOOC's near-term interests in Jiangsu.

Based on the current state of LNG projects, by 2010 China is likely to have about 9–10 million tonnes of LNG import capacity, providing another 11.5–13 Bcm of gas to China's national supply. In 2010, the government predicts that natural gas demand will reach 100 Bcm, up 14 percent per year on average from 55 Bcm in 2006, and expects that domestic production will increase by at least 11 percent per year on average over this period to 80 Bcm in 2010, leaving a gap of 20 Bcm of supply.[17] LNG alone does not appear to be able to fill the gap between demand and supply. This suggests that for the remainder of the decade, China's domestic gas market – even with potential higher 2010 production if Puguang gasfield production meets expectations – will remain primarily a seller's market, with probable negative impacts on the expansion of gas use in the power sector, and on the viability of China's gas-based fertiliser industry in the absence of government support.

Pipeline Projects

Infrastructure expansion is key to expansion of the domestic market for gas. Currently, over sixty cities are served with natural gas supply, and 210 more are targeted to be linked up by 2010. By mid-century, China forecasts that 65 percent of its cities will be served by a natural gas network.

Just two years after it was completed in 2004, the West–East Pipeline underwent modification with new compressor stations to raise the transmission capacity of the line from the original 12 Bcm to 17 Bcm. The expansion was completed in mid-2007 and will provide additional gas to the booming Shanghai market, where demand in 2007 is expected to reach 2.7 Bcm compared to the contracted volume of 2.2 Bcm. It will also support new direct-sale supplies from CNPC to large industrial users such as Bao Steel in Shanghai, for whom CNPC is constructing a spur supply line.

To satisfy this growing demand in the Lower Yangtze region, Sinopec began construction in mid-2007 on a 1675 km, 12 Bcm pipeline from the Puguang field in northeast Sichuan to Shanghai. The ¥65.4 billion (US$8.4 billion) project will provide gas to Chongqing, Hubei, Anhui, Jiangxi, Zhejiang, Jiangsu and Shanghai. Originally, Sinopec had proposed building the pipeline from Puguang to Jinan in Shandong province, one of the company's stronghold markets, but opposition from CNPC, operator of the two Shaan-Jing pipelines serving Beijing and north China and a direct competitor of Sinopec in the Shandong market, led NDRC to overrule Sinopec and redirect the pipeline to the Lower Yangtze valley.

In order to provide additional gas to the strong southern market, China is planning to build a second national gas pipeline, now slated for construction between 2008 and 2010. This pipeline would stretch from China's Northwest Xinjiang region to Guangdong 6500 km to the south east. With a planned capacity of 30 Bcm, this would be China's largest gas pipeline. Resource availability in Xinjiang, however, is insufficient to supply both the West–East Pipeline and the proposed new line, so CNPC is looking to Russia, Kazahkstan and Turkmenistan for the additional supply.

International Pipeline Supply

China has held numerous discussions with neighbouring countries on the prospects of importing additional supplies of gas to supplement domestic production. These discussions have been ongoing for more than a decade, but with the take-off of the domestic market in the early 2000s, the imperative to agree on one or more projects for the period after 2010 is becoming stronger.

Throughout the 1990s, particularly after the collapse of the Soviet Union, numerous proposals were put forth to build natural gas pipelines to China, including ones from Turkmenistan to China and onward to Japan, from Kazakhstan to Xinjiang to connect to the domestic pipeline system, west Siberia to Xinjiang to connect to the domestic pipeline system, Irkutsk to north or northeast China; and from Sakha and Sakhalin to northeast China, with or without Korea and Japanese extensions. Each proposal has different backers and promoters, but none progressed to a serious stage of feasibility assessment and planning (see Chapter 6).

China initiated many of the discussions that lie behind the proposals. As early as 1992, CNPC's chairman met with the Turkmen president to discuss the Turkmenistan pipeline scheme, and in 1994, CNPC signed a memorandum of understanding with the Russian Ministry of Fuel and Energy on east Siberian oil and gas exploration, including a pipeline to China.[18] At the time, CNPC, facing China's move from an oil-surplus to an oil-deficit market, was more focused on prospects for increasing and diversifying China's oil imports, particularly after 1994 when domestic oil price reform created a huge revenue surplus for the company. A substantial amount of these extra revenues was invested overseas in oilfields in Kazakhstan, Venezuela, Peru, Sudan and other countries, the first steps to secure overseas production assets to serve the

Chinese import market. At the same time, CNPC began focusing on developing potentially large oil reserves in the Tarim Basin of Xinjiang, and began planning a new oil pipeline that would run from Xinjiang to the major refining centres in the east.

Impetus for serious discussion of transnational pipelines came only after the settlement of outstanding political issues with potential suppliers. Russian relations were normalised with the signing of the 'Joint Statement on the Foundation of Mutual Relations between the People's Republic of China and the Russian Federation' between Presidents Yeltsin and Yang Shangkun in 1992. Border issues were largely resolved with the signing of 'The Agreement between Government of the PRC and Government of the Russian Federation on the Administration System of the Sino-Russian Border' between Premiers Chernomyrdin and Li Peng in 1994, followed shortly thereafter by a state visit by President Jiang Zemin to Russia, when additional protocols governing the border and a declaration of no nuclear first-use were signed.

On Yeltsin's return visit to China in 1996, China and Russia stopped short of signing a treaty of partnership, instead issuing a 'Joint Statement' establishing a 'strategic partnership of mutual coordination'. Later the same year, the heads of state of China, Russia, Kazakhstan, Kyrgyzstan and Tajikistan met to sign the 'Agreement on Confidence-Building in the Military Field in Border Areas of the PRC, the Russian Federation, Republic of Kazakhstan, the Kyrgyz Republic and Republic of Tajikistan'. This was a vital agreement for China (and established this grouping as the 'Shanghai Five'), as its resource-rich western regions are home to a number of Turkish-speaking minorities with cross-border relationships in the Central Asian republics. With separatist sentiments simmering in the region for a number of years, this agreement provided China with assurances that it would remain a domestic matter and that assistance or support from sympathetic kindred in Kazakhstan, Kyrgyzstan or Tajikistan would not be permitted. A second related agreement on border force reductions was signed by these countries in 1997.

After signing a joint declaration supporting the development of a multi-polar international system, indicating independent action from the pressures of other large powers, China and Russia signed numerous new statement and memoranda of cooperation in late 1997, including the 'Memorandum of Understanding between the Government of the PRC and the Government of the Russian Federation on the Basic Principles of the Projects to Lay a Natural Gas Pipeline from Eastern Siberia Area of Russia to China and to Exploit Russian Condensate Gasfield.' This agreement focused on the development of the huge

Kovyktinskoye field in Irkutsk with an estimated 1 Tcm in reserves and the potential for exporting up to 30 Bcm of gas to north or northeast China and Korea.

In 1999, Chinese Premier Zhu Rongji visited Moscow to sign a number of agreements to promote trade expansion. In the energy field, agreements included joint exploration of natural gas reserves in Kovyktino, Irkutsk, specifying future delivery of 20 Bcm of gas annually over thirty years (half for China and half for South Korea), construction of a gas pipeline from western Siberia to Shanghai by Russian Gazprom and CNPC, construction of an oil pipeline from Angarsk, Irkutsk to north China by Yukos and other Russian oil-transport companies, the Russian sale of two 1000 MW nuclear reactors for the Lianyungang nuclear power station in Jiangsu province, and the creation of an electricity export route from Siberia to China with export of 15–18 TWh of electricity annually.

Nine years of successive agreements, declarations, and joint statements bolstering bilateral ties, economic and military cooperation, transportation, communications, political ties, and international relations provided the foundation for the landmark 'Treaty for Good Neighbourliness, Friendship and Cooperation' signed by Presidents Putin and Jiang Zemin in mid-2001. This treaty, the first between the neighbours since Mao and Stalin's agreement in 1950, focused on five main areas, including opposition to US hegemonism, demarcation of the border, arms sales and technology transfer, energy and raw material supply, and the rise of Islamic militancy in Central Asia. The latter point extended an agreement that China and Russia had signed earlier that same year with the leaders of Kazakhstan, Kyrgyzstan, Tajikistan, and Uzbekistan establishing the Shanghai Cooperation Organisation to prevent the rise of Islamic militancy and promote economic cooperation.

These agreements and treaties provide China with a structured foundation for pursuing its interests in Russia and the Central Asian republics, in energy and other spheres. The magnitude of investment needed for realising proposed natural gas projects in these countries requires the minimisation of political risk, formalisation and routinisation of bilateral and multilateral ties, and a range of specific procedures and agreements on taxation, investment, resource development, trade, and other related areas. With these structures in place, China was able to move forward in considering the range of options available to supplement its domestic energy supply in consideration of its own interests in energy security, import diversification and energy diversification.

Energy relations with Russia are of particular importance to China in recognition of the potential large supplies of oil and natural gas available to its north. Russia, in turn, has focused more seriously on the potential of the Chinese market, with Gazprom moving quickly from asserting the westward orientation of its Siberian gas reserves to proposing a pipeline connection of its west Siberian reserves to China, stressing the existing infrastructure in Russia that would permit such a project to go forward at less cost than alternative schemes. Indeed, in early 2006, Presidents Putin and Hu signed an agreement at a summit in Beijing that Russia would build two natural gas pipelines to China within the following decade providing up to 80 Bcm of supply. At the same meeting, Gazprom signed an MOU with CNPC to undertake feasibility studies for two gas pipeline routes. Gazprom's leading proposal is a 30 Bcm pipeline to connect western Siberian gasfields from Nadym-Pur-Tazov in the northern part of west Siberia to Xinjiang, transversing the Russian Altai Republic, where surveys of possible routings have already begun. This routing would involve 2800 km in total of pipeline on the Russian side, and several hundred kilometres of pipeline in Xinjiang to connect with the West–East and second national pipelines. Although Gazprom's resource sufficiency is unclear for this project, entry into the Chinese market – particularly after having withdrawn from the West–East Pipeline project – may be a preferential alternative to selling the gas to Europe. Talks between Gazprom and CNPC have floundered, however, over the issue of price, with the Chinese continuing to insist on a cost-plus basis of around \$3.2/MMBtu and the Russians offering around \$6/MMBtu. Gazprom, nonetheless, announced in late 2007 that pipeline design and construction would begin in 2008 with the resolution of routing issues around a nature preserve in the Altai Republic, although no breakthroughs had been achieved on the pricing issue (see Chapter 6).

One of the eastern routes under consideration would bring Russian gas from the Sakhalin-1 project to Heilongjiang in northeastern China. Although Exxon, 30 percent stakeholder in Sakhalin-1, and CNPC signed a preliminary sales agreement for 8 Bcm per year in 2005, pricing was never settled. In addition, uncertainty over the ability of Sakhalin-1 to undertake exports has been called into question by Gazprom's assertion of full export rights for gas in the region, and moves to reprioritise it for domestic consumption instead of export. If exports to China are eventually approved, a pipeline would most likely follow the route of the existing Sakhalin–Khabarovsk pipeline, which opened in 2006.

One of the earliest pipeline proposals from Russia involved the sale of 20 Bcm of gas annually to China and 10 Bcm to Korea from the Kovykta field operated by TNK-BP in Irkutsk, following along a 4890 km pipeline linking Shenyang and Beijing in China through to an undersea pipeline to Korea. After eight years of study, Rusia Petroleum – the licence holder of the field – CNPC, and Korea Gas agreed to an international feasibility study for the pipeline in 2003. Early negotiations with China hit an impasse over its insistence on an alternate route from the field to northeastern China, by-passing land transit through Mongolia to Northern China, but significantly increasing the distance and cost. Although routing issues were not settled by the time that Presidents Putin and Hu met in Beijing in 2006 to sign the gas export framework, TNK-BP officials at the time were confident that any eastern route under consideration by Gazprom would include supply from the Kovykta field.

Gazprom's assertion of its monopoly on exports from Russia and refusal to allow pipeline access for production from Kovykta, prolonged the delay in moving the project forward. By 2007, following a trend of increased assertion of national interest in foreign-invested projects, the Russian Ministry of Industry and Energy threatened to withdraw TNK-BP's licence to develop Kovykta because of its failure to develop gas at a rate it had originally contracted. After months of negotiation with Gazprom, TNK-BP sold its 63 percent holdings in Rusia Petroleum to Gazprom in an agreement to set up an international investment group and an option for BP to repurchase 25 percent of the shares back by the end of the year. Although the inclusion of Gazprom into the Koyvkta project could accelerate progress on exports to China, the field is also a major source of helium, and the government has insisted on an integrated production plan that would maximise the value of the helium resources as well.

China has held discussions with Turkmenistan on possible exports to China since 1992, but agreement on how to proceed came only with a framework agreement in 2005, signed in Beijing by Presidents Nizayov and Hu. The second article of the agreement detailed the sale of 30 Bcm of natural gas annually for thirty years to China, effective on the date of the completion of a pipeline to the Chinese border. The gas for the project is in part to be sourced from new Turkmen gasfields in the Bagtyyarlyk region on the east bank of the Amu Darya river, containing an estimated 1.3 Tcm of reserves. In 2007, as a sign of increased cooperation with China, Turkmenistan granted China's CNPC a licence for exploration and production in this region, the first onshore rights granted to a foreign company. Presidents Hu and Berdymukhamedov further signed a thirty-year supply contract during Berdymukhamedov's visit to Beijing. Turkmenistan's support for this project was apparent in their announcement shortly after the signing that they had broken ground on the Turkmen section of the pipeline, in an inaugural ceremony attended by the president. Only 188 km of the proposed pipeline are in Turkmenistan, and it is unclear if full transit rights, tariffs and other issues related to the pipeline have been fully resolved with neighbouring Uzbekistan and Kazakhstan. In total, the pipeline would extend 2000 km to the Chinese border, to link with the second West–East Pipeline.

Turkmenistan's neighbour Uzbekistan, also vying to diversify away from Russian market dependency, signed an agreement with China in early 2007 for the sale of 30 Bcm of gas annually through a 530 km pipeline to China. While Uzbekistan noted that the China market would allow it to get a higher price for its gas than it is paid by Gazprom ($2.85/MMBtu), this contrasts with the reported negotiations on a promised $2.30/MMBtu price from Turkmenistan, lower than Gazprom's 2007 price.

In parallel with the Turkmen and Uzbek proposals, Kazakhstan is also considering a pipeline to Western China as an alternative to Kazakhstan's current complete dependence on exports to Russia. Discussions between CNPC and KazMunaiGas continued throughout the early 2000s, with an agreement to undertake a preliminary feasibility study in 2005, for completion in 2007. Two routes are under consideration. The northern line would run parallel to the existing 400,000 b/d oil export pipeline to China and the southern line through the more highly populated consumption areas in the southern part of the country. In total, the pipeline would be about 3000 km long, supplied by gas produced in the Karachaganak, Tengiz and Kashagan fields. Initial supply would be set at 10 Bcm per year, rising later to 30 Bcm as Kazakh production increased.

After first being proposed in the 1990s as transit routes from Central Asia to China and onward to Japan and Korea, the current pipeline proposals no longer include consideration of such transnational links beyond China. The rapid growth of China's domestic market, the growing gap between domestic supply and demand, and the long lead-times in finalising such a large-scale project have created a sufficient basis for domestic use of imported gas. Although China is interested in the lowest possible price for its imports, the low export price paid by Gazprom for Central Asian gas provides China with some negotiating flexibility, and provides the Central Asia suppliers with an alternative market to Russia.

In 2007, China trumped India and Korea in the award of the gas offtake agreement for the Shwe project in the Bay of Bengal in Myanmar (see Chapter 7). With Korea and India as the main operators of the fields, India had lobbied hard to get the award for supply of the gas to West Bengal, and Korea had proposed development of an LNG facility. Although India has moved on several diplomatic fronts to improve trade and security relations with Myanmar on its eastern borders, China's influence in Myanmar, including the issuance of soft loans, the construction of a power plant, and continued support for the regime in the UN Security Council against sanctions, swung the government's decision. China has planned to build a pipeline north to the southwestern province of Yunnan, where the gas supply could be integrated into the network connecting Sichuan, Yunnan, and Guangxi.

Filling the Supply Gap

Taken together, the projects described above would be sufficient to cover China's supply gap by 2020, but only a selected handful of the projects are likely to come to fruition during this time period. On the LNG side, total import capacity by 2020 – taking the major LNG terminals now in consideration, and assuming continued restriction on one terminal per province, except for Guangdong – could reach 35 million tonnes, or as high as 52 million tonnes if all proposed second phase expansions are implemented, and appropriate gas supply contracts can be secured (Table 2.14). This would provide 46–68 Bcm of total import supply, leaving a 12–54 Bcm shortfall on an import need of 80–100 Bcm.

LNG alone cannot provide China with all its import needs by 2020, nor can it serve to supply inland cities that are rapidly converting to gas based on initial supplies from the West–East and Sichuan/Chongqing pipeline networks. One or more international pipelines will be necessary.

Four of the proposed pipeline schemes (W. Siberia, Kazahkstan, Turkmenistan, Uzbekistan) enter China by the far west region of Xinjiang, far from the major consuming markets to the east. With only a 17 Bcm pipeline connection between Xinjiang and the east, now currently supplied by Xinjiang local gas production, and with a single 30 Bcm second national line mooted for the next five years or so, it appears that only one of the four schemes is feasible. The Uzbek pipeline is the shortest, but this project is only at initial stages of discussion. The Turkmen pipeline project is the most complex, relying on transit through two other countries, including Uzbekistan and Kazakhstan before reaching China, and a pricing agreement has yet to be reached despite top-level agreement on long-term supply. Nor is it clear that future Turkmen production is sufficient to supply

Table 2.14: Summary of Pre- and Post-2010 Supply Proposals

	By 2010	*After 2010 (initial capacities)*	*After 2010 (Phase II capacities)*
LNG			
Dapeng	3.7	5.7	5.7
Fujian	2.6	5.0	5.0
Shanghai	3.0	6.0	6.0
Zhejiang		3.0	6.0
Rudong		3.5	6.0
Zhuhai		3.0	6.0
Qingdao		3.0	5.0
Tangshan		3.0	6.0
Dalian		3.0	6.0
Total (Mtpa)	9.3	35.2	51.7
Total (Bcm)	12.2	46.3	68.0
Pipelines			
W. Siberia/Altai		30	
Sakhalin		10	
Kovykta		20	
Kazakhstan		10	
Turkmenistan		30	
Uzbekistan		30	
Myanmar		6–10	
Total (Bcm)		136–140	

Sources: Various press and company reports

all the commitments that Turkmenistan has proposed in the last few years. In the near term, Kazakh production would limit potential exports to about 10 Bcm, making this a much higher cost option. It is possible for both Turkmenistan and Kazakhstan to export to China in a single 30–40 Bcm pipeline, if agreement could be reached on volume allocations and transport fees. The Russian scheme backed by Gazprom to bring west Siberian gas to Xinjiang appears to be the most advanced in study and planning, but numerous hurdles, including new gas reserve development in western Siberia, geological, cost, and environmental challenges in bringing a pipeline from the north through the Altai mountains to China, and agreement on pricing and volume terms remain to be settled. China, however, appears to be placing more emphasis on developing its relationship with its Central Asia neighbours, particularly as an offset to Western interests in the region and, indeed, with Russian interests in the region. The limits on transport capacity from western China restricting the options to only one in the near term, the Central Asia option appears to be most likely.

In the east, new supply from either Kovykta or Sakhalin would require new pipeline construction in China, as the north and northeast remain without substantial long-distance gas pipeline capacity (except for the Shaan-Jing pipelines supplied from the Ordos gasfields). In northeast China, where the Daqing oilfield is increasingly concentrating on the production of natural gas to offset the long-term decline in oil production, construction began in 2007 to move gas south – in the first stage to Harbin – as part of the longer-term policy of 'Northern Gas to the South'. This could form the initial stage of a larger system supporting the import of 10 Bcm or more of gas from Sakhalin. Kovykta gas imports, however, could either connect to the nascent northeast pipeline system, if China is successful in rerouting it around Mongolia, or it could come directly south to north China, connecting with the 12 Bcm Shaan-Jing II pipeline, which is now linked to the West–East Pipeline system through the interconnector completed in 2005.

China and Russia have already removed many obstacles to deepening their commercial ties, particularly in the area of resource development and energy trade. The multiple bilateral agreements of the 1990s, the Friendship Treaty of 2001, and the framework agreement of 2006 elucidated the areas of concurrence in China and Russia's strategic, political and commercial interests, and both sides recognise the long-term benefit of closer trade and other ties. China would benefit from secure new supplies of natural gas (and oil) that would offset growing dependence on the Middle East, and from a closer military relationship that would offset the dominance of the United States and Japan in the Asia-Pacific region. Russia benefits from increased access to the Chinese market and an outlet for otherwise stranded natural gas resources, and from a stable Eastern relationship to balance its complex relationship with the EU. Russia's – in particular Gazprom's – exports to China would bring with it know-how, resources, and a long stable history of exports to other major markets. Both sides, however, face the challenge of financing the proposed large export projects and delivering the product to China's markets at a competitive price. This may require the participation of yet other major oil companies to share the risk and expand financing options, but it remains uncertain whether any of the projects is viable on a stand-alone basis without government participation to reduce market and financial risks.

By 2020, an 80–100 Bcm supply deficit could likely be filled by either two pipeline projects – 30 Bcm to the west and 20 Bcm to the east– along with 30–50 Bcm of LNG imports (23–38 million tonnes), or by a single 30 Bcm pipeline import project, supplemented by a stronger focus on expanding LNG imports to at least 40 million tonnes. The latter, however faces the larger challenge of competition from other Asia-Pacific region importers, including Japan, Korea, and Taiwan, and the United States. With natural gas output having passed peak in North America, California in particular is looking to one or more LNG terminals in the next decade to sustain its power and residential sectors. China's coastal consumers, however, are among the wealthiest in China, so the ability to pay in a tight LNG market may allow China to bid successfully for new LNG supplies. The lack of reform and rationalisation of the domestic pricing system has inhibited the willingness of China's oil and gas companies to commit to long-term high-cost imports, but as the higher price level of mid-decade for LNG appears increasingly to be a long-term structural change and not a short-term cycle, this inhibition is declining. PetroChina's 15–20 year agreement to buy LNG from Australia's Western Australia and Gorgon fields reflects such a trend (see Chapter 7).

Consumption Structure

Assuming that China reaches its target of about 200 Bcm of demand in 2020 and implements the range of current policies on urbanisation, coal substitution in the residential sector (including district heating), power generation expansion, alternative fuels in transport, and continues priority production of gas-based fertilisers domestically, the structure of demand may develop as shown in Table 2.15 and Figure 2.1. Concerned about the pollution impact in urban areas of a heavy reliance (over 100 million tonnes) on coal for residential use, China has planned to increase the number of cities using gas, with a target of 65 percent urban penetration by mid-century. By 2020, over 200 million

Table 2.15: Structure of Gas Demand to 2020. Bcm

	2000	*2005*	*2010*	*2015*	*2020*
Industry (including feedstocks)	17	29	42	51	63
Power Generation	2	3	13	25	37
Residential	3	8	26	37	52
Heat Supply	1	2	7	10	15
Transport	0	1	3	6	11
Other	1	3	10	15	22
Total	25	46	100	144	200

Source: Adapated from David Fridley et al, China Energy to 2020, Berkeley, CA: LBNL, October 2006.

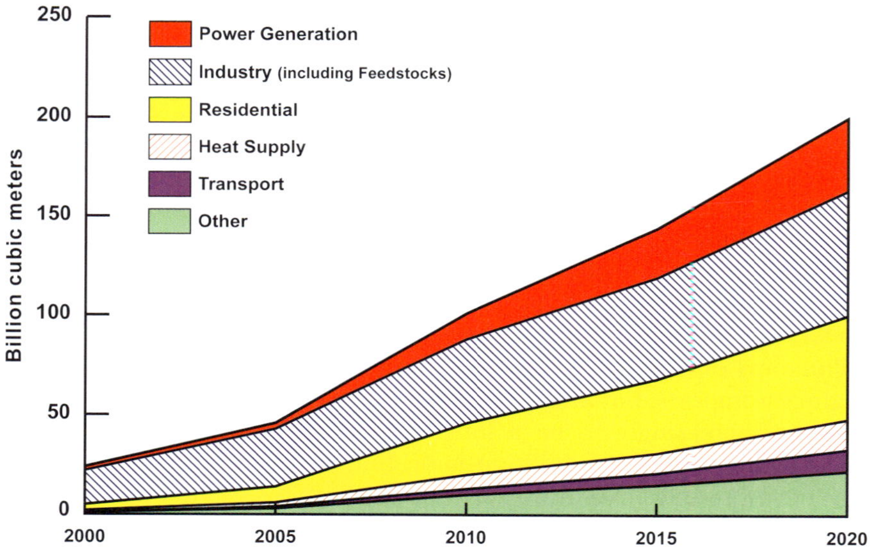

Figure 2.1: Structure of Gas Demand to 2020

people in 60 million new households are expected to join the urban population, as the urbanisation rate increases to about 55 percent. In the north, where district heating in winter is widespread, conversion of coal-fired boilers to natural gas is also proceeding apace. This use, however, is highly seasonal, and adds to the challenge of managing annual peak loads.

The outlook for industrial and power generation consumption presents several challenges. In the industrial sector, feedstock use trends will have a marked impact on the future direction of total industrial use, as currently about 30 percent of the gas used in the sector are feedstocks, particularly agricultural fertilisers and petrochemicals. The gas price hike of December 2005 had an immediate impact on the fertiliser producers, who historically have received very low prices in order to keep the price of fertilisers affordable. Chinese farmers, on average, use twice as much nitrogenous fertiliser per hectare as do their US counterparts, and future yield growth will depend mainly on the increase in fertiliser consumption. Although natural gas (including coal-bed methane production) is a favoured fuel for fertiliser production, China has also installed a number of coal gasification units to produce feedstock. If the rising price of natural gas creates economic difficulties for the fertiliser producers, the trend towards coal may accelerate and dampen gas demand growth. However, alternate subsidy approaches could be taken to maintain a reasonable price of fertilisers to farmers, sustaining gas use in this sector. Similarly, China is now planning development of some large coal-to-methanol plants to substitute for natural gas in petrochemical production, and the success of these pilots over the next few years will likely shape the role of gas as petrochemical feedstock to 2020. For under-the-boiler use of gas, industrial demand growth will also depend on local availability of gas and price differentials between gas and coal, which are subject to a range of potential policies, including those on climate change or carbon taxation.

China's nominal target for gas-fired power generation in 2020 is 70 GW. This represents about a doubling over current existing and planned power generation capacity, but would account for only about 5–6 percent of China's total installed capacity by 2020. Again, policies that change the relative cost of coal and natural gas would probably have the greatest impact on the penetration of gas into the generation sector, particularly in northern China. The full 70 GW of capacity could be entirely accommodated in southern China, but with an estimated gas demand of about 37 Bcm, southern supply would be insufficient until well into the next decade after additional LNG terminals are built or the second national pipeline reaches the south. If gas-fired power generation capacity growth falters, suppliers are likely to target the higher-priced residential sector.

Increasing scepticism and opposition has been voiced concerning the development of natural gas-based power generation. Citing China's low per capita natural gas reserve base, the large reserves of lower-cost coal, the risk of high gas import dependency and the high costs of LNG and pipeline imports, critics suggest that China's natural gas reserves are best directed to the production of chemical fertiliser for the vast agricultural sector and for residential use, while the power sector should continue to rely on low-cost abundant coal.[19]

The challenges to developing a national market of 200 Bcm in 2020 are substantial, but not insurmountable. Other energy sectors have undergone much larger transformations and expansion. Just in the six years since 2000, China's coal industry has more than doubled production from 1 billion to 2.3 billion tonnes, and capacity is expected to reach 3 billion tonnes by 2010. Similarly, refining capacity more than doubled between 1990 and 2006. An expansion of the natural gas market is similarly feasible, at least on a technical and infrastructure basis, but a schlerotic pricing system unreflective of a quickly changing natural gas market remains a serious challenge. As was the case with both coal and petroleum, rapidly rising demand and increasing import dependency put enormous pressure on the state-regulated pricing scheme, to the point that coal prices have now essentially been deregulated, and oil pricing controls have been narrowed to just diesel fuel and gasoline. Reforms in pricing in each case have been for the benefit of the producers, and similar changes can be expected in the natural gas sector as well.

Given this national commitment to natural gas development and the multiple national policy goals it serves – economic development of the West, diversification of energy supply, reducing emissions, strengthening national energy companies, and technology transfer, among others – it is reasonable to imagine this political commitment translating into the wide array of reforms and policy development necessary not only to expand the domestic market with domestically produced gas, but also to lay the foundation for a move into large-scale pipeline projects with China's neighbours.

Limits and Implications

Since 2000, China has been promoting a planning goal of raising the proportion of cleaner natural gas in its energy mix to 10 percent by 2020. As with most planning goals, the target is not likely to be met with exactitude; it is better understood as an indication of the level of government priority for further expansion of the natural gas market. These priorities in turn guide the development of market, investment, pricing, international relations, and other policies that underpin growth in exploration, production, imports, and market development. Since 2000, natural gas production and use have expanded tremendously, having more than doubled. Transport and distribution infrastructure has grown nation-wide, including the construction of a cross-country natural gas trunk line, and LNG imports have begun. Despite early challenges of high prices and expansion of the consumer base, China now sees demand outstripping supply, and expects the tightness in the market to extend beyond this decade. The future of natural gas, then, is likely to be driven more by supply issues than the state of domestic demand and infrastructure.

At the same time, however, consumption of all forms of energy has been growing rapidly since 2000, so the proportion of natural gas in the primary energy mix has barely risen, from 2 percent in 2000 to just 3 percent in 2005 (Table 2.16). In terms of final energy, growth in electricity consumption – at 13.1 percent per year between 2000 and 2005 – is second only to that of natural gas, at 13.8 percent per year over this period, but electricity demand is expanding from a much higher base. Over 90 percent of China's thermal electricity generation is coal-based, and this steady high growth in electricity consumption has been a major driver in the surging demand for coal. This unexpectedly high growth rate of coal use has slowed the rise in the share of natural gas in the total.

Looking out to 2020, a key uncertainty in determining the degree to which a successful natural gas expansion programme could expand its role in the total primary energy mix is the outlook for coal demand; other energy forms are unlikely to grow substantially beyond what is projected in Table 2.16 because of a range of other constraints, including resource availability and construction timeline requirements. The scenarios in Table 2.16 then, are less forecasts of demand growth than an assessment of the outlook for each energy form under the circumstances that would permit China to achieve a 200 Bcm level of natural gas consumption, including no or limited constraints on resource availability, a growing domestic economy, and a stable international and national political climate under which long-term decisions and investments can be made with some confidence.

Table 2.16: Scenarios of China's Energy Mix to 2020 (million tonnes of coal equivalent)

	2000	*2005*	*2010*	*2015*	*2020*
Coal ('Controlled Growth')	943	1,548	1,751	1,835	2,028
Coal ('Industry-Led Growth')	943	1,548	1,929	2,183	2,469
Petroleum	321	465	595	754	910
Primary Electricity	97	182	208	229	352
Natural Gas	33	62	125	192	273
(Bcm)	(25)	(46)	(94)	(144)	(205)
Total ('Controlled Growth')	1,393	2,257	2,680	3,010	3,562
Total ('Industry-Led Growth')	1,393	2,257	2,857	3,357	4,004
Natural Gas as Total					
'Controlled Growth' (percent)	2	3	5	6	8
'Industry-Led Growth' (percent)	2	3	4	6	7

Source: Lawrence Berkeley National Laboratory China 2020 Study

Nuclear power (included as primary electricity here), for example, is assumed to reach the 40 GW capacity target in 2020. This development alone requires the completion of an average 2.5 GW of new capacity each year to 2020, a pace that some observers do not find credible. Similarly, the balance of primary electricity will come from hydropower expansion, and it is assumed as well that China will meet its target of 300 GW of installed capacity in 2020, nearly triple the current level.

Most of China's remaining large hydropower resources, however, is located in southwestern China, far from consumption centres, and already environmental opposition, both local and national, has slowed development. The consequence of not meeting these 2020 capacity goals is, of course, a greater reliance on coal or natural gas for power generation.

Petroleum demand is also subject to a variety of uncertainties and risks that limit further upside expansion by 2020. Although demand growth today is strong, primarily from growth in the transportation and petrochemical sectors, China has nearly reached the level of 50 percent import dependency for oil. The demand projection for 2020 of 910 million tonnes coal equivalent (or 12.7 Million b/d) would bring China to over 80 percent import dependency. With non-OPEC production likely to reach a peak before 2015,[20] and the circumstances of OPEC's long-term production capacity uncertain, combined with rapid domestic demand growth in the major world oil exporters, a continued low-price environment for sustained rapid growth in Chinese oil consumption to 2020 appears unlikely. Even at this level of projected demand, China may face challenges both on price and supply availability.

Thus, China's domestic coal reserves will be called upon to provide the bulk of its energy demand growth over this period. Two scenarios are presented here, differing primarily in the degree to which China can keep its heavy industrial growth – a main driver of direct coal and coal-based electricity demand growth – under control. Currently, industry consumes about 70 percent of China's electricity production and the majority of its direct-use coal, and the explosive growth in heavy industrial production, such as iron and steel, cement, and non-ferrous metals has not only supported China's emergence as the 'manufacturer to the world', but has also significantly raised the trajectory of its energy demand path above what was projected at the beginning of the decade. Cement production, for example, exceeded 1.2 billion tonnes in 2006 and is 54 percent higher than was projected in 2001. Similarly, glass production has exceeded projections by 82 percent, and iron and steel production by 42 percent. In nearly all major heavy industrial sectors, output in 2006 has already exceeded the level projected by the end of the 11th Five-Year Plan in 2010.[21] This runaway growth in demand for coal and coal-based electricity prompted the imposition in 2006 of the '2010 Target' to reduce the energy intensity of the economy by 20 percent by 2010, with a major focus on improving the efficiency and reducing the growth of consumption in China's top 1000 enterprises, which together account for 30 percent of the country's energy use. Further, China's planners hope that after 2010, when the current round of extensive infrastructure development is largely completed, production of both cement and iron and steel will peak and begin to fall, thus reducing the energy stresses on the economy from these two energy-intensive sectors. If such controls – both on efficiency and production levels – can be implemented successfully and maintained, then the trajectory of China's coal demand should slacken and fall from the blistering 10 percent per year average rate from 2000 to 2005, to as little as 1.5 percent per year average growth after 2010. Reductions in industrial coal and coal-based electricity demand would thus offset continued expansion of coal use in the power sector for the residential and other sectors.

If, however, such controls are not completely successful and heavy industry output continues to expand after 2010, even at a slowing rate, then the demand trajectory for coal could attain a stronger 2.5 percent average annual rate through 2020. Just this 1 percent difference in annual growth after 2010 adds about 440 million tonnes additional coal demand to be satisfied in 2020 compared to the controlled growth scenario, highlighting the government's current imperative to manage this demand growth.

Under both scenarios, however, natural gas fails to reach its planning target of 10 percent penetration of the primary energy mix; in the controlled-growth scenario, China would need to import an additional 70 Bcm of natural gas by 2020 to reach a 10 percent level of natural gas use, and under the industry-led growth scenario, an additional 100 Bcm of imports would be needed. This would require the construction of two to three additional world-scale natural gas import pipelines.

Could natural gas play a larger role beyond the 200 Bcm of consumption projected in 2020? One motivation to attempt a move in this direction is China's growing concern over global warming, particularly in light of their becoming the world's largest CO_2 emitter in 2007, primarily because of the explosive growth in coal consumption since 2000. Chinese scientists have already raised alarms over the accelerated melting of the Himalayan glaciers, source of all of China's major river systems, and the severe impact warming will have on the country's ability to feed itself over the next few decades. Natural gas is the least carbon intensive of the fossil fuels and the higher efficiency of natural gas combined-cycle power plants could offset emissions – by up to 35 percent – from coal-based power generation.

The challenges to accelerating the consumption of natural gas beyond that projected for 2020 highlight a range of issues arising from the current market development path and the limited options China has to respond to global warming in the context of its energy structure. Initially, China had promoted the use of natural gas for power generation to reduce the growth of coal use in that sector. In the first half of this decade, however, gas-fired power generation expansion remained weak while uptake in the residential and industrial sectors grew strongly. By 2005, the consumption of coal in households had decreased by over 100 million tonnes from its peak in the late 1980s, replaced largely by natural gas, coal gas, and LPG. In northern cities such as Beijing, where urban pollution is legend, district heating plants were largely converted to natural gas. Natural gas also penetrated the transportation sector, with fleets of buses and taxis rolled out in Beijing, Xi'an, Qingdao, Guangzhou, and other cities.

This type of market development shifted the focus away from high volume (and low gas priced) single consumers such as power plants to dispersed, small volume (and high gas priced) users requiring great investment in distribution infrastructure and equipment conversion. This subjected the gas grid to much greater seasonal variation than had been anticipated when the first major supply pipelines were built. Indeed, Beijing found itself unable to meet peak demand in the first few years after gas arrived until appropriate storage could be built to offset seasonality. Only natural gas from LNG terminals was directed in significant volumes to power generation, but the slowing pace of LNG construction has hampered growth in gas power generation from this source as well.

Urban pollution remains a serious problem in China – over 200 cities are considered to have moderate to severe air pollution – and although Beijing has spent billions of dollars to clean up the air in the city in time for the 2008 Olympics, other cities in China may well show the world at that time the degree to which degraded air has become the consequence of China's industrialisation and rising urban wealth and energy consumption. These persistent environmental problems could spur a concerted effort to accelerate the import of natural gas as a way to combat the rising toll from polluted air. However, even such a programme, expanding natural gas consumption beyond the 200 Bcm expected by 2020, would in all likelihood only have a noticeable environmental impact if the bulk of the gas were directed to power generation, where displacement of coal would be the highest. Indeed, pollution in cities is coming increasingly less from residential sources as coal continues to be backed out and, in the case of particulate matter, less from transportation as well. Nonetheless, even if China could achieve an additional 100 Bcm in imports by 2020, and the entire volume were dedicated to power generation, the additional capacity could replace only about 15 percent of the growth in coal-based capacity expected by 2020 and displace about 220 million tonnes of coal, or just 5–7 percent of total coal consumption projected for that year.

Such limited capacity to decrease the use of coal and diversify the energy mix in the face of the need to control emissions growth, and the growing probability that China itself will face physical limits to further expansion of coal production shortly after 2020[22] highlights the bind that confronts China's leadership in developing policies to ensure an expanding energy system delivering growth to the economy. Energy security policies, for example, underpin the need to rely largely on China's domestic coal resources and avoid high import dependency on other energy forms. Environmental policies suggest that diversification away from coal would be beneficial. Electricity pricing policies, however, are concerned with keeping prices as low and stable as possible. Such policies would need to be rationalised to achieve a substantial growth in natural gas use beyond current trajectories. Widespread gas uptake in the power sector remains hindered by gas pricing policies that put gas at a disadvantage to coal in much of the country, by electricity tariffs that give priority to lowest-priced fuels, and by the widespread lack of enforcement of policies to clean up emissions from power plants. Still today, with acid rain covering over one-third of the country, few power plants operate desulphurisation units because of the higher costs. Without internalisation of the costs of environmental pollution from coal-fired power generation, greatly expanding use of natural gas in the sector will be challenging. Policies such as a carbon tax on coal, and required retrofit of all coal plants with emission controls, reformed electricity tariffs and possibly feed-in policies for natural gas and renewable electricity would substantially raise the incentive for increased natural gas use in the sector.

China's natural gas industry has developed remarkably in the last decade and has laid the foundation for continued strong growth over the next decade. Maximising the benefits of this valuable energy form, however, will require reforms to offset the historical domination of coal in a wide range of policy thrusts. The exigencies to shift directions to cleaner fuels are strong, but the limitations are equally challenging.

Notes

1 National Bureau of Statistics, *China Statistical Yearbook 2000*, Beijing: Statistical Publishing House, 2000.

2 Li Mingzhai, Feng Sanli, Hu Aimei, Pan Jun, 'Chinese Coalbed Methane Industry Development', *Zhongguo Nengyuan (Energy in China)*, vol. 5, 2001, p. 10.

3 *China Petroleum Industry Yearbook* 1997. Beijing: Petroleum Industry Publishing House, p. 443.
4 China National Bureau of Statistics, *China Statistical Yearbook 2006*, Beijing, 2006.
5 The State Development Planning Commission was responsible for review and approval of all large investment projects; it has since been merged into the National Development and Reform Commission.
6 'China Natural Gas Vehicle Overview', International Association for Natural Gas Vehicles, *http://www.iangv.org/content/view/80/100/*
7 Interfax-China, 4 April 2007.
8 Energy Research Institute, Power Economic Research Center, *Policy Study: Gas-Fired Power Generation in China*, Beijing, March 2006.
9 Marie-Françoise Chabrelie, 'LNG: A Commodity in the Making,' *Panorama, 2006*, online at *http://www.ifp.fr/IFP/en/events/panorama/IFP-Panorama06_10-GNL-VA.pdf*, accessed June 2007.
10 Even before the creation of Sinopec, refining assets also lay with other ministries, such as those for textiles and agriculture. The creation of Sinopec and CNPC did not completely divide upstream and downstream assets. CNPC, for example, retained control of numerous oilfield refineries, and many small 'teakettle' refineries – essentially oil-processing plants rather than refineries – remained under the ownership of local governments.
11 Estimated proven reserve figures provided by Sinopec and CNPC in investor prospectuses, as of end-1999.
12 'Oil Giant Wants in on Gas Project', China Internet Information Center, online at *http://www.china.org.cn/english/BAT/18165.htm.*
13 Erica Downs, *The Brookings Foreign Policy Studies Energy Security Series: China*, Washington DC: Brookings Institution, 2006, p. 19.
14 'Meiguo daji Zhongguo jingji, xu jingti dangqian diyoujia fengxian', *Shanghai Securities News*, 23 January 2007, reposted at *http://www.wyzxsx.com/Article/Class4/200701/14569.html*
15 'Central Asia: SCO Leaders Focus on Energy, Security, Cooperation', *www.eurasianet.org/departments/insight/articles/pp081607.shtml*, 16 August 2007
16 Development Research Centre, *Research on National Energy Comprehensive Strategy and Policy of China*, Economic Science Press, 2005.
17 *2007 Energy Blue Book*, Social Sciences Press: Beijing, April 2007.
18 For details of China's contacts and discussions with potential gas supplies, see Lan and Paik, *China Natural Gas Report*, Xinhua News Agency and Royal Institute of International Affairs (London, 1998), pp. 103–13.
19 See, for example, 'Natural Gas Market Development and Existing Problems', Zhong Yan Wang, April 2007, online at *http://www.china5e.com/news/oil/200704/200704180032.html*
20 See, for example, Michael R. Rodgers, PFC Energy, 'Recent Trends in Exploration Results and the Implication for Future Petroleum Liquids Supply, October 2006, online at *http://www.aspo-usa.com/fall2006/presentations/pdf/Rodgers_M_Boston_2006.pdf*
21 Lin, Jiang, Nan Zhou, Mark Levine, David Fridley, *Achieving China's Target for*

Energy Intensity Reduction in 2010: An exploration of recent trends and possible future scenarios, Environmental Energy Technology Division, LBNL, December 2006.

22 See, for example, Tao Zaipu, Li Mingyu, 'What is the Limit of China's Coal Supply? *Energy Policy*, 35 (2007) 3145–3154, January 2007; *Asian Utilities and Coal*, UBS Investment Research, September 2007; Energy Watch Group, *Coal: Resources and Future Production*, EWG-Series No 1/2007, July 2007 (rev).

CHAPTER 3

NATURAL GAS IN INDIA

Sunjoy Joshi and Najeeb Jung

Introduction

The history of oil in India has closely followed the history of oil itself. It began with the drilling of the first oil wells in Assam as early as 1865 and the Digboi discovery in Upper Assam in 1886. Natural gas, however, has been the foundling child of the energy industry, its share in the country's energy basket being a mere 8.5 percent against a global average of 24 (Figure 3.1). While the EIA's International Energy Outlook (IEO 2006) projects that India and China will be the fastest

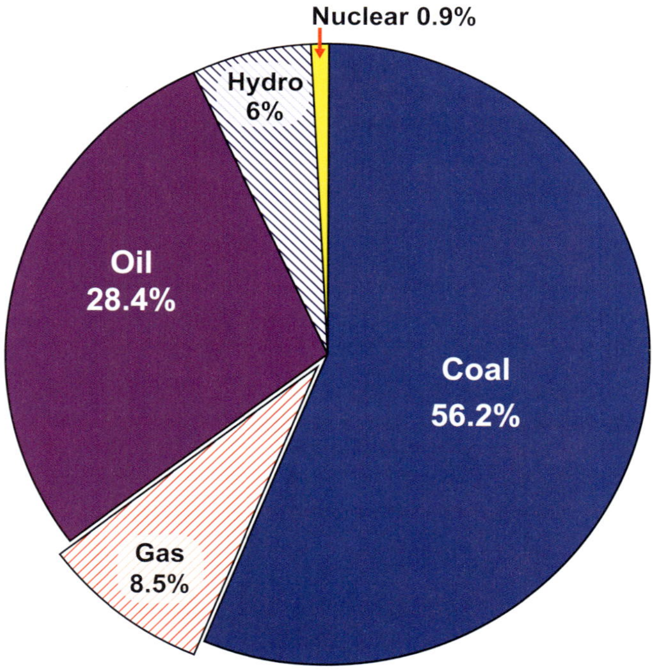

Figure 3.1: Primary Energy Consumption in 2006

Source: *BP Statistical Review of World Energy* 2007.

growing gas markets among emerging non OECD Asian economies it must be remembered that even with a booming economy, which has been persistently registering GDP growth rates of 9 percent, the development of natural gas in India continues to be constrained by its availability. Paradoxically Asia itself represents the largest part of the world LNG market accounting for two-thirds of global LNG trade, Japan being the largest importer and Indonesia the largest exporter. However, in India, unlike other gas deficient countries, the uncertainty engendered by price controls has shackled growth in investments and infrastructure for gas imports and so further restricted the availability of domestic gas resources.

Yet, since the turn of the century, in tandem with policies purportedly announcing a gradual phasing out of price controls, there has been renewed interest in natural gas. The first LNG terminal made its appearance on the West Coast of the country in 2004. Simultaneously, transnational pipelines from the East and West have become subjects of active diplomacy. But perhaps more significant has been the spate of gas finds announced since 2002 as exploration under the New Exploration Licensing Policy started yielding results. The year 2006 also saw the adoption of the Petroleum and Natural Gas Regulatory Board Act; the Board was set up the following year, marking an important milestone in the growth of gas markets in India. The extent of this growth may, however, depend on the shape of corresponding reforms in other sectors of the economy.

This section provides a historical background to the development of the gas sector in India with particular focus on the evolution of pricing policies leading to the existing hybrid gas market. It looks at the emerging regulatory environment and the slow and uneven pace of reforms in the user industries at a time when large investments from both the private and public natural gas sectors are poised to receive their first taste of success.

Overview

The Development of Oil and Gas in India

While India is a Federal country, the Constitution places the responsibility for the management and regulation of hydrocarbon resources upon the Union Government. Ownership of oil and gas resources on land, however, lies with the state governments but the Union Government owns the resources in offshore areas (the territorial waters, the continental shelf as well as the Exclusive Economic Zone are classified as offshore). Accordingly while the legal framework for exploration and exploitation of hydrocarbon resources is set out in a piece of legislation framed by the Union Government (the Oil Fields (Regulation and Development) Act of 1948 and the Petroleum and Natural Gas Rules, 1959, framed under it), licences as well as mining leases for exploration and exploitation of oil and gas are issued by the Union or state governments according with their geographical jurisdictions. However, even where the licences are issued by state governments, awards as well as their conditions are determined by the Union Government.

After independence, in the 1960s, state-owned companies and joint ventures between the government and private oil and gas companies dominated the oil and gas sector in India. In the 1970s the Indian oil and gas industry was largely nationalised, the government taking over the operations of IBP, Esso, Caltex and Burmah–Shell. Oil India Ltd (OIL), a Joint Venture Company between Burmah Oil Company and the Government of India, also became a national oil company (NOC), wholly owned by the Government of India, in 1981. Following nationalisation, only state-owned enterprises were allowed to participate in the oil and gas industry and all aspects of the industry were highly regulated. Licences for oil and gas were given by nomination to government entities such as the Oil and Natural Gas Commission (ONGC) and OIL. With the discovery of the large Mumbai High offshore oilfield in 1974, the Oil and Natural Gas Commission (which subsequently became a national oil company in 1993) emerged as the largest producer of both oil and gas in India, dwarfing Oil India Ltd, which continued to operate mostly in the north east of the country. Even in 2006–07, ONGC was producing 76.7 percent of the country's oil and 70.7 percent of its natural gas.

Natural Gas: administered prices leave a legacy of mismatched demand and supply

The supply of gas was initiated by OIL in 1959 in Assam. Five years later ONGC began to supply Gujarat with largely associated gas; more than a third of the production (a paltry 2.3 Billion cubic metres (Bcm) in 1980) was flared. After the discovery and development of the giant Bassein Gas field of ONGC in the western offshore the severe lack of infrastructure to handle gas resources became evident in the early 1980s. In 1984 the government carved out the Gas Authority of India Ltd (GAIL) from ONGC giving it a mandate to plan and build the necessary gas transportation infrastructure. In Assam this task remained with the

Assam Gas Company (also a government undertaking). In 1986 GAIL began work on the 2688 km Hazira–Bijapur–Jagdishpur (HBJ) pipeline linking ONGC's western offshore gas fields with fertiliser and power plants in the west and north of the country. Its convenience as a fuel source along with low controlled prices encouraged the development of a large number of industries along the route of the HBJ pipeline. Production increased to 19 Bcm in 1994–95 with just 2 Bcm being flared. By 2003 gas consumption had reached almost 30 Bcm. However, as a percentage of total domestic energy consumption, it remained relatively flat over this period, increasing from 6.4 percent for the year ended 31 December 1993 to 7.8 percent for the year ended 31 December 2003, which was well below the world average of 23.9 percent.[1]

Table 3.1: Gas Production and Imports in India. Bcm

	2000–01	*2001–02*	*2002–03*	*2003–04*	*2004–05*	*2005–06*	*2006–07*
ONGC	24.02	24.042	24.238	23.584	22.970	22.576	22.440
OIL	1.861	1.618	1.744	1.887	2.010	2.272	2.260
Pvt/JV	3.596	4.054	5.407	6.491	6.782	7.358	7.039
Total	29.477	29.714	31.389	31.962	31.762	32.206	31.742
LNG				0.072	3.230	6.120	8.310
Total	29.477	29.714	31.389	32.034	34.992	38.326	40.052

Source: Ministry of Petroleum & Natural Gas, Government of India *www.petroleum.nic.in*

Although domestic gas production continues to be dominated by ONGC, the share of private and joint venture companies has more than doubled since 2001. However, in spite of acute shortage of gas, imports in the form of LNG could only begin in 2004 after Petronet LNG[2] set up the Dahej terminal on the West Coast. In 2006 the total gas available (after flaring and consumption by the industry itself) was 34 Bcm, of which 20 Bcm came from ONGC and OIL and 6.1 Bcm was re-gasified LNG. Private and joint ventures[3] supplied the remaining 7.4 Bcm (Table 3.1). 70 percent of India's gas is produced from offshore fields (Table 3.2).

On the demand side the major consumers of gas are the power and fertiliser industries (40 percent and 29 percent respectively), while city gas accounts for just 4 percent (Figure 3.2).

Even though the power sector is the biggest consumer of gas, what is more significant is that, only 16 percent of thermal power generation capacity of 86,015 MW, and just 10 percent of the total installed power capacity of 132,330 MW is gas based (Figure 3.3).[4] Coal accounts for

Table 3.2: Production of Natural Gas by Region 2005–06. Bcm

Region	*ONGC*	*OIL*	*Pvt/JVs*	*Total*
Offshore	16.375	-	5.909	22.284
Andhra Pradesh	1.525	-	-	1.525
Arunachal Pradesh	-	0.029	0.006	0.035
Assam	0.498	2.011	0.016	2.525
Gujarat	2.185	-	1.109	3.294
Rajasthan	0.017	0.225	-	0.242
Tamil Nadu	1.130	-	-	1.130
Tripura	0.520	-	-	0.520
Total	22.250	2.265	7.040	31.555

Source: *www.petroleum.nic.in*

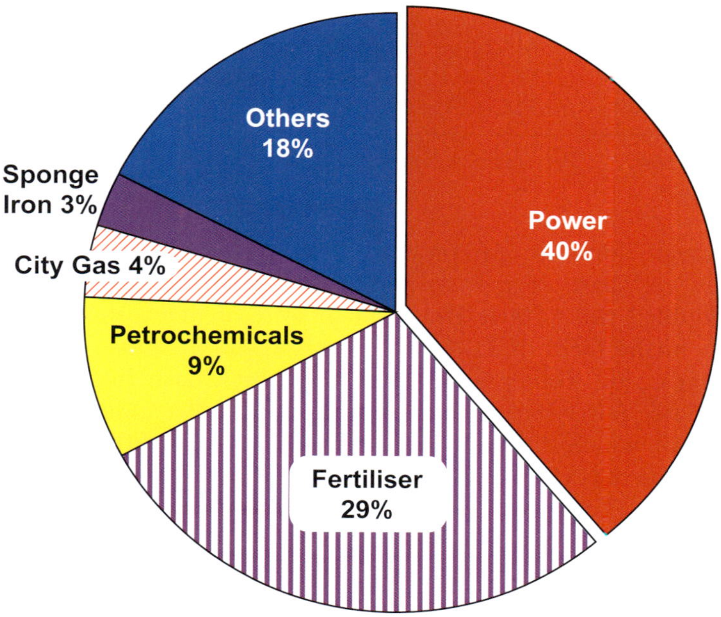

Figure 3.2: Sectoral Shares of Gas Consumption 2006

Source: Ministry of Petroleum and Natural Gas, Government of India

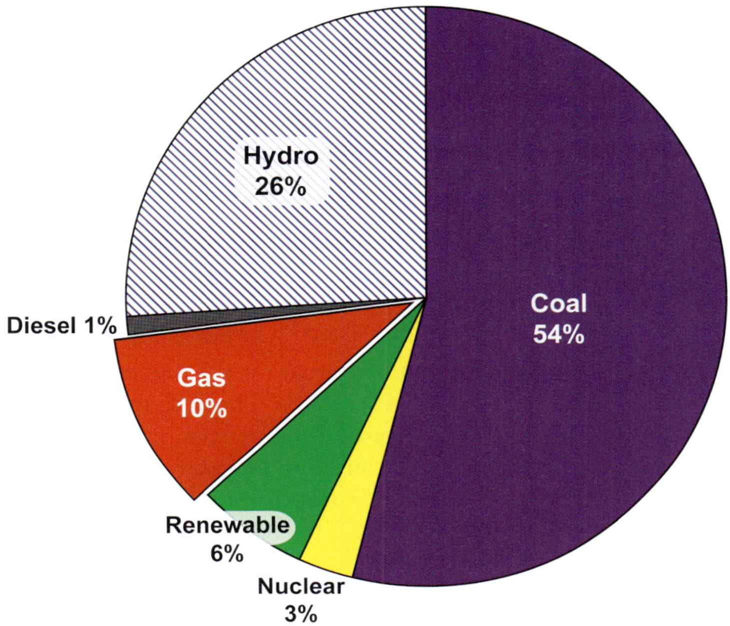

Figure 3.3: Installed Generating Capacity March 2007, total = 132330 MW

Source*:* Central Electricity Authority, *www.cea.nic.in*

over 50 percent of the country's primary energy supply and still dominates thermal power with a capacity share of 83 percent. The existing gas-based power plants have a demand of 25 Bcm which, at the levels of availability of the mid-2000s, was not met. During 2005–06 this shortfall of gas is estimated to have resulted in annual lost generation of almost 7700 million KWH.[5]

Natural gas is the preferred feedstock for urea, the major fertiliser consumed in India. At present only 67 percent of the installed capacity for urea is based on natural gas. A new pricing scheme introduced for this sector in 2003 was to be implemented in three stages;[6] the second of these established energy and efficiency norms for urea units and was completed in March 2006. The new policy encourages the conversion of units running on expensive feedstock (naphtha and high and low sulphur fuel oil) to gas, so that the subsidy burden on the government becomes lower. However, availability of reliable sources of natural gas has remained the major constraint. In 2004–05, against a demand of 12 Bcm from existing fertiliser units the actual gas supply was 8.8 Bcm. The following year, with the commencement of LNG imports, the situation had eased and the supply increased to 10.4 Bcm. According to the Report of the Working Group on Fertilisers set up by the Planning Commission in 2006, the total requirement at the end of the Eleventh Plan in March 2012 would be around 28 Bcm.[7]

Uncertainties about market behaviour, regulation, the cost of supply and the absence of a proper transportation and distribution infrastructure have led different agencies to give widely differing projections[8] about gas demand in India. In its 1999 report 'Preparation of Natural Gas Development Plan India' the Asian Development Bank projected a demand of 95 Bcm by 2012. According to the EIA's International Energy Outlook 2006 with a 'high'[9] GDP growth rate of 6.4 percent India's gas consumption was expected to grow at 7.5 percent annually to reach 57 Bcm in 2012 and 182 Bcm by 2030. The Indian government's Hydrocarbon Vision 2025 report, counting on gas prices to be one of the main drivers, projected the demand to increase from 20 Bcm in 1998–99 to 79 Bcm at $4/MMBtu import prices and 114 Bcm at $3/MMBtu import prices by 2011–12.[10] The corresponding projections by 2024–25 at these levels were 118 and 143 Bcm. Against these, an analysis by Petrofed[11] projects a demand of 130 Bcm by 2020 and 243 Bcm by 2030 assuming a GDP growth of 6 percent. However, should the country grow at the projected high growth rate of 8 percent this study places demand at 180.5 Bcm by 2020 and 406 Bcm by 2030.

It is also instructive to look at short-term projections for gas demand done by various Ministries in the course of formulating their Eleventh Plan proposals. Projections done by the Ministry of Petroleum are shown in Table 3.3.

It is evident from Table 3.3 that not much change is projected in the relative consumption of the various consuming industries. For the

Table 3.3: Gas Demand Projections Eleventh Plan 2007–12. Bcm

Year	*2007–08*	*2008–09*	*2009–10*	*2010–11*	*2011–12*
Power	29.1	33.3	37.5	41.7	46.2
Fertiliser	15.0	15.7	20.4	27.8	27.8
City Gas	4.4	4.7	5.0	5.4	5.8
Industrial	5.5	5.9	6.3	6.7	7.2
Petrochemicals	9.3	9.9	10.6	11.3	12.1
Sponge Iron/Steel	2.2	2.3	2.5	2.7	2.9
Total	65.4	71.8	82.3	95.7	102.1

Source: Report of the Working Group on Petroleum and Natural Gas for the Eleventh Plan, Ministry of Petroleum and Natural Gas, Government of India, Nov 2006

fertiliser sector it has been assumed that all non-gas based urea plants would convert to gas over a three-year period. For city gas it has been assumed, rather arbitrarily, that the coverage would grow from the present ten cities to forty by the end of the plan, presuming an annual growth rate of 8 percent. Petrochemicals and sponge iron sectors have been presumed to grow at 7 percent annually. More significantly, for the power sector it has been assumed that the Power Ministry's target of setting up an additional 70,000 MW generation capacity would be met and that 12,700 MW of this would be gas based.

However, the Power Ministry in its own projections for the same exercise does not foresee more than 4700 MW gas-based capacity being added even after including 2600 MW from the disputed[12] National Thermal Power Corporation's (NTPC) projects at Kawas and Gandhar. The reasons cited for these conservative estimates by the Ministry of Power are 'uncertainty about the availability, period of availability, and price of gas'.[13] In accordance with these estimates just 32.5 Bcm of gas would be required for power by 2012. The Power Ministry foresees the bulk of this capacity addition (47,000 MW) coming from coal with hydro contributing 16,000 MW and nuclear another 3200 MW.

The underlying assumption in many of these reports has been that India's domestic gas production will either stagnate or decline and that no new major gas finds would be forthcoming. The country's ability to be able to contract any significant amount of gas imports, at 'affordable' prices, either through LNG or transnational pipelines has also been in doubt. At the root of all these varying projections lies the legacy of a gas consumption that has been severely limited by supply constraints. By not allowing gas markets to develop prices set by supply and demand, price controls have discouraged investment decisions in the development of either LNG terminals or transnational pipelines for the import of gas. In the absence of price incentives, the major gas producer, ONGC, has stated (for good commercial reasons) that domestic gas production would start declining in the absence of significant investments in secondary recovery from existing fields, or the development of new reserves. Given this background, it is hardly surprising that announcements of new gas finds have tended to be greeted with a lot of popular scepticism. The upstream regulator, the Director General of Hydrocarbons (DGH), and companies including ONGC have on occasion publicly sparred over the quantification of new discoveries and declaration of reserves.

It is against this backdrop of limited and declining gas availability that the government had initiated a process of allocating gas to consumers through the Gas Linkage Committee (GLC) under which subsidised gas at administered prices was allocated amongst 'priority' users with the bulk going to the power and fertiliser sectors. With increasing demand for cheap gas at administered prices, the GLC, by 2002, had already allocated 44.5 Bcm of gas against a possible supply of just 23.7 Bcm by GAIL. Consequently the allocation mechanism, having become inappropriate, lapsed into disuse. After June 2005, in the face of declining availability, gas supplies under the Administered Price Mechanism (APM) were restricted to preferred customers, that is to power and fertiliser companies, to consumers covered under judicial directives[14] and to small consumers (below 0.05 MMSCMD). In November 2005 the GLC itself, with no gas left to allocate, was finally wound up.

In a way these difficulties were not just inevitable but symptomatic of the constraints and constrictions in related sectors of the economy such as power, fertiliser and coal. The sale of power to politically sensitive consumers such as agriculture continues to be heavily subsidised. Added to this is an electricity transmission and distribution system managed by state utilities, which is highly prone to heavy transmission losses on account of theft. A worsening power situation did lead to an enthusiastic call for change in this sector in the early nineties. However, due to the offering of irrational investment incentives to Independent Power Producers (IPPs),[15] the first flush of reform only addressed the short-term goal of generating more power without seriously looking at the supply side. As a result IPP power ended up being the most expensive power available to the already tottering state electricity utilities. The cash-starved Maharashtra State Electricity Board in 2001 had to withdraw from its power purchase agreement with Enron's Dabhol power generation project and this led to a contentious arbitration. Finally in 2005 the power project along with the Dabhol LNG terminal was taken over by a special purpose vehicle of GAIL and NTPC named the Ratnagiri Gas and Power Private Ltd.

Analysing the problem from the point of view of cost of generation, the IEP report referred to earlier finds that as long as the price of coal remains at or below US $2.27/MMBtu (i.e. $45 per tonne of imported coal of a calorific value of 6000 kcal/kg), natural gas, even at a price of $4.5/MMBtu, would not be the preferred alternative even for peak power loads.[16] The IEP report therefore concludes: 'Any supply strategy over the coming decades will have to emphasize India's major resource, i.e. coal.' As far as the IEP report is concerned, therefore, natural gas does not emerge as a major fuel in any of the scenarios projected and its share remains below 11 percent. As the report explains, 'Even when gas is pushed for power generation, only

16 percent of the power generated comes from gas.' Of course this would be contingent upon the government undertaking major reforms by way of privatising the coal sector and successfully managing to implement policies that increase domestic coal production well above present levels.

While the single major consumer, power is still perceived to be very sensitive to the price of gas, it is widely believed that, under existing conditions, gas would have to compete with coal as a fuel source. However, for the required expansion in coal-based generation capacity, large quantities of coal would have to be imported as demand would far outpace domestic output. The IEP report estimates that by the year 2031–32, India's coal imports could be in the range 72 MMt to nearly 462 MMt. This would mean an import dependence of between 11 and 45 percent.

However, notwithstanding the IEP report, there are well reasoned concerns about the country's capacity to adequately expand coal-based electricity supply due to limits on the expansion of either coal production or transportation (both inland rail for domestic supplies and port handling capacities in case of imports). Hardening of imported coal prices would also have a corresponding impact in the form of an increased demand for gas. The Tenth Plan (2002–07), for instance, had projected a total capacity addition of 41,110 MW of which, at the end of the Plan period (March 2007) only 21,180 had actually been commissioned. This had included a coal-based thermal target of 20,553 MW against which just 9510 MW was actually installed. Consequently, judging by past performance, there remains a big question mark over whether the projected thermal coal-based capacity expansion can take place over the next five years. What can be said with a fair degree of certainty is that as the power supply situation worsens, increasing gas supplies, even at higher market prices, will certainly promote a shift to gas-based generation at least for captive or merchant use.

Given that imported coal prices have increased substantially over the past few years, coal futures prices in Europe are expected to average $70/Mt over the period (Q1 2007 until the calendar year 2010). Assuming long-term contracts for imported coal can be entered into at $50 per ton fob, the cif price works out to $60/ton. This would translate into a delivered price of $80/ton or $3.2/MMBtu after adding duties, handling and other charges.[17] Since gas-based units have a conversion efficiency which is 35–40 percent more than coal-based plants[18] and come at a lower capital cost, a delivered gas price of $6/MMBtu could well be competitive against imported coal prices of $80/ton and above. Simultaneously, should there be changes in tariff policy such as the introduction of differential peak-time tariffs there could be a larger gas-based demand for power.

At the same time should increasing global, regional or local environmental changes bring about a gradual and conscious policy shift to encourage cleaner fuels, a higher proportion of gas-based generation would become inevitable. The rapid transition to a gas-based public transport system as a result of environmental action by the judiciary in the city of Delhi (see Box 1) illustrates the kinds of pressures that would force a greater role for gas in the future energy baance.

Winds of Change

Rising oil prices, a deteriorating balance of trade with increasing dependence on oil imports, and the accompanying downturn in industrial production, led to a balance of payments crisis in 1991. The government had to embark on a process of structural adjustments to satisfy conditions of loans advanced by the World Bank, International Monetary Fund and the Asian Development Bank. Needing to attract both investments and technology in the oil and gas sector, the

Box 1

In 1998 the Supreme Court, in a Public Interest Litigation seeking to clean up Delhi's polluted air mandated the conversion of all of Delhi's public road transport to CNG beginning in September 2001. In spite of initial resistance by Government agencies, by 2006 Delhi, with 125,000 vehicles serviced by 144 CNG stations, had the largest fleet of buses running on gas anywhere in the world. Under the Supreme Court's directives similar steps are to be taken in twelve other cities. This CNG segment, though still in its infancy, has been growing strongly over the past two years in the states of Gujarat, Andhra Pradesh, Delhi, Maharashtra and Uttar Pradesh. In 2006–07, CNG sales increased at an annual rate of 21.6 percent over the previous year, reaching 690 thousand metric tonnes (0.95 Bcm) with Delhi alone accounting for almost half this share. By March 2007, 408,000 CNG vehicles were being serviced by 356 stations in these four states. A number of new players have been trying to enter the CNG and city gas distribution services but the real constraint so far has been the availability, rather than the price, of natural gas.

government began a process of reforms aimed at gradual deregulation in order to rationalise the industry, improve efficiency, and reduce the mounting burden of government subsidies. The reform measures sought to de-license refining and allow private as well as foreign companies to enter exploration and production as well as marketing and retailing of products. To attract much needed foreign exchange as well as modern technology along with best practices, bids were invited for farming out ONGC and OIL's discovered fields as Joint Ventures which, between 1994 to 1996, led to the signing of the first few Production Sharing Contracts for the Panna–Mukta, Tapti and Ravva fields as joint ventures between private companies and ONGC. While the government retained the right to nominate a buyer for the gas, prices under these contracts were fixed on the basis of a pre-agreed formula linked to a basket of fuel oils. Bids were also invited for smaller fields and exploration blocks.

However, it was only in 1999, with the coming of the New Exploration Licensing Policy (NELP), that the upstream sector finally became truly deregulated. Under NELP, the practice of assigning exploration acreages to NOCs by nomination came to an end. Henceforth, NOCs were required to bid for acreage, either alone or in consortia with others, on equal terms under open international competitive bidding. In the case of already existing nomination acreages they were also required to complete committed exploration programmes within a stipulated time-frame, after which these would have to be relinquished if no fresh discoveries had been made. Most importantly, NELP for the first time provided that the contractors under the PSCs would be allowed full freedom to market the oil and gas within the country at prices determined on arm's length principles. Government gave up its right to nominate buyers for the oil and gas produced. This was significant because a parallel process of reforms in the downstream had by this time already created a massive growth in refining capacity in the private sector.[19]

The Maturing of the New Exploration Licensing Policy (NELP)

India has 26 sedimentary basins extending over 1.39 million sq km onshore, and 1.75 million sq km offshore (of which 1.35 million sq km are in deep waters beyond 400 m). While 80 percent of Indian basins is still either unexplored or poorly explored, since 2000 six rounds of NELP have brought 1.36 million sq. km under active exploration with the award of 162 blocks. So far nine of these blocks have been relinquished.

When first announced, NELP had been greeted with a fair degree of scepticism about the prospectivity of Indian exploration acreages. However, between 2000 and 2007, as exploration programmes progressed, various operators began announcing a string of oil and gas discoveries (Map 3.1). The most significant of these have been the Dhirubhai series of gas discoveries made by Reliance Industries Ltd in the Krishna Godavari (KG) basin, beginning 2002, as well as the company's significant finds in the adjoining Mahanadi, the North East Coast (NEC), and very recently, the Cauvery deep water basins. On the basis of already completed appraisals it is estimated that 1797 MMt of oil and gas reserves have been added during the Tenth Plan (2002–2007) which is double what had been anticipated (Figure 3.4). More significantly, it is operators other than the NOCs who have added the bulk (58 percent) of these reserves.[20] By April 2007, the country's established gas reserves were estimated at 1116 Bcm, up by 47 percent from 760 Bcm in 2000. The Gujarat State Petroleum Corporation Ltd as well as ONGC[21] also announced very promising results on the east coast. In the first flush of excitement both ONGC and GSPC have given estimates of over 600 Bcm each, however, in the absence of proper appraisal, the Directorate General of Hydrocarbons has insisted that it is too early to place a reserve figure on these finds, significant though they may be.[22]

Meanwhile the development plan of Reliance Industries Ltd (RIL) for 300 Bcm of gas from its Dhirubhai-1 and -3 discoveries, at a water depth of up to 1200 metres in the KG-DWN-98/3 Block, is already being implemented and the first gas is scheduled to flow in June 2008 and the scheme is expected to attain a production level of 29 Bcm by 2011. If this plan stays on schedule, RIL will have brought this gas into production within six years of discovery. This development alone means that the country's domestic gas supply is set to more than double to 57 Bcm per year, and RIL to emerge as the biggest producer of natural gas in the country. Subsequently, RIL has announced twelve more discoveries (including oil finds) in the area and current estimates are that these would add at least another 150 Bcm to reserves. In addition RIL seems quite confident of establishing more reserves in its acreages in the Mahanadi and North East Coast, and so the facilities at Kakinada for block D-6 development are being designed with enough capacity to handle 44 Bcm per year. Development plans for the other gas finds have yet to be submitted by the contractors. The good news, however, does not seem to be ending here. RIL recently announced the D-35 discovery in the Cauvery deepwater basin on the east coast of India. This is the first discovery in this deepwater basin and holds much promise for the future.

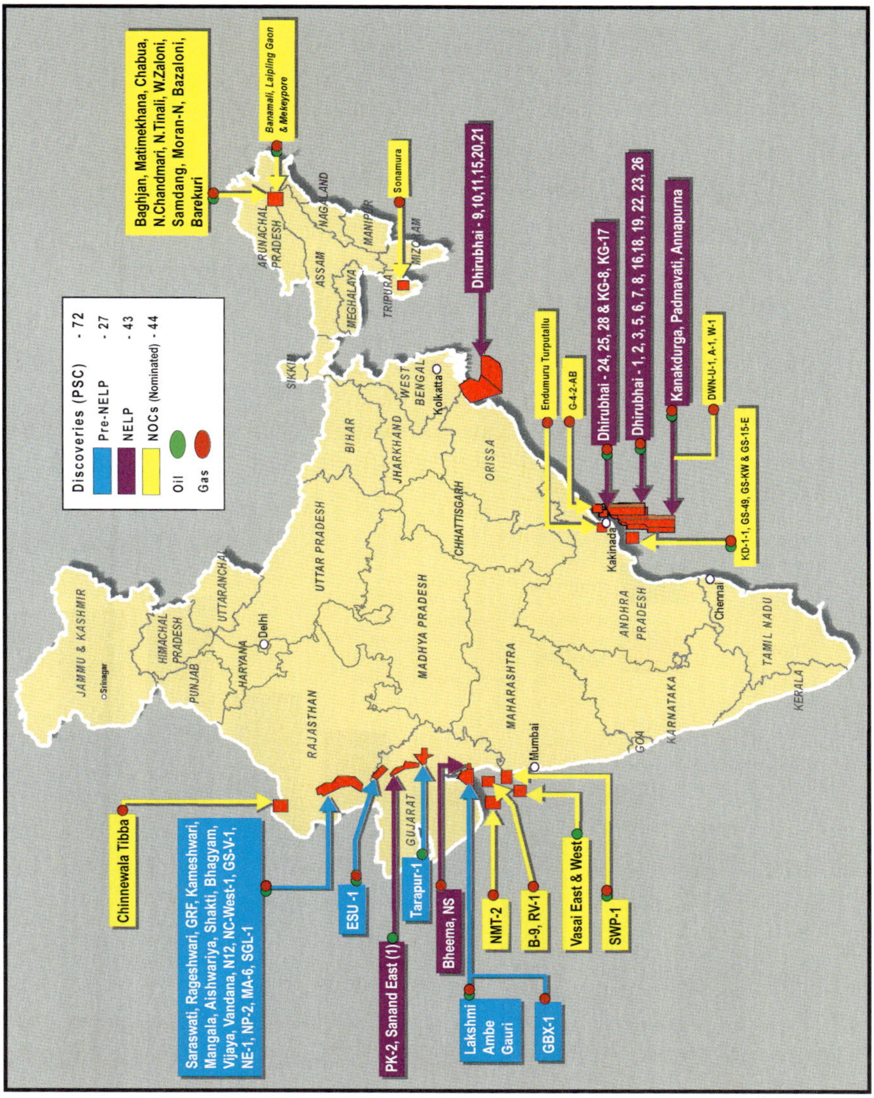

Map 3.1: New Discoveries

(Map reproduced courtesy Director General of Hydrocarbons, Government of India)

Meanwhile, in Oct 2007, ONGC submitted a consolidated appraisal of its finds in the KG offshore, which it hopes to develop together with its discoveries in the KG-DWN-98/2 block (see footnote 22). This places the reserves estimates in the area from recent, as well as earlier, discoveries (some in adjacent nomination acreages) at 180 Bcm. ONGC now hopes to produce an additional 9 Bcm of gas per year from this area from 2013.[23]

Gujarat State Petroleum Corporation's (GSPC) discoveries (Deendayal series), still under appraisal, are remarkable because they were the first significant discoveries in the Cretaceous in this geographical region.[24] They thus opened an entirely new geological region for explorers to run their models in the hope of even larger finds on the east coast, and perhaps, elsewhere in the Indian basins.[25] What is still unknown is the speed with which GSPC can complete the appraisal and find the partners or technology to develop gas from these high pressure and high temperature reservoirs.

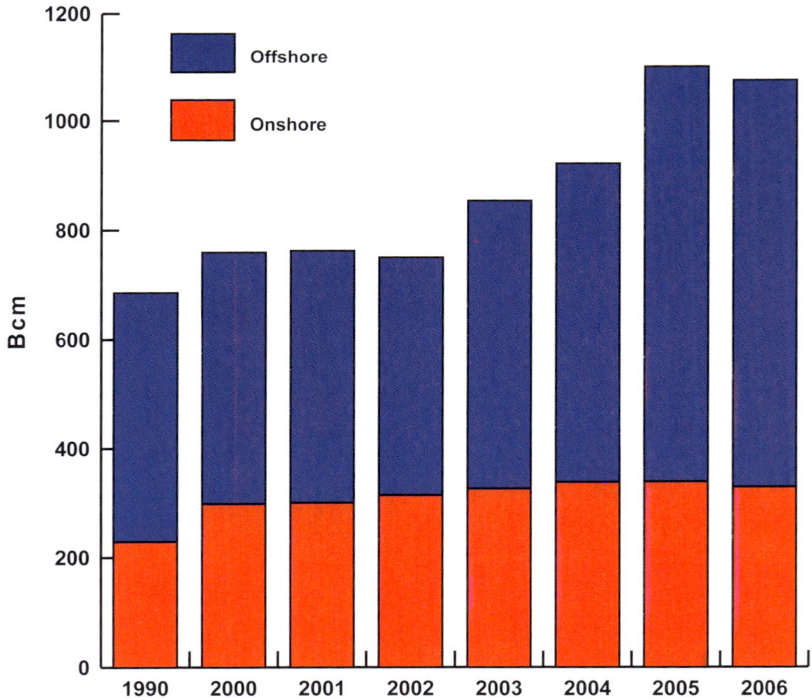

Figure 3.4: Gas Reserves (Bcm)

Source: Ministry of Petroleum and Natural Gas

The first five years of NELP have led to investments of around 1.7 billion USD of which 1.3 billion has come from the private sector. They have also led to the addition of 625 Bcm of gas reserves. In spite of these successes, however, the acreage holding pattern under NELP shows that ONGC continues to dominate, holding more than 50 percent of the total acreage. However, the Indian private sector company RIL, a new entrant in the exploration and production business, follows with more than 30 percent of the total. The major international oil companies (IOCs) are still almost totally absent from Indian basins. The reason for this absence is not so much lack of participation, as very aggressive bidding by Indian NOCs as well as private companies. British Petroleum has taken progressively greater interest in the last two rounds. It has also been keen on linking up with ONGC for exploration in the Kutch basin[26] on the west coast, where positive synergies can develop with BP's exploration efforts in the Pakistan offshore. ENI has chosen to partner ONGC in some deep offshore blocks on the east coast as well as in the Andaman seas. However, on the whole, IOCs have fought shy of committing major risk investments in the upstream both because of the absence of a well defined gas market with clarity on gas pricing (in spite of contractual provisions under NELP) and because of the lack of infrastructure and of a clear policy on its development and regulation. The Regulatory Board created by the Petroleum and Natural Gas Regulatory Board Act of 2006 is at an early stage of development (see below). However, Indian companies like Reliance Industries Limited (RIL), ONGC and the Gujarat State Petroleum Corporation (GSPC) have already got a head start which others may find difficult to keep pace with. Buoyed by the new discoveries, GAIL, which was hived off as a gas transportation company from ONGC in 1984, Reliance Gas Transportation Infrastructure Ltd (RGTIL),[27] a Reliance Industries Ltd group company, and GSPC have all announced ambitious plans for pipelines from the east coast, Andhra Pradesh to the west as well as to the north and south. For a list of planned pipelines see Appendix 3.2 and Map 3.3.

The Commercialisation of Coal Bed Methane

India, being the world's third largest producer of coal with the fourth largest proven reserves holds interesting prospects for the recovery of coal bed methane (CBM). On the basis of exploration (up to a depth of 1200 m) the cumulative total of coal resources estimated in the country at the start of 2006 was 253 billion tonnes. Based on this quantum of resource the Director General of Hydrocarbons, quite optimistically, places the CBM potential gas reserves in the country between 1400 to 2600 Bcm. However, it was only in 1997, after the adoption of the Coal Bed Methane Policy, that CBM operations began to be looked at as something more than R&D efforts. Even though all matters relating to coal fall within the jurisdiction of the Ministry of Coal, the Ministry of Petroleum and Natural Gas has become the nodal agency for CBM. Since 2001 CBM blocks have been carved out and open international competitive bids invited for contracts with the Government of India on the basis of a Production Level Payment (PLP) system for the exploitation of CBM. As in the case of NELP Production Sharing Contracts, contractors under CBM are also allowed freedom to market the gas within the country.

So far three rounds of bidding under the CBM policy have been held involving 26 blocks. These cover an area of 13,600 sq. km. Because of co-ordination issues between the diverse agencies involved in CBM, the pace of these rounds has not been as encouraging as those for oil and gas. Simultaneous coal mining and CBM operations in the same area can still become causes of conflict between agencies and lead to operational delays. Both the Ministries of Coal and Petroleum have now started drawing up detailed guidelines to avoid such possible conflicts. Meanwhile, exploratory efforts have established a potential of 178 Bcm in four blocks. Two of these blocks are operated by RIL, while ONGC and the Great Eastern Energy Corporation Ltd (GEECL) operate one block each. Encouraged by these results, CBM III saw the active participation of a number of foreign companies such as BP, Arrow (Australia) and others in the round. It is estimated that potential CBM reserves in the blocks awarded to date may be 1400 Bcm. So far the Directorate General of Hydrocarbons (DGH) has approved the commercial development plan for GEECL's block in Ranigang, West Bengal. Latest estimates put the original reserves of gas in place in this block at 54 Bcm; GEECL commenced commercial production in July 2007 and expects to produce at the rate of 0.5 Bcm per year from the block by 2008. RIL's development plan for the Sohagpur East and Sohagpur West blocks in Madhya Pradesh has been submitted to the DGH. The plan is believed to propose a steady production of 1.8 Bcm per year of gas from both blocks over a fifteen-year period and the two blocks are believed to hold 103 Bcm of gas. However, since recovery rates have yet to be established, credible availability estimates for CBM will have to await commercial production.

Transnational Natural Gas Pipeline Projects

Iran–Pakistan–India

A Memorandum of Understanding to transport natural gas from Iran to India was signed between the two countries in 1993. The proposal was to transport 50 to 72 MMSCMD (13.35 to 26.30 Bcm) of natural gas through a pipeline landing at Kutch in India via the shallow waters offshore route passing through the extended economic zone of Pakistan. The Government of Pakistan was unresponsive to this project and no further work was done. Nonetheless, the project stayed on the radar as one of the possible pipeline projects that could at some stage come to fruition to bring natural gas to India. The main stumbling block was the strained political relationship between India and Pakistan. While India remained unsure about the safe passage of gas through Pakistan, the latter did not press the matter because its own Sui fields were adequate to satisfy domestic demand.

In 2004, after a thawing of relations between the two countries and a realisation within Pakistan of its own declining reserves and need to import natural gas from 2010 onwards,[28] there has been serious dialogue between Iran, India and Pakistan to advance the project. The Petroleum Ministers of the three countries have met several times and even the three Heads of State have met and expressed a desire to move ahead. Committees have been established at the bureaucratic level and today the outline seems clearer.

The Iran–Pakistan–India (IPI) natural gas pipeline is being planned to bring natural gas from the South Pars fields in southern Iran through an on-land route to Pakistan and India (Map 3.2). The design capacity presently being planned is to transport 150 MMSCMD (54.75 Bcm) of which 60 MMSCMD (21.9 Bcm) would be for Pakistan and 90 MMSCMD (32.85 Bcm) for India. Iran has recently proposed to take up the project in two phases: in the first phase, it proposes to deliver 21.9 Bcm of gas through a 56 inch diameter pipeline. The remaining 32.85 Bcm would be supplied through a second pipeline. The capital cost of the project for the first phase is expected to be around $7 billion. The pipeline would run 1880 kms from Iran to the Sui Northern Gas company's compressor station at the mid-country location of Bhong in Pakistan, and then an additional 70 kms to the Indian border at Fazilka.

Despite apparent progress over the past three years doubts have continued to plague the project. The initial concern had been whether, in the light of the US sanctions on Iran, India and Pakistan would in fact move beyond the consultation stage into actual project implementation. Repeated statements by leaders[29] in the two countries indicated that there was a fair degree of commitment towards the project despite US reservations. Indian leaders claimed to have made their position clear to the US State Department indicating that, in view of her burgeoning energy requirement, India had little choice but to proceed. They have also said that the US government has not proposed an alternative plan to meet Indian energy needs leaving India with no choice. The same reasoning perhaps holds even more true for Pakistan. According to the Asian Development Bank the country will require imports of natural gas from 2010 onwards. Starting from a relatively small deficit ranging from 2.2 to 8.4 Bcm in 2010, the need for imports in Pakistan is expected to grow to 27.4 Bcm by 2015 and could be as high as 35.4 to 57.7 Bcm by 2020.[30] This demand has been calculated assuming a growth of about 6.5 percent GDP per annum. Consequently, some analysts believe that political pressure from the United States against the pipeline will not work beyond a certain point. This had been tested when both India and Pakistan developed and exploded nuclear weapons in 1999, much against the wishes of the US government, and were able to withstand sanctions from the World Bank and other multilateral funding agencies without any significant impact on their respective economies. As far as India is concerned, the security of the pipeline over the troubled areas of Baluchistan in Pakistan has been the more worrisome aspect of the project.

Iran had all along been negotiating to enter into separate gas sales and purchase agreements with Pakistan and India. In January 2007 Pakistan and Iran mutually agreed on a pricing formula that fixed the price of Iranian gas as follows:

- when JCC=$30–70, US$/MMBtu=0.063 * JCC($/barrel) + US$1.15;
- when JCC <US$30, US$/MMBtu=0.05 * JCC($/barrel) + US$1.54;
- when JCC >US$70, US$/MMBtu=0.05 * JCC($/barrel) + US$2.06

With this pricing formula Tehran has proposed to sell gas to India and Pakistan at $4.93 per MMBtu at a JCC price of $60 per barrel.[31] Add to this transit fees of $0.49/MMBtu (Pakistan had been seeking $0.943/MMBtu) and transportation charges of $1.57/MMBtu and the gas at India's border would cost almost $7/MMBtu.[32] For India it is a matter of serious debate whether GAIL would be in a position to find buyers at this price. The fact that transit through Pakistan is a contentious issue means that it is better strategy for India to be seen as a buyer of gas at its borders rather than a partner in the project. This would help it to pass on both supply side risks, particularly disruption threats, and pipeline security to the supplier and to some extent insulate GAIL from price risks in an overly cost sensitive gas market which is likely to have increasing access to cheaper domestic gas. More importantly it would free India from the pressures of having to negotiate the politically sensitive issue of the transit fee with Pakistan.

The other serious concern is whether Iran has the ability to deliver natural gas within the agreed time-frame. History does not favour Iran for it has never entered into contracts and delivered on a major project such as this. On the political side too there are concerns coming out of Iran's stand on the development of its nuclear reactors and the potential results of that. India seems to have become far more cautious in recent times and appears to be veering towards enhanced participation in the Turkmenistan–Afghanistan–Pakistan pipeline project.

Turkmenistan–Afghanistan–Pakistan–India

In 2002, the Presidents of Turkmenistan, Afghanistan and Pakistan signed a framework agreement that proposed to develop a pipeline supplying natural gas from the Daulatabad gas field in south-east Turkmenistan to Afghanistan and Pakistan. They invited the Asian Development Bank (ADB) to develop the project and India to join in it. A Steering Committee comprising the Petroleum Ministers of the three countries was constituted with ADB as convenor for the meetings. The ADB completed the project report in mid 2004 but India did not join in. The proposal was for a 1680 km pipeline, 56 in. in diameter, passing through Herat in west Afghanistan, crossing Kandahar and landing at Quetta in Pakistan, with a spur line going to Karachi (Map 3.2). The project was reported to be financially viable with an internal rate of return of 17.5 percent.

While considering the project report, the Steering Committee asked Turkmenistan to conduct an independent audit of the Daulatabad field to assess the gas reserves in the field and the condition of the existing production facilities and to address the question of possible upgrading of the facilities. Subsequently Turkmenistan appointed Degolyer and McNaughton to audit the field and its facilities but the Government of Turkmenistan did not publish the findings of this audit. Therefore doubts continued on the issue of whether Turkmenistan and/or Daulatabad actually had enough gas to meet commitments towards this project. Nevertheless, India began to attend Project Steering Committee meetings as an observer.

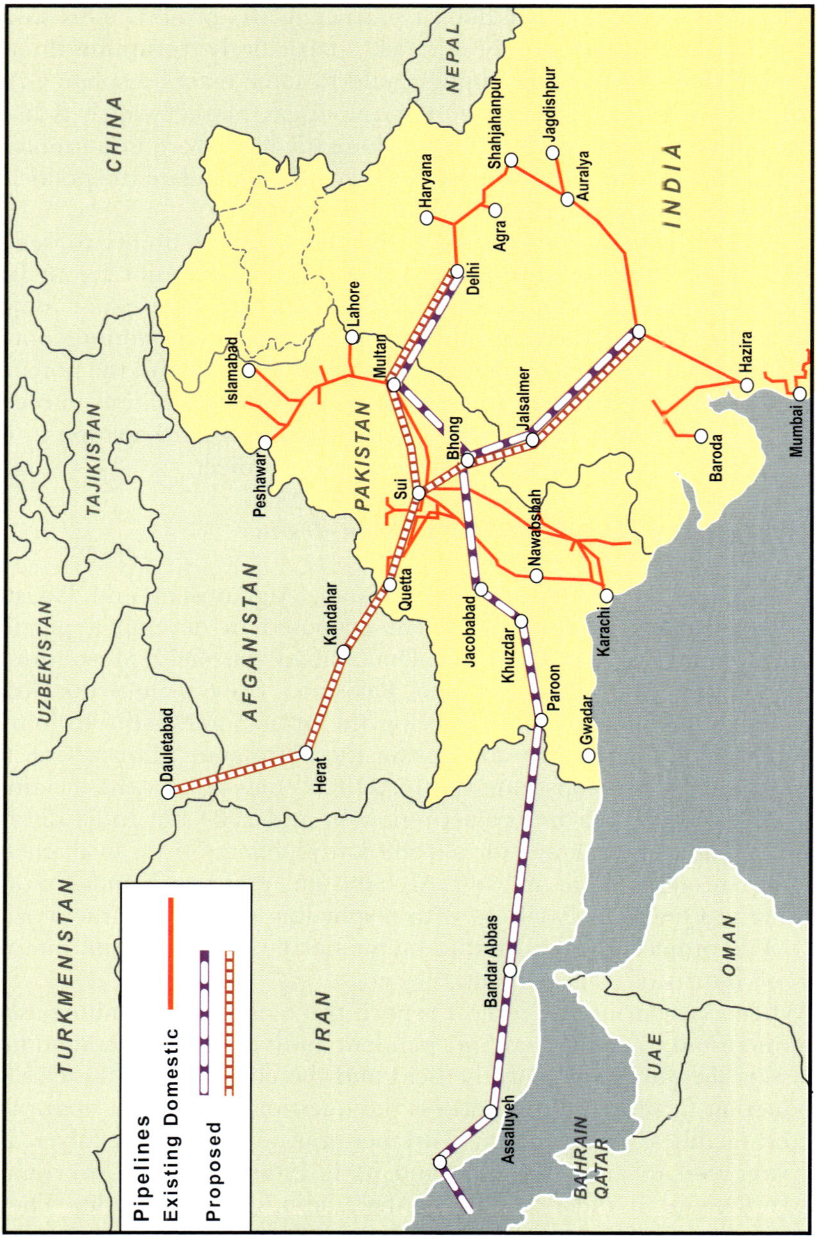

Map 3.2: Transnational Pipelines

Source: Asian Development Bank, *http://mea.gov.in/srec/internalpages/tapi.pdf*

Despite the keenness initially shown by Turkmenistan, Afghanistan and Pakistan there are serious doubts about the viability of this project. A major issue is that of civil rights in Turkmenistan. While in 2007 the USA appeared to favour this project (seemingly because of the progress made by the Iranian pipeline to which it objects), in the earlier phase of project development it had serious concerns about civil rights issues. President Niyazov, a known dictator, had eliminated all opposition in the country. Elections after his death have not been democratic, although President Berdimuhammedov may not have the same level of autocratic power as his predecessor. The situation raises big questions about the political future of the country.

Serious doubts still exist about the remaining reserves of natural gas in the Daulatabad field. Moreover, Turkmenistan has signed a major gas sales agreement with Russia that will put enormous pressure on the availability of natural gas exports to other countries. These were some of the questions in the mind of the other members of the TAP consortium (Afghanistan and Pakistan) when they asked Turkmenistan for an independent audit of the field. This audit was completed in 2005 but the Turkmens have not revealed its conclusions to the Steering Committee. Doubts over the availability of gas therefore remain but there have been recent new finds in Turkmenistan which are believed to be very significant, and it is possible that gas for the pipeline could be sourced from these new fields.

When the project was initially conceptualised and great interest shown by the three participating countries, there was also euphoria and excitement over the future of Afghanistan: the Taliban had melted away, the new government under President Karzai appeared energetic, and considerable development activity was seen in Kabul. Four years later, there is a perceptible change in the Afghan environment. Much uncertainty exists over the state of old landmines that were laid earlier by the Mujahedeen fighters fighting the Russians, and others subsequently laid in Taliban times. Moreover, the Talibs have regrouped and are a major force in the Kandahar region through which the proposed pipeline would pass. It is doubtful whether the consortium partners will be able to attract private capital or private participation in the prevailing environment. The Afghan Ministry of Mines and Natural Resources, which looks after the oil and gas sector, had a good reputation under the first Minister, Joma Mohamadi, a former World Bank engineer, who had returned to take up the assignment at the behest of Mr Karzai. Mohamadi died in a mysterious air crash in early 2003 and since then the Ministry has never regained its earlier momentum or capacity to plan and execute a major project.

In 2007, the project seemed to be at a standstill with ADB, the lead development partner, trying but failing to make progress. Therefore, it would not be prudent for India to rely on this as a live project.

India–Bangladesh

There has been discussion about importing natural gas from Bangladesh to India since the early 1990s. The country's natural gas resources have seemed promising and raise the prospect of the country emerging as an exporter of natural gas. However, Bangladesh has not been able to formulate a clear policy on exports. In 2002, the Asian Development Bank forced the issue with the Government of Bangladesh proposing a pipeline from the Bibiyana gasfield into India through West Bengal and Bihar to Uttar Pradesh in the North, joining the Hazira–Vijaypur–Jagdishpur (HVJ) at Dadri. This proposal made no progress. Although it is widely suggested that Bangladesh may not be willing to sell gas to India but would sell to other foreign companies, this proposal has also come to nothing and potential buyers have lost hope.

One part of the argument against exports rests on the question of whether Bangladesh has adequate proven and recoverable reserves to meet its own growing demand over a fifty-year time period. There are indeed varying estimates of reserves in Bangladesh. A study jointly carried out by the Hydrocarbon Unit of the Energy and Minerals Division of the Energy Ministry and the Norwegian Petroleum Directorate published in January 2002, has been accepted as the basis for the government's decisions. The government had also set up an independent committee which estimated reserves in the range of 420–460 Bcm compared with cumulative gas demand during the period 2002–2030 of around 700 Bcm even with modest GDP growth of 5 percent per annum.

One major reason for the government not agreeing to export gas to India is the lack of political consensus on this matter among the political parties of the country. Successive governments, led by Mrs Khalida Zia and Shaikh Haseena, have been unable to decide the matter for political rather than economic reasons. The Government of Bangladesh has been aggrieved with India on the grounds that the import–export balance between the two countries is heavily skewed in favour of India. India has done little to improve this situation and continues to be perceived as a big brother which does not merit grand friendly overtures. So much political emotion surrounds this issue that any government which proposed such initiatives would risk being accused of selling out Bangladeshi interests to India.

India–Myanmar

GAIL has proposed a 18 MMSCMD (6.57 Bcm) natural gas pipeline bringing natural gas from Myanmar via Bangladesh into India. This pipeline would extend 1573 km from the border of Myanmar to Gaya in Bihar. This gas would come from blocks A1 and A3 for which the Government of Myanmar invited bids in 2006 and in which GAIL and ONGC had acquired a 30 percent participating interest. However, the Government of Myanmar accepted an offer from the China National Petroleum Corporation to build a pipeline connecting the A1 and A3 gas fields to the Chinese border, a distance of 900 km. This announcement has for the moment put an end to India's ambitions of acquiring gas from Myanmar.

LNG Projects

Until 2002, there was much discussion about developing a number of LNG projects. The Tenth Plan had envisaged establishing at least four LNG terminals in the country with the expectation that LNG imports could reach almost 18 Bcm by 2007. This was the time when both the public as well as the private sector were looking at LNG as a panacea for meeting the growing gap between demand and supply.

Petronet LNG Ltd began as a unique experiment started on the initiative of the Government of India when the four major public sector companies, the ONGC, GAIL, IOC and Bharat Petroleum Corporation formed a joint venture company to import LNG. A number of other projects were also announced, the most important being BG's terminal at Pipavav (Gujarat), Shell's project at Hazira (Gujarat), Reliance Industries Ltd at Jamnagar (Gujarat), India Natural Gas Company at Trombay (Maharashtra), Dabhol (Maharashtra), Ennore (Tamil Nadu), and Kakinada (Andhra Pradesh).

When Reliance Industries announced a natural gas discovery that could have reserves of over 300 Bcm (see above), most projects went back to the drawing board as LNG import prospects were reviewed. Around the same time improving relations with Pakistan allowed talks to begin on the much anticipated Iran–Pakistan–India natural gas pipeline project.

In 2007 a number of LNG projects are operational: Petronet LNG already has a tie up with RasGas of Qatar for 5 MMtpa for Dahej and another 2.5 MMtpa has been contracted from RasGas for Dahej; Shell is operating a terminal of 2.5 MMtpa expandable to 5 MMtpa at Hazira; a 5 MMtpa terminal is under construction at Dabhol; and a 2.5 MMtpa terminal with a provision to expand up to 5 MMtpa is being developed at Kochi.

India will thus have LNG import capacity of about 22.5 MMtpa by 2011. However, with the exception of the agreement already reached with RasGas for Dahej, secure sources of LNG to feed the other terminals, including the nearly complete Dabhol have been near impossible to obtain because of the extremely tight supply position in the international LNG market. None of the companies involved seem to have any clear long-, or even medium-term strategy to source LNG. Added to this is the confusion due to the nascent and underdeveloped state of the country's gas market with potential investors and suppliers being unsure of gas demand at international LNG prices.

With production from the Reliance fields in the KG basin coming onto the market by mid 2008, and the uncertainty surrounding future gas demand, new LNG plants or even plans for expansion of existing ones will have to await a fresh assessment of the market situation. Once Reliance and the other companies producing gas in the KG basin bring their gas onto the market, the requirement for LNG will become clearer.

Gas Pricing

Gas pricing in India is still in the process of evolving from a strictly controlled administered pricing regime to one of market-determined prices. In tracing the history of gas pricing it must always be remembered that until very recently ONGC was producing over 90 percent of all the gas in the country. Before the 1970s an expert committee fixed natural gas prices for consumers. In the early 1970s ONGC, the dominant producer, negotiated gas prices separately for each consumer. The arrangement led to arbitrary price variations across consumers. The mid-1970s saw prices based on the thermal equivalence of substitute fuels and opportunity costs to the consumer. This led to price variations not just across consumers but also across different uses by the same consumer. This practice of determining gas prices on the basis of negotiation between ONGC/OIL and gas consumers continued till 1986. For gas supplies in the industrially underdeveloped north east prices were charged in accordance with the government's orders at special subsidised rates, even before uniform administered prices came to be introduced across the country in 1987.

In 1987, three years after the creation of GAIL, the government began setting uniform prices for natural gas on the cost to producer (cost-plus) principle, which was called the Administered Pricing Mechanism (APM). Under this arrangement producers were allowed a specified return on their investment and cost of production. Gas prices were accordingly fixed at Rs1400/Mcm ($1.21/MMBtu)[33] at the transfer (landfall) point with a postage stamp tariff of Rs850/Mcm ($0.74/MMBtu) being levied on customers along the Hazira–Vijaypur–Jagdishpur pipeline. For the north east states, with little industrial development but substantial gas availability, consumer prices were fixed at a concessional Rs1000/Mcm (0.87/MMBtu) as an incentive to encourage development of gas-based industries. In 1992 these prices were revised following the recommendations of the Kelkar Committee. The consumer price was fixed at Rs1550/Mcm ($1.35/MMBtu)[34] to be raised annually by Rs100/Mcm until it reached Rs1850/Mcm ($1.60/MMBtu) in 1995. Producers were paid Rs1500/Mcm ($1.30/MMBtu). The difference between consumer and producer prices was credited to a Gas Pool Account which was set up to compensate both ONGC and OIL for selling gas below the cost of production (Rs1000/Mcm) in the north east. The prices quoted above do not include royalty, taxes, and transportation. A table showing these prices in $/MMBtu (inclusive of royalty) at the actual prevailing rupee dollar exchange rate is given in Appendix 3.3.

During the mid 90s, the government entered into the first ever Production Sharing Contracts (PSCs) for select discovered fields of ONGC and OIL in the hope of bringing in foreign investment and technology. These contracts would not have been possible on the basis of controlled prices. Consequently, in the case of medium-sized fields, while all gas was to be bought by a government nominee (GAIL), gas prices were linked to a basket of fuel oils with a designated floor and a ceiling. For small, isolated and marginal fields, in the absence of any developed transportation network, producers were given the freedom to find their own local markets as well as prices. In 1996, given that domestic gas availability only expected to reach a maximum of 28 Bcm by 2001–02 against a demand of 65 Bcm already registered with GAIL, the need for either LNG or pipeline imports became more and more urgent. The government appointed the Sankar Committee to deliberate on alternative pricing strategies for gas which, while accepting the need to move toward market determined prices, left intact the principle of separately setting consumer and producer prices.

The government considered these recommendations and in October 1997 it linked consumer prices equal to 55 percent of the price of a basket of fuel oils and then increasing by 10 percentage points each year up to 75 percent parity by 1999–2000. For the north-east states, consumer prices were fixed on the same principle but at a lower percentage of the same fuel oil basket, starting at 30 percent to be increased by 5 percentage points each year until it reached 45 percent. This consumer price was linked to a calorific value of 10,000 kcal/cu m. and could vary proportionately for gas with higher or lower calorific content. Using this formula, consumer prices were reset every quarter by GAIL. However, they were subject to a floor of Rs2150/Mcm and a ceiling of Rs2850/Mcm.[35] For the north east, the floor and ceiling were fixed at Rs1700 and 1200/Mcm respectively.

Simultaneously, by the late 1990s, private gas started flowing from the fields auctioned for joint venture operations under Production Sharing Contracts (PSCs) (see footnote 3). Small producers, working isolated and marginal gas fields, particularly in Gujarat, started developing their own gas markets while joint venture gas from the medium-size fields began flowing to GAIL. Contractors in these joint ventures began getting prices in accordance with the pricing formula in the PSCs[36] even as users continued paying GAIL the administered consumer price as notified by GAIL under the pricing formula declared by the government in 1997. Out of the consumer prices collected, GAIL was allowed to retain the amount required to compensate it for the higher cost of gas purchased from the joint venture companies and charge its transportation tariff.[37] Consequently the price realised by the producer ONGC was determined out of the net revenues from gas sales after paying for more expensive joint venture gas and after deducting the transportation tariff as well as Rs2.5 billion as an annual contribution to the Gas Pool Account (to compensate producers for supplying gas at lower prices in the north east). This was known as the net-back method of determining producer prices. In effect this meant that ONGC subsidised the difference between fixed consumer prices and contractually determined higher joint venture gas prices.

The intention of the government was to gradually phase out price controls and eventually bring in market-determined prices based on full parity with fuel oils. The October 1997 gas pricing order was to be effective until March 2000. But with increasing cost pressures being felt by the power and fertiliser sectors which were still subject to price controls, successive governments could not muster the courage to revise the ceiling upward and continue to pursue the reforms. The result was that consumer prices reached the ceiling of Rs2850 (at 75 percent parity) in October 1999 and were forced to remain frozen at this level until June 2005 even while crude oil prices were rising dramatically. Under the net-back method of determining producer prices this translated into revenues of $1–1.3/MMBtu to producers such as ONGC.

The stranglehold of administered prices, of course, owed a lot to the uneven pace of reforms in the end user sectors particularly fertiliser and power, the major consumers of all natural gas sold in India (see Figure 3.2). Since these sectors were unable to charge their customers market prices, they were in no position to accept market prices for gas and continued to lobby hard for artificially low prices of fuel and feedstock. Looking at diminishing and then negative returns from its producing gas fields ONGC saw little, if any, incentive to invest in exploration, development or improved recovery of domestic gas. The disparity vis-à-vis crude oil became even more pronounced after April 2002 when administered prices for domestic crude oil produced by ONGC and OIL ended and they began receiving import parity prices for all crude produced.[38]

Succeeding years saw a process of ferment with several developments taking place simultaneously. In an entirely unrelated development, as crude prices continued their relentless march towards $50 a barrel and beyond, the price formulae under the Tapti and Panna–Mukta PSCs became due for contractual revision in June 2004 and February 2005 respectively. Simultaneously, in 2004, Petronet LNG set up its 5 MMtpa terminal at Dahej and markets started receiving re-gasified LNG at non-administered prices. Shell's LNG terminal at Hazira, which followed shortly after, was based on spot LNG rather than long-term contracts. Consequently, Indian gas markets, initially reluctantly, got their first exposure to international gas prices. Gas allocations having far outstripped supply, the country launched into a phase of hyperactive diplomacy to develop transnational pipelines which either came from or passed through the territory of recalcitrant neighbours. Whatever the results of this diplomacy, gas prices at other than administered levels started to be actively considered within the domain of public policy.

The contractual provisions under the Panna–Mukta and Tapti PSCs had fixed the price of gas at 90 percent of the designated fuel oil basket, with a floor and ceiling of $2.11 and $3.11/MMBtu respectively. Seven years after these contracts were signed (that is in June 2004 and February 2005 respectively) the ceiling was due to be raised to 150 percent of the average gas prices prevailing during the previous year calculated at the same 90 percent of the designated fuel oil basket. The revised ceiling ($5.73/MMBtu for Panna–Mukta and $5.57/MMBtu for Tapti) meant that gas from these fields would have to be bought by the government nominee (GAIL) at over $5/MMBtu. This was in effect even higher than existing market prices (for gas from the small fields) in the region as well as the ex-Dahej price for re-gasified LNG being imported into India. Since the contractors were clamouring for a hike in prices under the contractual provisions, the only solution was a compromise under which the joint venture partners were given marketing freedom for a limited 1.75 Bcm[39] of gas per year. The remaining 2.2 Bcm per year was allocated to GAIL but its prices, for the first time, were benchmarked at $3.86/MMBtu until April 2006. This was the ex-Dahej price for LNG being imported by Petronet LNG from Ras Gas, Qatar under a 25-year contract.[40] Meanwhile the PSC contractors (which included ONGC as a joint venture partner) were asked to enter into independent price discovery by inviting bids from potential buyers in the open market. This process pegged gas prices from this field at $4.75/MMBtu. From April 2006 this became the market determined price for GAIL's share as well. Thus was the first successful transition to a market linkage for gas prices in India achieved for limited quantities of gas. Taking a cue from this, gas prices on the East Coast from the Ravva PSC were revised to $3.5/MMBtu[41] from July 2005, and Ravva satellite gas, which had been priced at $3.3/MMBtu until September 2006, was revised to $4.30/MMBtu in January 2007. The experience of the Panna–Mukta and Tapti gas price revisions followed by Ravva demonstrated for the first time that bulk consumers in India were willing to accept market prices.[42] By 2007–08 Petronet LNG itself was negotiating for as many as 24 spot cargoes of LNG at international prices and seeking assistance from Shell for use of its facility at Hazira.

When Shell first began importing spot LNG cargoes in 2005 at its Hazira terminal the general perception had been that there would not be a market for such high priced LNG. Shell rarely managed to operate its terminal at more than a third of its capacity until 2006. However, customers (particularly fertiliser and petrochemicals as well as some power plants) have increasingly turned to imported gas in order to replace more expensive naphtha. In 2007 the Hazira terminal may well operate at over 60 percent capacity. Already Petronet LNG and some others have been trying to negotiate with Shell for use of its facilities. This may be a lesson for those who feel that India may not have a market for gas at higher prices. Gas markets in India, on the whole, have so far proved themselves to be more flexible than policy makers had asumed.

As far as APM gas was concerned consumer prices were revised in July 2005 to Rs3200/Mcm ($1.98/MMBtu).[43] For the north east states, consumer prices were fixed at Rs1920/Mcm ($1.18/MMBtu). Joint venture gas having already made the transition to market-related prices, the supply of APM gas was reserved against existing allocations to power and fertiliser companies and for small consumers having a demand below 0.05 MMSCMD.[44] All other consumers would pay market-related prices. ONGC currently gets paid the consumer price (Rs3200 plus 10 percent royalty) after deducting Rs1.5 billion annually for the Gas Pool account to compensate for sale of subsidised gas to north east customers. After May 2006, the government having accepted the recommendations of the Tariff Commission on transportation, GAIL adopted cost of service transmission tariffs for the integrated HVJ–DVPL pipeline. The effective tariff based on the current compression cost, inflation, tax rates and volumes of gas transported, works out at Rs960–970/Mcm ($0.60/MMBtu).

Meanwhile in October 2005 the controversial Dabhol Power Corporation, owner of the 2150 MW Dabhol power plant as well as the partially completed 5 MMtpa LNG terminal, was taken over by Ratnagiri Gas and Power Private Ltd (RGPPL) – a SPV comprising NTPC, GAIL and institutional lenders. Its LNG terminal not yet being operational, the power plant was forced to run on expensive naphtha at under 30 percent capacity, which translates to a power cost of Rs5.10 a unit. In the meantime, GAIL as well as Petronet LNG has been sourcing LNG for the plant on spot basis at around $9/MMBtu. This price was considered too high for Dabhol power, so the government decided on a uniform pooled price for all LNG being supplied by Petronet irrespective of the source. Under this arrangement so long as RasGas LNG is available at the fixed price of $3.86 ex-Dahej (see note 41), Dabhol would get LNG at a delivered price of $5.83/MMBtu. This would reduce the cost of power being bought to Rs3.08 a unit, at least until December 2008 when RasGas prices themselves become due for revision. By then the Dabhol terminal should be operational and the power plant will either have to source its own LNG, or alternatively seek domestic gas which will probably become available from the KG basin (provided the dispute between RIL and NTPC in gas supplies gets resolved). In July 2007 with the completion of the Dabhol pipeline, GAIL started supplying LNG to Dabhol. However, the pooled price mechanism itself has been challenged in the courts by consumers who had already contracted for RasGas LNG, as they now have to bear the burden of higher pooled prices.

Table 3.4 shows estimates of domestic natural gas production on the basis of current trends in production and of the development plans being implemented.

Almost all the planned increase in production shown[45] is expected to come from Reliance's D-6 field alone, which at a production level of 29.2 Bcm per year would, by 2009–10, account for 47 percent of all domestic gas in the Indian market. This is in addition to the 5 Bcm which flows at present from the Panna–Mukta and Tapti fields in

Table 3.4: Projections of Domestic Natural Gas Production

Year	*2007–08*	*2008–09*	*2009–10*	*2010–11*	*2011–12*
ONGC	22.10	22.53	22.77	22.99	22.00
OIL	3.13	3.21	3.25	3.28	3.56
Pvt./JV	8.55	22.55	36.71	36.07	41.67
Total (Bcm)	33.78	48.29	62.73	62.33	67.23
Total in MMSCMD	92.2	132.3	172	171	184

Source: Ministry of Petroleum and Natural Gas, Government of India

which RIL has a 30 percent share. Simultaneously ONGC's production from its nomination acreages is set to decline from the current level of 20 Bcm, and this is the most significant anticipated change in market behaviour. As the share of APM gas declines and more and more gas starts flowing in from the PSCs, APM supplies to the favoured sectors will need to be reduced. Any additional gas will have to be bought at market related prices even from the NOCs under PSC arrangements. By 2009, market priced PSC gas is expected to contribute 54 percent of all available gas of which almost 80 percent will flow from RIL's D-6 field (Figure 3.5).

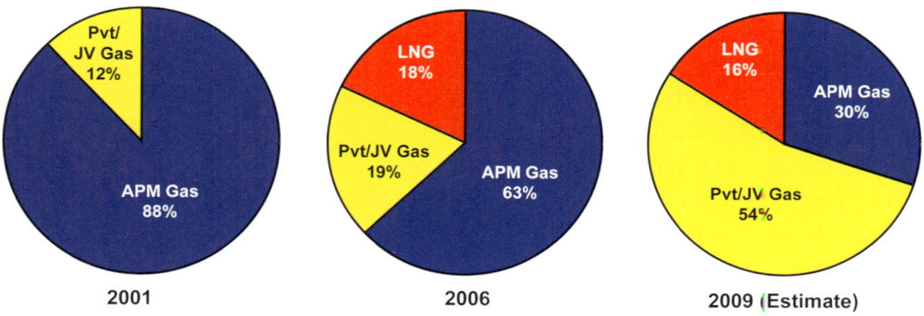

Figure 3.5: The Declining Share of APM Gas

Just 30 percent of the total gas supply will then be available from ONGC and OIL at administered prices. Already by July 2007, the variety of gas prices prevailing in the country (Table 3.5) presented a confusing picture with transition to market-related prices happening at a rather slow and uncertain pace.

As APM gas volumes decline and gas from NELP fields increases, it would normally be expected that the price of gas would correspondingly

Table 3.5: Gas Prices in July 2007: A mixed bag

Gas source	*Region*	*Bcm/yr*	*$/MMBtu*
APM Gas	North East	2.04	1.18
APM Gas	Rest of India	18.98	1.98
PSC Panna–Mukta, Tapti	West Coast	5.08	4.08–4.75[46]
PSC Gauri–Lakshmi	West Coast	0.47	3.68–4.46
PSCs Hazira etc	West Coast	1.39	5.00
PSC Ravva	East Coast	0.76	3.5–4.30
Petronet LNG	West Coast	6.57	3.86[47]
Spot LNG	West Coast	1.46	6 – 9

* 1 MMSCMD = 0.365 Bcm per year

settle at market levels allowing for a smooth transition to a market determined gas pricing regime for all sectors. While allowing for market prices, NELP restricts oil and gas sales exclusively to those within the country and consequently these prices would be expected to settle below international ones. However, fertiliser and power sector reforms were still quite far away and APM gas still accounted for 63 percent of gas supplies in the country; when the first major gas price under NELP came up for approval there was significant pressure on the government to re-introduce cost plus pricing for gas. Contractual Provisions notwithstanding, even such an authoritative document as the IEP Report prepared by an Expert Committee under the Planning Commission, although it accepted market related prices for all fuels, recommended cost plus pricing due to limited supply. The IEP document argues: 'While requiring that no new gas capacity (for power generation) be built without firm and bankable gas supply agreements, effort should be made to allocate available domestic gas supplies to the fertiliser, petrochemicals, transport and power sectors at prices that are regulated to yield a fair return to domestic gas producers.'[48]

Given that the fertiliser and power sectors alone account for 74 percent of the total gas consumption in the country the outlook for the future growth of these sectors is certainly going to have an impact on the movement and control of gas prices. While the power requirement in 2004–05 was 691 billion kWh, the supply was 625 billion kWh. The total installed power capacity in the country is 127 GW; but, assuming a GDP growth rate of 8 percent per annum, the country would need 220 GW of capacity by 2012. The IEP report mirrors the apprehensions and fears in this regard: 'as long as there is shortage of natural gas in the country and the two major users of gas, namely fertiliser and power, work in a regulated cost plus environment, a competitive market determined price would be highly distorted. Such distortions would get further amplified by the prevailing regime of fertiliser subsidies and power sector subsidies and cross subsidies. In such a situation price of domestic natural gas and its allocation should be independently regulated on a cost plus basis including reasonable returns' (IEP Report, p. 127). The Report of the Working Group on Power reiterates 'as long as there is shortage of natural gas and the two major users of gas fertilizer and power work in a regulated cost plus environment, price of domestic gas and its allocation should be independently regulated on cost plus basis'.[49] It seemed that few lessons had been learnt from the experience of introducing cost plus regimes for private participation in the power sector (through IPPs) and the fertiliser sector.

These arguments were fully rehearsed when RIL tried to discover a market price for KG D-6 gas. Its first attempt to find a market price for NELP gas from RIL's KG-DWN-98/3 block began unfortunately. In 2003 RIL bid a price of $2.9/MMBtu (on Gross Calorific Value basis)[50] for supplying 12 MMSCMD (4.8 Bcm) of gas to the public sector National Thermal Power Corporation (NTPC) once the gas came on stream in 2008. But subsequent market changes mired the deal in controversy with NTPC dragging RIL into the courts to enforce this price. Meanwhile RIL itself had to break up its power and oil and gas components into two separately controlled and managed entities. As a part of the de-merger agreement, gas supplies were committed for power production projects to be undertaken by the de-merged Reliance Natural Resources Ltd (RNRL). However, the government refused to approve RIL's proposal for supplying gas to RNRL at the same prices as the NTPC offer. When RNRL took RIL to court over the issue of gas supplies, the two companies were directed to re-negotiate the contract length, quantity and price terms in their gas supply agreements.

While these disputes were still pending in court, RIL submitted a fresh proposal for the approval of the government, linking the price of KG D-6 gas to crude oil under a formula[51] in which 55 percent of the price component is fixed and 45 percent linked to the price of Brent crude. On the basis of quotes received from buyers, RIL proposed a gas price which, when Brent prices were at or above $65 a barrel, worked out to $4.33/MMBtu. Both the fertiliser and power ministries lodged strong protests with the government, insisting that no free and fair markets existed for these large volumes of gas in the country due to shortage of gas, and sought a policy on pricing and utilisation before approval was given for the proposed gas prices. Upstream investors protested against the stand taken by this powerful group of gas consumers and sought to maintain marketing freedom, as promised under the

PSCs. This move along with the government's stand on pooling LNG prices to bail out Ratnagiri Power, made the pricing of gas from two major market-based sources a contentious issue. Eventually, in spite of PSC provisions, the RIL tangle was referred to a Group of Ministers (GOM) which was to consider RIL's proposal in the broader context of a policy for allocation and pricing of gas. The GOM finally agreed to the proposal in broad terms with the following modifications:[52]

(i) The ceiling was pegged at a crude price of \$60/bbl in place of the \$65/bbl proposed by the contractors.
(ii) The biddable component C in the formula was brought down to zero.
(iii) The price was to be denominated in US \$ rather than Rs.
(iv) The fixed component of Rs. 112.5 per MMBtu in the formula was converted to US \$2.5 using an exchange rate of US\$1=Rs45.[53]

These modifications effectively reduced the formula to $2.5 + (CP–25)^{0.15}$ with CP moving in the range \$25–60/bbl. At crude oil prices well above \$60/bbl, this had the effect of marginally reducing the proposed gas price of \$4.33 to \$4.20 per MMBtu and effectively creating a ceiling at this price. Other contractors, who are already selling contracted quantities of gas well above this price have become somewhat apprehensive about this proposed cap and the extent to which the government is prepared to interfere in the operation of gas markets.

The GOM decision has also ensured that the base price in NELP contracts does not fall below \$2.50 per MMBtu, a level which could provide some comfort to investors. While the government has asserted its intent to stand by existing contracts it has, however, required that for all future contracts, the government's prior approval would have to be obtained for the formula on which bids are invited by contractors. Whether this would lead to amendments in the terms of offer for new blocks and the provisions of future PSCs remains to be seen. The government's response to the pricing issue and its intervention, though symbolic, also highlights the political sensitivities which exist on gas pricing issues. It still remains to be seen whether, after the KG D-6 deliberations at the highest level, the government would accede to demands from politically sensitive groups of consumers and bring about changes in the terms for forthcoming rounds of NELP. At stake, however, would be the runaway success that NELP has had in attracting investments which have led to new finds. Simultaneously, there is now increasing pressure on the government to take its share of gas from the Production Sharing Contracts in kind, in order to allocate it to priority sectors, such as power and fertilisers, at subsidised rates. If that happened it would certainly affect the future growth of the gas market in India and might even slow the pace of new investments in exploration and development in what seems to be a very promising frontier area in this part of the world. How investors would view these developments would only become apparent from the outcome of forthcoming licensing rounds.

Regulation and the Development of Infrastructure

Until very recently all aspects of the gas sector, including pricing, allocation and usage, were regulated by the government. Between 1987 and 1989 GAIL commissioned the country's first major gas transportation pipeline, the 1700 km long Hazira–Vijaipur–Jagdishpur (HVJ) pipeline with a capacity of 18.2 MMSCMD (6.64 Bcm). By 1997 the length of this network had increased to 2800 km and the capacity to 33.4 MMSCMD (12.19 Bcm). With the coming of Petronet LNG's Dahej terminal, GAIL commissioned the 610 km Dahej–Vijaipur line (DVPL) linking Dahej to the HVJ network. Besides GAIL, Gujarat State Petronet Ltd, set up in 1998 as a company of the Gujarat State Government-owned GSPC group, has built 1090 km of pipelines transporting about 18 MMSCMD of gas from Hazira to Kalol. Gujarat is also serviced by the Gujarat Gas Company Ltd which is promoted by the BG group (with a 65.2 percent share) and operates 1800 km of pipeline network linking Hazira to Ankleshwar.

As Table 3.6 shows, the gas transportation network, both in terms of length as well as capacity is at present dominated by GAIL. In a highly skewed pattern of distribution, the northern, western and southern states dominate demand having a combined consumption of 90 percent of available supply. One reason for this has also been the fact that the eastern region is richer in resources such as coal. However, following the development of new gas finds, the pipeline network is set to expand from about 9400 to more than 19,000 km creating a pan-Indian grid. City gas distribution (CGD) networks would be installed in a number of cities. An estimated $28 billion investment would be needed to build the proposed pipelines which would eventually result in a pan-Indian gas grid connecting the domestic, commercial and the transportation sectors in about 150 cities in the country. Consequently city gas demand, although the infrastructure is yet to be laid, is anticipated to increase six-fold to 5.8 Bcm by 2012. Subsequent growth, once greater clarity has emerged on regulatory issues, would be far more rapid.

Given that the market is now supplied by gas production available

Table 3.6: Pipeline Capacity (2006)

Company	*Network (Region)*	*Length (Km)*	*Capacity Bcm/yr*
GAIL	HVJ (including spur lines)	3190	12.2
	DVPL	610	8.7
	Assam (Lakwa)	9	0.9
	Tripura (Agartala)	60	0.8
	Gujarat region	745	7.1
	Maharashtra (Mumbai)	125	8.6
	Andhra Pradesh (KG Basin)	835	5.8
	Tamil Nadu (Cauvery Basin)	255	3.2
	Total	5829	47.3
GSPL	Gujarat	1090	6.6
GGCL	Gujarat	1800	1.2
Assam Gas Company	Assam	600	2.2

Source: MOPNG, GGCL and GSPL

under open market pricing from PSCs as well as by imports of LNG by both Petronet LNG and Shell, the government has come under increasing pressure to establish an agency, and a policy, for gas industry regulation. The Petroleum and Natural Gas Regulatory Board Bill was finally passed by Parliament in March 2006 and, after more than a year, the Board was created. The government has also announced its pipeline policy for shipping natural gas in December 2006 (see below).

Even while these processes are continuing, Reliance Gas Transportation Infrastructure Ltd (RGTIL), having obtained all clearances, has already commenced work on its ambitious 48 inch 1386 km East–West Kakinada–Hyderabad–Ahmedabad pipeline to transport KG basin gas from Kakinada in Andhra Pradesh to Bharuch in Gujarat (Map 3.3). Once in Gujarat, RIL plans to use both GAIL's and GSPL's existing transport network in the western part of the country to reach its customers. It has already entered into agreements with both GAIL and GSPL to make this possible. RGTIL has also received authorisation for the Kakinada–Vijaywada–Chennai pipeline. Meanwhile GAIL and RGTIL, separately, with the approval of the Ministry of Petroleum have in (March 2007) invited Expressions of Interest for 9500 km of additional pipelines (for details see Appendix 3.2).

While the Regulatory Board takes shape, the emerging gas pipeline transportation network will be dominated by GAIL and RGTIL. This situation is quite different from that of a few years ago when GAIL

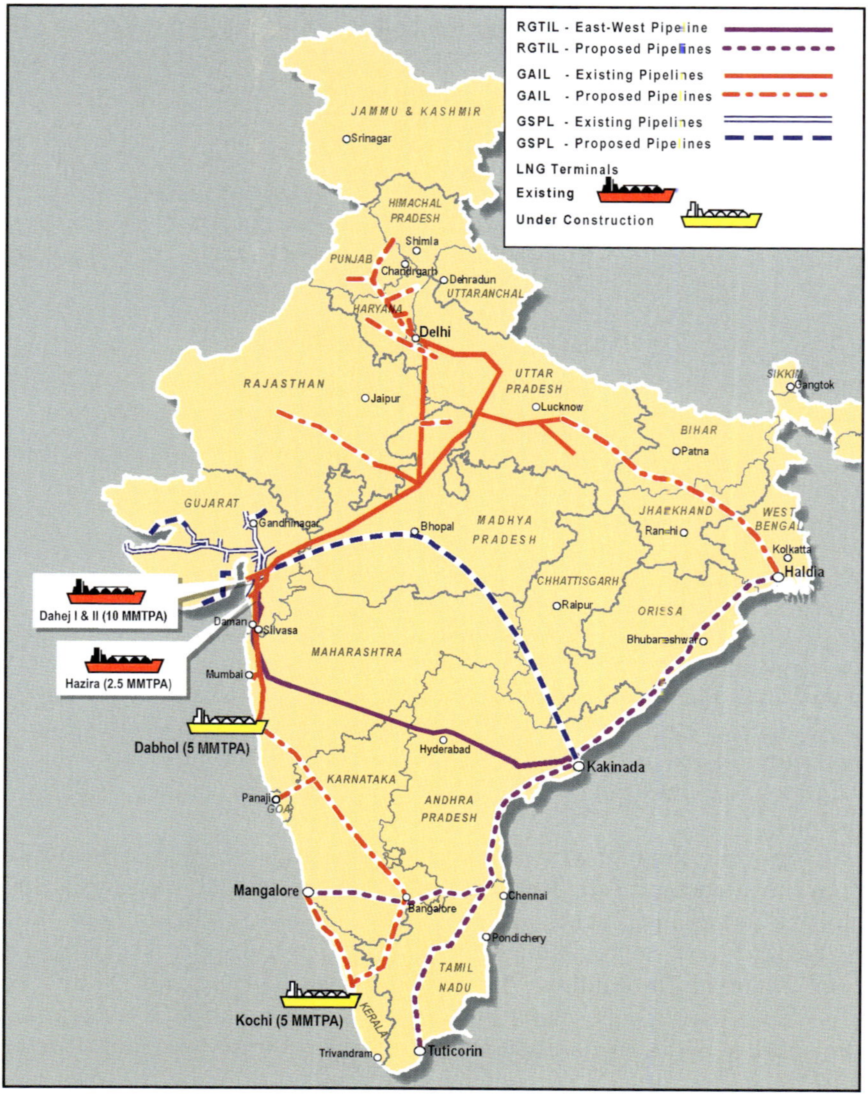

Map 3.3: Pipeline Infrastructure – Existing and Future

had been seeking exclusive control and rights to lay its own 7890 km gas grid across the country. Since the government procedures for seeking necessary approvals for laying pipelines make it mandatory for gas transportation companies to prove firm gas supplies before being considered for a grant of permission, GAIL (which does not own any major gas supplies) would have been at a disadvantage in obtaining necessary approvals for expanding its pipeline networks. But now that an understanding has been reached between GAIL and RIL, it seems that the two are prepared to co-operate with each other to lay the gas grid, RIL agreeing to give GAIL access to its gas.

If gas from the new discoveries as well as the proposed LNG terminals is to be brought to the market, India will be laying at least 10,000 km of fresh pipelines during the period 2008–12. The investments proposed by GAIL and RIL to make these pipelines a reality, are expected to reach $18 billion or more. A substantial share of this investment, particularly the expansion of the city gas distribution network across the country will depend upon various regulatory issues being resolved. Expansion of the city gas network across 150 cities when it does take place will cost another $10 billion.

The PNGRB Act of 2006 is the enabling legislation for setting up a Petroleum and Natural Gas Regulatory Board which would essentially regulate the midstream and downstream. The upstream continues to be regulated by the government with the technical assistance of the Director General of Hydrocarbons. The following functions have been assigned to the PNGRB:

1. Protecting consumer interests by fostering competition and promoting fair trade.
2. Authorising and granting registration to entities for:
   (a) marketing notified petroleum and petroleum products and natural gas
   (b) establishing and operating LNG terminals.
3. Authorising entities so that they can:
   (a) Lay, build, operate or expand common carrier or contract carrier pipelines.
   (b) Lay, build, operate or expand city or local gas distribution networks.
4. Declaring pipelines as common carriers or contract carriers.
5. Regulating:
   (a) Access to common carrier or contract carrier pipelines in order to ensure fair trade and competition.
   (b) Transmission tariff rates for common carrier and contract carrier pipelines.
   (c) Access to city or local natural gas distribution networks so as to ensure fair trade and competition amongst entities.
6. In respect of notified petroleum, petroleum products and natural gas:

(a) Ensuring adequate availability.
(b) Ensuring the display of information about the maximum retail prices fixed by the entity for consumers at retail outlets.
(c) Monitoring prices and taking corrective measures to prevent profiteering.
(d) Securing equitable distribution.
(e) Providing and enforcing retail service obligations for retail outlets and marketing service obligations for entities.
(f) Monitoring transportation rates for natural gas by common carrier or contract carrier pipelines as well as the prices charged by local distribution companies to prevent profiteering.

Following the PNGRB Act, the Gas Pipeline Policy notified by the government on 20 December 2006 makes it mandatory for any company laying a pipeline that is not intended for a dedicated customer to provide 33 percent capacity in excess of its own requirements. This 33 percent capacity is to be made available on a non-discriminatory and open access common carrier basis to any third party at transportation rates laid down by the Regulator. The stipulation for mandatory excess capacity has come in for its fair share of criticism because of the cost which it adds to projects.

Companies seeking to set up a city gas distribution network will be selected on composite criteria including tariffs, capital investments, IRR, CNG volumes and the number of small commercial and industrial users.

The other salient features of the policy are:

1. All natural gas and city or local distribution pipelines are to be laid in accordance with the authorisation granted by the Regulatory Board. Dedicated pipelines, however, will not need any authorisation.
2. The regulator shall develop a comprehensive set of technical and safety standards as well as a code for grid connectivity.
3. There should be progressive unbundling of gas marketing and sales from city and local gas distribution networks and common carrier transmission pipelines.
4. Periods of exclusivity for distribution networks and the marketing of gas therefrom will be decided by the Board.
5. Transmission tariffs as well as the manner of determining such tariffs will be laid down by the Board.
6. Foreign direct investment up to 100 percent would be permitted in laying NG pipelines under the automatic approval route.

Under the stated policy, the government was also meant to prepare a long-term plan for creating a gas pipeline network in consultation with the Regulator. However, as described in the first part of this section, a substantial part of this pipeline infrastructure seems to be well on its way even as the new regulation system is being created.

One of the key areas of dispute in the future will be the Board's position on unbundling of activities down the entire chain. Because gas transporters are required to indicate firm sources for gas before receiving authorisation to lay pipelines, the playing field is tilted in favour of producers with firm development plans. This becomes even more pronounced because of the limited availability both of LNG supplies and of infrastructure for imports. GAIL, with its strong foothold in the distribution and marketing network, would be the only other company which could partially offset the inherent advantage enjoyed by producers. However, as long as LNG supplies remain tight, GAIL will have to link up with producers such as RIL and subsequently GSPC and ONGC to ensure its supplies.[54]

Conclusion: The Future of Reforms in the Gas Sector

The future growth of the gas sector is inexorably linked to changes in other parts of the economy, most particularly the pace of reforms in the power sector. As long as the product prices of the prime gas consumers remain regulated, the price of gas as an input cost will be a matter of serious concern for the exchequer and susceptible to interventionist pressures from interest groups. It will be some time before the pricing mechanisms of agricultural inputs like energy and fertiliser embrace markets, so it is a foregone conclusion that certain subsidies will be inherent in a system where income distribution is glaringly unequal. In a rapidly globalising economy, policy makers thus are faced with a difficult choice, to which there are no easy answers. One cannot therefore foresee many of the present subsidies disappearing in the short or even medium term. The focus in public discourse is at present shifting more to questions related to the distribution and administration of subsidies rather than questioning the need for subsidies as such. The questions being raised are: do these subsidies really need to come in the form of controlled costs and restrictive allocation mechanisms for inputs like natural gas, which ultimately only create oligarchic access on the one hand and a bureaucracy of control and rent seeking on the other? Does not this process put a premium on inefficiencies leading to high costs for the very goods and services the state is spending precious resources to subsidise? Would state support mechanisms not be better and more transparently directed if they addressed the ultimate end users as beneficiaries? Thus the real political choices that will need to be made in the future are between the relative merits of transparent subsidies on the one hand and hidden or disguised subsidies and cross subsidies on the other.

Ultimately, the actual driver for this reform process may not necessarily be concern for increasing productivity and efficiency, but may rather evolve more as a response to crises. The severe and progressively worsening deficit in power being faced by the country may just constitute the trigger for change. Up to the present for policy makers at least, an important question has revolved around gas as a substitute fuel in power generation where it obviously needs to compete with coal. But any calculations on the relative desirability of these fuels are valid only if they take into account the amount of power needed to fuel the country's accelerated rate of growth. India is already going through a severe and worsening crisis relating to the availability and the quality of power. In 2005, some states reported a peak shortage of 25 percent or more; for the whole country it was 11.7 percent.[55] Since the existing generating capacity does not have the mix capable of handling peak, intermediate and base loads, the gap between demand and supply for power will grow. Transmission and distribution investments have lagged far behind additions to generating capacity, so the problem of transferring power from surplus to deficit areas is a major one. According to the IEP report there may already be 50,000 MW of captive and standby capacity existing in the sector. Under these circumstances the question of supply should be increasingly viewed from the viewpoint of the power user. The IEP report itself acknowledges that, calculated on the basis of purchasing power parity, power tariffs for commercial users in India are perhaps the highest in the world. This is in large part due to cross subsidies leading to high amounts of theft and transmission and distribution losses. The comparisons given in the IEP document itself are striking: 'For example, in 2002, industries in India paid 47 US cents per unit as opposed to 20 cents in China, 17 cents in Brazil, 12 cents in Japan, 5.5 cents in US and 5 cents in Germany in terms of PPP.'[56] Bulk industrial and commercial users, accustomed to paying higher differential power tariffs[57] and facing increasingly uncertain supplies of power, would prefer to switch over to captive as well as merchant power. Such supplies should be based on perhaps more expensive, but more secure, domestic or even imported gas, should these become available in the coming years.

While the role of the new Regulatory Authority is going to be critical in determining the pace at which gas markets develop and build in the coming years, increasing supplies would force the pace of development of city gas distribution. During 2005–06, 10.5 million tonnes of liquefied petroleum gas (LPG) were consumed as domestic fuel in Indian households. This translates to a potential demand for 12.8 Bcm of natural gas. Even though LPG continues to be subsidised,[58] the local distribution plans of all major gas players look upon this consumer group as an area of high growth and maximum value addition in the immediate future. Natural gas at market prices is expected to compete with and replace subsidised LPG wherever it can reach these customers. This would reduce the burden of subsidies on this fuel which constitutes the country's only significant imported petroleum product. As the gas grid develops and the distribution network extends its penetration across the country, increasing numbers of domestic, commercial and industrial consumers would be brought within its expanding fold.

The gas sector in India, even though it has still not emerged from an era of severe price-led supply constraints, is at a major turning point. Should the promise of east coast gas be fulfilled, imported gas would certainly face severe price competition from domestic gas. In recent years, private investors have been shy to commit to investments in LNG for this reason. Already, supplies from the RIL's D-6 block alone are expected to double the availability of domestic natural gas by 2009. Now that LNG imports are underway, Petronet LNG is going ahead with the Kochi terminal and active diplomacy is taking place on transnational gas pipeline imports, India is closer to being integrated into the international gas market than at any time in the past. Since the huge volumes of gas from Reliance's KG D6 are being allowed into the market at prices more or less in line with RIL's proposal, both LNG and imported pipeline gas will have to find markets at prices that will compete with the sudden increase in available volumes of domestic gas. The expectation that gas production will double very soon as this single major development plan is completed, means that the urgency for both LNG and transnational pipelines has to some extent receded. While already committed LNG investments may see the light of day, further plans will necessarily be on hold until the gas markets move to the next level of growth. Given the success of NELP, this growth may come more from future development plans of domestic gas rather than gas imports. The pace of this growth and integration will of course be determined to a large extent by reforms in allied sectors such as power, coal and fertilisers. Meanwhile, the availability of gas will create and build its own markets and may even, for example in the case of city gas, drive rather than follow the reform process.

APPENDIX 3.1: Demand Projections for Natural Gas in India by Different Agencies

Table A3.1.1: Demand Projections for Natural Gas in India by Different Agencies. Bcm

Year	*EIA (2004)*			*IEA (2007)*	*IHV (2000)*	*India Vision 2020 (2002)*		*Projections by Power and Energy Division*	*IRADe and PWC**	
	Reference Case	*High Case*	*Low Case*	*AP case*		*BAU*	*BCS*		*BAU*	*HOG*
Base Year (Bcm)	2001 (22.6)	2001 (22.6)	2001 (22.6)	2005 (35)	1999–2000 (40)	1997 (21.5)		2001–02 (29.5)	2003–04 (31)	
2004–05	27.0	28.1	27.0		71.2	32.5	31.8	35.8	33.9	34.7
2009–10	33.9	36.9	33.9		101.1	42.0	40.5	48.9	52.9	59.9
2014–15	45.3	48.2	39.8	57	120.1	54.4	51.8	66.8	82.5	104.0
2019–20	56.6	62.4	48.2		130.7	70.8	64.6	90.9	129.9	179.9
2024–25	71.2	82.1	56.6		142.7	94.2	82.5	119.0	178.1	269.4
2029–30				107				157.0	243.5	405.5

EIA: US Energy Information Administration
IEA: International Energy Agency
IHV: India Hydrocarbon Vision 2025
IRADe: Integrated Research and Action for Development
BAU: Business as Usual
BCS: Best Case Scenario
PWC: Price Waterhouse Coopers
HOG: High Output Growth

Source: Integrated Energy Policy: Report of the Expert Committee, Government of India, Planning Commission, New Delhi, 2006, p. 27.

APPENDIX 3.2

Pipelines under construction

GAIL:

(1)	Dahej–Uran–Dabhol	620 km	completed in July 2007
(2)	Jagoti–Pithampur	180.5 km	
(3)	Vijaypur–Kota	192 km	

RGTIL:

(1)	Kakinada–Hyderabad–Uran–Ahmedabad (East–West pipeline)	1,385 km

Pipelines for which expressions of interest have been invited

GAIL:

(1)	Dadri–Bawana–Nangal	590 km
(2)	Chainsa–Gurgaon–Jhajjar–Hissar	310 km
(3)	Jagdishpur–Haldia	876 km
(4)	Dabhol–Bangalore	730 km
(5)	Kochhi–Kanjrikkod–Bangalore–Mangalore	862 km

RGTIL:

(1)	Kakinada–Basudebpur–Howrah	1,100 km
(2)	Chennai–Tuticorin	670 km
(3)	Chennai–Bangalore–Mangalore	660 km
(4)	Vijayawada–Nellore–Chennai	445 km

GSPC:

(1)	Kakinada–Ahmedabad	1,851 km

APPENDIX 3.3: Dollar equivalence of Consumer Prices for APM Gas

Table A3.3.1: Dollar equivalence of Consumer Prices for APM Gas

Year	*Region*	*Rs per thousand cu m*	*Royalty @ 10%*	*$ per MMBtu*	*Reference exchange rate for $*
1987	All india	1400	140	3.31	13.00
	NE	1000	140	2.45	13.00
1992	All india	1500	150	1.78	25.92
	NE	1500	150	1.78	25.92
1993	All india	1500	150	1.47	31.44
1994	All india	1500	150	1.47	31.37
1995	All india	1500	150	1.42	32.42
1996	All india	1500	150	1.30	35.43
till Sep 97	All india	1500	150	1.27	36.32
97–98*	All india	2150	215	1.60	41.27
98–99	All india	2150	215	1.53	43.06
99–00	All india	2850	285	1.95	44.94
00–01	All india	2850	285	1.86	47.19
01–02	All india	2850	285	1.80	48.60
02–03	All india	2850	285	1.88	46.58
03–04	All india	2850	285	1.93	45.32
04–05	All india	2850	285	1.99	44.10
till Jun 05	All india	2850	285	1.98	44.26
July 05**	All india	3200	320	2.22	44.30

* Calorific value for APM gas, effective October,1997 was increased from 9,000 kcal/SCM to 10,000 kcal/SCM

** prices applicable only to power and fertiliser consumers

Note: Consumer Prices shown here in dollars include royalty but not transport and taxes.

Source: compiled by authors from various sources

APPENDIX 3.4: List of Common Acronyms used

APM	Administered Price Mechanism
DGH	Directorate General of Hydrocarbons
GGCL	Gujarat Gas Company Ltd
GSPC	Gujarat State Petroleum Corporation
GSPL	Gujarat State Petronet Ltd
HVJ	Hazira–Vijaipur–Jagdishpur pipeline
IEP	Integrated Energy Policy (a report by the Planning Commission, published in Aug 2006)
KG	Krishna Godavari Basin
MMSCMD	Million Standard Cubic Metres per day
NELP	New Exploration Licensing Policy (launched in 1999)
NTPC	National Thermal Power Corporation
OIL	Oil India Limited
ONGC	Oil and Natural Gas Corporation Ltd
PSC	Production Sharing Contract
RGTIL	Reliance Gas Transportation Infrastructure Ltd
RIL	Reliance Industries Ltd
RNRL	Reliance Natural Resources Ltd (de-merged from RIL in 2005)

Notes

1 *BP Statistical Review of World Energy*, June 2006.
2 A Joint Venture Company established by the Government of India expressly for the import of LNG in which the Public Sector Undertakings ONGC, GAIL, Indian Oil Corporation (IOC) and Bharat Petroleum Corporation Ltd (BPCL) hold 12.5% each. While Gaz de France International (GDFI) and the ADB hold 10% and 5.2% respectively, the remainder is held by the public.
3 The joint ventures, Panna-Mukta, Tapti, Ravva and Kharsang resulted from the earlier PSCs where ONGC and OIL's medium size discovered fields and exploration acreages were auctioned to private companies, while the NOCs were allowed to retain a working interest of 40%. Fields having small size and marginal discoveries did not carry this stipulation and are held exclusively by private companies. Pre-NELP exploration blocks also gave NOCs an option to take an initial 10% stake which could be increased up to 40% after discovery and declaration of commerciality.
4 This is exclusive of about 14,600 MW of grid connected captive power capacity under the captive power policy. All figures are taken from the Central Electricity Authority (CEA).
5 Source: Central Electricity Authority.
6 Department of Fertiliser, Government of India.
7 Report of the Working Group on Fertilisers, Department of Fertiliser, Government of India, 2006 (*http://planningcommission.gov.in/aboutus/committee/wrkgrp11/wg11_fertiliser.pdf*).
8 A compilation of various projections done in the Planning Commission's Integrated Energy Policy (IEP) Report is in Annex 1.
9 Recently Indian GDP has been persistently growing at 8 percent or more.
10 India Hydrocarbon Vision 2025, Ministry of Petroleum and Natural Gas, Government of India.
11 'Fuelling India's Growth Vision 2030' Petroleum Federation of India, New Delhi, 2005.
12 RIL had bid for supply of 12 MMSCMD of gas at $2.90 to NTPC in 2003 from production anticipated from its KG basin finds. This has now become a matter of legal dispute between the two parties.
13 Report of the Working Group on Power for the XI Plan, Ministry of Power, Government of India, New Delhi 2006, p. 6.
14 The Supreme Court has directed that CNG supplies should be made to 12 identified cities with high pollution levels.
15 These included a guaranteed 16% return on equity in $ terms, five-year tax holiday and counter guarantees by the Union Government.
16 See IEP report p. 46.
17 Import duty is 5%; there is a countervailing duty (CVD) of 16%; and handling charges are $5/ton.

18 Assuming that the average conversion efficiency of gas to power is 1600 to 1750 kcal/kwh while coal to power is 2300 to 2400 kcal/kwh.

19 Refining capacity rose from 62 MMtpa in April 98 to 113 MMtpa by April 2000. In April 2007 it has touched 149 MMtpa.

20 Report of the Working Group on Petroleum and Natural Gas for the Eleventh Plan, Ministry of Petroleum and Natural Gas, Government of India, Nov 2006.

21 ONGC in 2006–07 announced gas discoveries in block MN-OSN-2000/2 in the Mahanadi Basin and KG DWN 98/2 in the Krishna Godavari Basin.

22 Subsequent appraisals have so far confirmed proven reserves of 40 Bcm in GSPC's block KG-OSN-2001/3 and 59 Bcm in the discovery by ONGC.

23 *Hindustan Times*, New Delhi, Oct 18, 2007.

24 The Cretaceous is a geological period which relates to rock strata and depths at which hydrocarbons may be located.

25 RILs latest Cauvery discovery, D-35, is also reported to be in the Cretaceous.

26 BP has been trying to farm into ONGC's nomination block GK-DW-1 but could not get government approvals as post NELP the only available route is through open bidding.

27 Set up as a gas transportation infrastructure company to lay down and operate pipelines from RIL's fields.

28 Asian Development Bank, *Natural Gas Options for India and Pakistan: Market Outlook and Transportation Economics*, November 2004.

29 See *Financial Express*, 26 September 2007: 'IPI to be reality if price is Agreed on: India' also available at *http://www.financialexpress.com/news/IPI-to-be-reality-if-price-is-agreed-on-India/221360/*.

30 Presentation on Turkmenistan–Afghanistan–Pakistan (–India) Natural Gas Pipeline Project by Dan Millison, P.E., ADB South Asia Energy Division November 2006, available at *http://mea.gov.in/srec/internalpages/tapi.pdf*.

31 JCC or Japan Crude Cocktail is the reference point from which the price of LNG (and much other gas) traded in Asia is calculated.

32 *Financial Express*, 10 April 2007.

33 For ease of comparison the reference exchange rate for the period 1987 to 1995 has been taken to be flat at $1 = Rs32.

34 Until the 1990s because of restrictive exchange rate policies followed by the government comparison of these prices in dollar terms at rates prevailing at that time does not reflect the true dollar price of gas. It was only in 1992 that the rupee was partially floated and linked to an exchange rate more representative of the market. However, for ease of reference a chart comparing these prices at the actual prevailing exchange rates over the years has been included in the appendix to this Chapter.

35 Taking $1 = Rs45 as the average rate of exchange 1998 to 2005 this meant a floor and ceiling of $ 1.32 and $ 1.76 per MMBtu respectively.

36 Perhaps, in retrospect, it is here rather than anywhere else that the seeds of transformation of gas markets in India will really be found.
37 The transportation tariff is linked to a calorific value of 8500 kcal/cu m.
38 In effect this arrangement was rather short lived. International crude prices on fire and two successive governments were unable to raise consumer prices for petroleum products; ONGC and OIL were thus forced to give subsidies by way of price discounts on crude supplied to public sector refiners.
39 In common gas industry practice in India volumes are computed in terms of millions of standard cubic metres per day (MMSCMD). Thus the gas volume reserved for GAIL was 6 MMSCMD while the joint venture partners were allowed to market 4.8 MMSCMD.
40 Petronet LNG's contract with Ras Gas Qatar for supply of 5MMtpa commenced in April 2004. Price is linked to JCC under an agreed formula. However, the fob price until December 2008 has been agreed at $2.53/MMBtu which translates to a LNG price of $3.86/MMBtu ex Dahej.
41 Ravva Gas is being bought by GAIL on the basis of a formula with a floor and ceiling of $3.50 and $4.30/MMBtu for Ravva field associate gas and Ravva Satellite field non-associated gas respectively.
42 This freedom given to the Panna–Mukta Tapti contractors was however, short-lived. In November 2007, with GAIL lobbying for more gas it was re-designated the government's nominee under the contract for the entire quantity of gas, albeit at the higher contractual ceiling price of $5.7 per MMBtu.
43 Post 1998 the exchange rate has been assumed at $1=Rs45.
44 These small consumers would also have to move to market related prices over the next 3 to 5 years. In June 2006 prices for small consumers were raised to Rs3840/Mcm ($2.37/MMBtu).
45 These projections are strictly on the basis of approved development plans. Since ONGC and GSPC development plans for their discoveries in the KG basin have not yet been approved, they are not included in these projections.
46 Additional 6–7 MMSCMD contracted at $4.6–$5.7/MMBtu from September 2007.
47 Fixed price valid until December 2008.
48 p. xv, IEP Report, Planning Commission, Government of India, August 2006.
49 p.48,, IEP Report, Planning Commission, Government of India, August 2006.
50 $ 3.17/MMBtu on NCV basis.
51 Gas Price (Rs/MMBtu) = $112.5*K + ER(CP-25)^{0.15} + C$, where ER is the Rs/dollar rate of exchange, and the value of K is 1 for any ER between Rs25 and Rs65 to the dollar, CP = Price of Brent for the previous year, with a floor and ceiling of $25 and $65 per barrel, C is a biddable premium to be quoted by the customers in Rs per MMBtu.
52 *The Economic Times,* New Delhi, 13 September 2007.

53 This was not significant because at the time it was done the rapidly appreciating rupee was already trading at Rs40 to the dollar compared with Rs45 when the formula was formulated.

54 Once Petronet LNG implements its 2.5 MMtpa LNG terminal at Kochi it could supply about 9 MMSCMD to meet the gas demand along the Kochi–Bangalore route by 2012. But Petronet LNG is still to tie up the LNG for this terminal.

55 IEP, Report of the Expert Committee set up by the Planning Commission, New Delhi, 2006.

56 IEP Report, Planning Commission, August 2006 p. 5.

57 Power tariffs are structured so that industrial and commercial users subsidise agricultural and domestic consumption. Industries in India paid 47 cents a unit as opposed to 20 in China, 17 in Brazil and 5.5 in the USA in 2002.

58 The under-recoveries to public oil companies on account of subsidised prices for 2005–06 are estimated to be over USD 2.4 billion.

CHAPTER 4

NATURAL GAS IN JAPAN

Akira Miyamoto

Introduction

The global energy market has changed drastically in the 2000s and with it the environment surrounding Japan's natural gas market. For Japan, with its high degree of dependence on overseas energy supply, these changes appear to create significant difficulties for the future. At the same time, however, several positive factors for the development of natural gas have emerged in the domestic energy market. Whereas six years ago when the chapter was written for the first edition of this book the growth of gas was affected by a wide range of issues in a variety of areas, the debate now seems to have narrowed to a smaller number of key points. This has happened because the gas market has grown steadily over this period, creating increasing confidence in its future development. Nevertheless there are still critical issues lacking simple solutions, and these need to be overcome if natural gas is to become a principal axis of Japan's energy supply. This chapter examines the long-term market potential for gas by focusing on recent changes in the domestic energy situation, and at the same time, identifies issues essential for further development.

Changes in the External Environment

As the first step towards gaining a deeper understanding of the changes to Japan's natural gas market over the past five years, and for its medium- and long-term outlook, consideration needs to be given to the following changes in the general economic and policy environment.

Economic Recovery

After the collapse of the 'bubble economy' in 1991, Japan's economy entered a long period of recession, which depressed growth in energy demand. When the previous edition of this book was written, the

Japanese economy was deep in that recession. In April 2002, however, it at last started to show signs of recovery, entering a period of prosperity that has continued through 2007, making it the longest stage of continuous expansion in the post-war era, even surpassing the 'Izanagi boom'.[1] Increases in energy consumption have not matched overall economic growth, due to a combination of energy conservation and high energy prices. Average growth in GDP from 2000 to 2005 was 1.34 percent, but energy consumption growth over the same period was only 0.28 percent, representing extraordinarily low energy-GDP elasticity of 0.21.

Japan's population peaked in 2005, and the globally unprecedented ageing society and low birth rate are expected to accelerate, raising the spectre of declining vitality on a national scale. Nevertheless, the long-term economic outlook on a 2020 horizon is for firm growth with an annual rate of around 1.4–1.5 percent.[2]

Energy Policy

The second point is the change in Japan's energy policy. After the 9/11 terrorist attacks in 2001, countries around the world recognised once more the importance of ensuring national security, and this resulted in substantial changes in the fundamental framework for determining energy policy. The approach of Japan's energy policy up to that point had been to achieve simultaneously three basic targets, namely economic growth, environmental sustainability, and supply stability, and a start had been made on promoting energy market liberalisation in line with the global trend at that time. However, after the 9/11 incident, policy shifted to give top priority to energy security. Following the publication of the New National Energy Strategy,[3] the Basic Energy Plan was revised in March 2007 for the first time in three years.[4] Specifically, an increased emphasis was placed on nuclear power through the Nuclear Energy National Plan, and on other measures including securing energy resources and further conservation, while energy market liberalisation was effectively put on hold.

High Oil Prices

The high crude oil prices since the second quarter of 2004 resulted in a general rise in prices for LNG and other forms of energy. With regard to energy market trends, it is important to note that this series of energy price rises brought about drastic changes in the relative competitiveness between energy supplies, which had a substantial impact on the demand for different energy sources. For natural gas, the effect was enhanced competitiveness relative to oil products, but at the same time, reduced competitiveness against electricity. Generally speaking, market conditions in 2007 were conducive to growth in demand for natural gas, but it is not clear whether this situation will continue.

International LNG Markets

The impact of a tighter supply/demand balance in the Asia-Pacific LNG market in the mid to late 2000s, has had a significant effect on the gas market. The period 2005–07 has seen extraordinarily high prices for spot LNG transactions. If this trend continues into the early 2010s, international LNG market conditions could be a significant constraint on Japanese gas demand

Global Warming

Finally, there is the issue of global warming, which is basically a positive factor for growth in natural gas consumption. In 2005, Japan's greenhouse gas emissions were 7.8 percent above 1990 levels. In particular, CO_2 emissions attributable to energy consumption had reached 1203 Mt-CO_2, 13.6 percent above the 1990 benchmark. The corresponding policy objective under the Kyoto Protocol is 1056 Mt-CO_2, and achieving this objective would require, for example, more than a 10 percent improvement in energy efficiency by 2020, but this objective is now extremely difficult to achieve. Consequently, in order to meet numerical targets in the period 2008–2012, new policy initiatives such as environmental taxes or domestic emissions trading would be required. At the same time, since there is a strong likelihood of more rigorous policies being implemented regarding voluntary targets, particularly by industry,[5] it is necessary to consider the potential effect of this on the natural gas market up to the target date of the Kyoto Protocol as well as in a longer-term perspective.

Natural Gas Demand and Supply

Recent Trends

As stated previously, despite economic recovery the growth of energy consumption since 2002 has been relatively small. Examination of final energy consumption at the sectoral level reveals that demand is generally flat or even declining in industry. The position is similar in the transportation sector, where there is greater use of small cars as fuel costs rise; demand in the commercial and residential sectors is increasing slightly. Overall the trend is virtually flat.

Table 4.1 examines trends by energy source in terms of primary energy during the period 2000–2005. From 2002 onwards, demand for oil, LNG, and coal has been affected by a number of electric power utilities temporarily shutting down some of their nuclear power plants. For this reason, the underlying trends – a marked decline in oil consumption and a steady increase in coal and natural gas consumption – are not immediately apparent. The share of natural gas in primary energy slightly increased from 13.7 percent in FY2000 to 14.4 percent in FY2005.

Table 4.1: Primary Energy Supply. 10^{12}kcal

FY	*2000*	*2001*	*2002*	*2003*	*2004*	*2005*
Oil	2,892	2,711	2,757	2,781	2,748	2,815
Coal	1,002	1,021	1,071	1,109	1,205	1,166
Natural gas	734	720	743	790	784	783
Hydro	193	187	184	212	210	174
Nuclear	692	688	634	516	607	655
Others	73	70	73	75	73	73
Total	5,587	5,397	5,462	5,483	5,627	5,666
(%/year)	(1.8)	(-3.4)	(1.2)	(0.4)	(2.6)	(0.7)
Real GDP*	505.6	501.6	507.0	517.7	527.9	540.4
(Trillion Yen)	(2.6)	(-0.8)	(1.1)	(2.1)	(2.0)	(2.4)
CO2 emissions** (million t-C)	317.2	312.2	320.1	326.9	331.9	332.3

* Trillion Yen at constant price based on calendar 2000

** CO_2 emissions in FY1990 were 287.9 million t-C

Source: EDMC *Handbook of Energy & Economic Statistics in Japan 2007*, The Energy Conservation Center, Japan

The supply/demand balance for natural gas is shown in Table 4.2. From FY2000, total supply of natural gas grew substantially to 65.8 million tonnes (LNG equivalent) in FY2006. However, a breakdown of the balance shows that there was virtually no growth in supply for power generation, while supply for the gas distribution sector increased significantly. The figures for individual years also show an interesting trend: while supply for city gas followed a trend of steady growth, the volume of gas supplied for power generation fluctuated considerably. These trends in natural gas consumption over the past five years

Table 4.2: Supply/Demand Balance for Natural Gas. 1,000 tonnes LNG equivalent**

FY	*2000*	*2001*	*2002*	*2003*	*2004*	*2005*	*2006*	*Annual growth rate*
PNG								
Domestic Production***	1,878	1,853	2,068	2,115	2,223	2,360	2,480	4.7 %
For generation	399	374	437	427	395	385	391*	-0.3 %
For others	1,479	1,479	1,631	1,688	1,827	1,973	2,089*	5.9 %
LNG								
Imports	54,157	54,421	55,018	58,538	58,018	57,797	63,308	2.6 %
For generation	37,844	38,175	37,914	39,063	37,170	34,640	37,773	2.7 %
For others	16,313	16,246	17,104	19,475	20,848	23,157	25,535	4.3 %
Total								
Supply	56,035	56,274	57,086	60,653	60,241	60,157	65,788	2.7 %
For power generation****	38,243	38,549	38,351	39,490	37,565	35,025	38,164	0.0 %
For others	17,792	17,725	18,735	21,163	22,675	25,132	27,624	7.6 %

* estimated by author
** The calorific value of LNG is 13,000 kcal/kg.
*** The calorific value of domestic natural gas is 9,771 kcal/m^3.
**** Supply for power generation: supply to electric utilities and wholesalers
City gas consumed by electric utilities is not classified as being for power generation.

Source: '*Denryoku Jyukyu no Gaiyou,2006*' [*Overview of Power Supply & Demand fiscal 2006*], Agency for Natural Resources and Energy, METI; Annual Report on Energy Production and Supply/Demand Statistics, METI; Annual Report on Gas Industry Statistics, METI.

incorporate essential factors that will play a part in determining the trends for natural gas demand over the long term, so the details will be examined in the next section.

Demand for Power Generation[6]

From FY 2000 to FY 2005, demand for electricity grew at an annual rate of 1.2 percent. The power industry, however, was hit heavily during this period by the need to temporarily shut down nuclear plants,[7] which resulted in major changes to the composition of fuels used for power generation. Keeping these factors in mind, the following conclusions can be drawn about the trends in natural gas consumption for the power generation sector.

Power generated in Japan by energy source is shown in Table 4.3. In FY 2003, when the load factor of 59.7 percent for all nuclear plants was at its lowest during the past five years, around 80 billion kWh of electricity were generated by alternative sources, predominantly thermal power. The shortfall in electricity supply was actually made up by additional supplies from adjacent power utilities or by utilising the utility's own spare capacity.[8] As a result, LNG consumption for power generation grew to 39.06 million tonnes in FY2003, exceeding the supply plan for the year by some 2.6 million tonnes, but then declined significantly to 34.64 million tonnes in FY 2005 (when the load factor for nuclear power had recovered to 71.9 percent). It needs to be borne in mind that these fluctuations in LNG consumption are also affected to some extent by changes in the composition of power sources, such as increases in coal-fired generating capacity (Table 4.4), but the temporary shutdown of nuclear plants was the major reason.

The second point relates to the impact of high oil prices on natural gas demand in the power sector. Changes in price competitiveness among fuels will be discussed in detail in the next section, but the following considerations can be derived from the perspective of how electric utilities generally manage different power sources. In general, LNG-fired capacity has retained its role in providing middle- or peak-load capacity, and data on power source composition indicate that coal-fired generation capacity is increasing (Table 4.5), while government policy for electric power calls for greater emphasis on nuclear power. Judging from these circumstances, LNG is increasingly used for peak-load capacity, with the effect that it is probably competing more frequently with oil-fired power generation. The current situation, however, is that at least the LNG being purchased on long-term contracts can be procured less expensively than oil, so using LNG-fired capacity should

Table 4.3: Changes in Power Generation of 10 Incumbent Electric Power Utilities by Energy Source. 10^9 kWh

FY	*2000*	*2001*	*2002*	*2003*	*2004*	*2005*	*2006*
Load Factor for Nuclear Power Plants	81.7 %	80.5 %	73.4 %	59.7 %	68.9 %	71.9 %	69.9 %
Power Source							
Nuclear	302.5	301.3	275.5	220.5	262.5	287.0	287.1
Thermal	429.4	411.6	446.6	466.0	457.8	462.2	468.0
Coal	98.2	106.9	118.2	127.4	139.3	147.5	NA
LNG	255.1	251.2	261.9	269.1	255.4	243.6	NA
Oil	63.9	40.5	53.1	55.5	53.4	65.6	NA
Hydro	66.5	64.7	63.3	72.4	70.9	60.0	65.9
Generation Total	798.4	777.6	785.3	759.0	791.2	809.2	821.0
Purchased Power	134.6	139.5	152.2	168.3	168.3	167.2	160.8
Electricity Sales	837.9	824.1	841.5	834.3	865.4	882.6	889.4

Sources: Data until FY 2005: '*Denryoku Jyukyu no Gaiyou,2006*' [*Overview of Power Supply & Demand fiscal 2006*], Agency for Natural Resources and Energy, METI; Data for FY 2006: Interim report published on the web-site of The Federation of Electric Power Companies of Japan

Table 4.4: Fuel Consumption by Electric Utilities

FY		*2000*	*2001*	*2002*	*2003*	*2004*	*2005*	*2006*
Coal	10^3 tonne	34,367	37,429	41,350	44,557	48,229	50,565	50,892
Fuel Oil	10^3 kl	10,135	6,302	7,554	9,106	8,314	9,876	7,371
Crude Oil	10^3 kl	6,864	4,640	7,527	6,303	6,050	7,674	6,033

Source: as Table 4.3

Table 4.5: Installed Generating Capacity of Power Plants. Excluding private power generation. GW

FY	*2000*	*2001*	*2002*	*2003*	*2004*	*2005*	*2006*
Hydro	44.9	44.9	44.9	45.2	45.2	45.7	45.8
Nuclear Power	44.9	45.7	45.7	45.7	47.1	49.6	49.6
Thermal	139.4	139.7	142.8	143.8	145.2	143.6	143.2
Coal	29.2	30.5	33.8	35.8	37.8	37.7	37.4
LNG	57.2	58.8	59.3	60.4	59.9	58.7	60.1
Oil & others	53.0	50.4	49.7	47.6	47.5	47.2	45.7
Total	229.1	230.3	233.5	234.7	237.6	238.9	238.4

Source: '*Dengen Kaihatsu no Gaiyo, FY2006*' [*Overview of Generation Capacity Development fiscal 2006*], Agency for Natural Resources and Energy, METI

be more efficient than using oil-fired capacity. Supporting this proposition is the fact that in FY 2006, although the load factor for nuclear power plants and the total volume of power generated were more or less the same as the preceding year, natural gas consumption for power generation increased considerably year on year by approximately 3.1 million tonnes. Over the same period, consumption of coal and coal-based fuels remained largely unchanged, whereas oil-based fuels (fuel oil and crude oil) declined in FY 2006. It can be inferred from these figures that in FY 2006 there was a shift from oil-fired to LNG-based capacity, possibly due to high oil prices.[9]

The proportion of LNG in power generation has therefore fluctuated considerably over the last five years, partly because of the effect of changes in nuclear power and partly because of changes in oil prices. These are some of the most important premises on which medium- and long-term projections in natural gas consumption in the power sector are based.

Demand in the Distribution Sector

In contrast to the consumption of natural gas for power generation, the consumption for other purposes grew by 55 percent (to over 9.8 million tonnes) from FY 2000 to FY 2006, growing at an annual rate of 7.6 percent. Most of this increase stems from the gas distribution sector, as city gas demonstrated a similarly high annual growth rate of 6.1 percent, including a very high rate of 9.6 percent in industrial consumption.

LNG consumption by region (Table 4.6) indicates that the greatest increase was in the Tokyo Metropolitan area, which accounted for 55 percent of the total increase. A major factor is the strong growth in Tokyo Gas' sales within its own service area and sales to adjacent city gas utilities, but wholesale sales and direct sales by Tokyo Electric Power Company (TEPCO) are also growing. Non-power gas sales handled by TEPCO are expected to reach a level of 1 million tonnes (LNG equivalent) in FY 2008.[10]

Table 4.6: LNG Imports by Region 2000–06 (excluding supplies for power generation).* Thousand tonnes

Regions	*FY 2000*	*FY 2006*	*Incremental volume*	*Annual growth (percent)*	*Share in total (percent)*
Tohoku	390.5	657.9	267.4	69	3
Kanto	6367.3	11285.5	4918.2	77	57
Shizuoka	284.4	607.3	322.9	114	4
Chubu	1812.2	2869.7	1057.5	58	12
Kansai	5550.6	7964.1	2413.5	44	28
Chugoku					
Mizushima Iwakuni	8.9	259.8	250.8	2804	3
Hiroshima	134.0	359.3	225.3	168	3
Kyushu					
Northern prefectures	1186.1	1416.5	230.5	19	3
Kagoshima	69.3	115.1	46.0	66	1
Total	15803.4	25453.9	9650.5	61	

* Imports by region minus LNG supplied to incumbent power utilities

Source: Trade Statistics of Japan, Ministry of Finance, The Federation of Electric Power Companies of Japan website.

The region with the second largest increase was the Kansai area, which accounted for 27 percent of the total growth nationwide. However, since the share of city gas in the industrial and commercial markets was already high in comparison with other regions, the rate of growth in gas sales by Osaka Gas was relatively low among the large major gas utilities. Yet, during the same period, there was a large increase in gas sales by Kansai Electric (KEPCO), which had been actively developing its gas business as a result of market liberalisation. It is reported that the total gas sales by KEPCO reached about 700,000 tonnes in FY 2006.[11]

After the Kansai region the next largest increase was in the Chubu region, reflecting the extremely high growth in sales for industrial use that Toho Gas had achieved for several years, as well as the general strong performance of the regional economy. Chubu Electric's gas sales in the region in 2006 were 360,000 tonnes.[12]

As described, LNG sales volumes for city gas usage have increased in the major urban areas. There has also been substantial growth in natural gas consumption in rural regions. High growth rates in LNG imports by city gas utilities in Shizuoka, Hiroshima and others have been driven predominantly by sales to industry and wholesaling to neighbouring city gas utilities (Table 4.6). Over the same period, there was progress in a number of regions, upgrading supply capacities in response to growing demand for natural gas as described in the next section.

As these situations demonstrate, natural gas consumption in the gas distribution sector is growing steadily throughout Japan and the following points can be identified as background factors: first of all, the growth in consumption is partly attributable to regional city gas companies switching to high calorific value gas (IGT 2010) which was noted in the previous edition of this book,[13] and the proportion of city gas feedstock accounted for by natural gas is rising, along with growth in natural gas service areas. In addition to city gas utilities, electric utilities have played an important role, particularly in wholesaling to medium-size and small city gas utilities, and direct sales to large users not only in major urban areas but also in areas not previously supplied with natural gas.

The second point relates to a rapid increase in industrial demand and wholesaling to other regions as shown in Table 4.7 and Figure 4.1. With a few exceptions, both figures show consistently high growth from FY 2002 onwards. This is due to manufacturing industry switching to fuels with lower CO_2 emissions in advance of the Kyoto Protocol deadline for achieving their own greenhouse gas emissions targets. Energy suppliers also need to meet environmental compliance requirements, and regional expansion of gas has made it easier to install energy-efficient appliances such as cogeneration systems. Furthermore, the figures from FY 2004 onwards are influenced by high oil prices producing a substantial enhancement in the price competitiveness of natural gas over other fuels.

The above statistics demonstrate firm growth in demand for natural

Table 4.7: Gas Utility Sales by Sector and Region 2002–06.* (year-on-year growth %)

Sector	*Company*	*2002*	*2003*	*2004*	*2005*	*2006*
Industrial	Tokyo	23.6	14.8	11.0	5.7	5.4
	Shizuoka	9.5	19.3	4.1	14.7	28.9
	Toho	8.8	21.5	16.3	21.0	14.1
	Osaka	1.5	1.3	3.5	4.8	7.5
	Hiroshima	128.4	13.6	23.1	21.0	15.1
	Saibu	12.8	22.5	16.1	18.7	17.3
Wholesale	Tokyo	16.4	8.0	13.5	19.9	8.7
	Shizuoka	-	-	177.7	57.9	39.7
	Toho	11.1	-54.6	14.6	20.3	9.9
	Osaka	24.4	8.3	67.6	17.7	15.9
	Hiroshima	-	-	-	-	-
	Saibu	567.2	58.8	4.6	104.8	33.7
Total	Tokyo	13.5	6.8	6.8	6.9	1.3
	Shizuoka	6.7	17.5	9.6	15.2	22.9
	Toho	7.8	7.2	10.3	15.8	8.1
	Osaka	2.8	1.0	3.7	4.9	3.4
	Hiroshima	32.1	6.6	11.5	11.6	8.4
	Saibu	5.0	4.9	9.8	13.3	7.7

* Fiscal years except Hiroshima gas where figures are calendar years, wholesale did not start until 2006

Source: Company's annual reports and other public data.

gas over the past few years, but there are also some negative factors concealed by this growth. Firstly, electric utilities are strongly promoting all-electric homes, and this is becoming an increasing threat to residential demand for city gas. Secondly, where city gas is used for cogeneration and/or air-conditioning in the industrial or commercial sectors, oil and LNG price increases are damaging the price competitiveness of city gas relative to electricity, and competitiveness is actually being lost. This point will also be described in more detail in the next section.

Changes in LNG Procurement and Prices

Growth in Spot Transactions

As described above, demand for LNG in the power and gas distribution sectors has exceeded expectations. This has meant that supplies under long-term contracts were insufficient to meet demand and resulted in some utilities increasing spot LNG purchases. As can be seen in

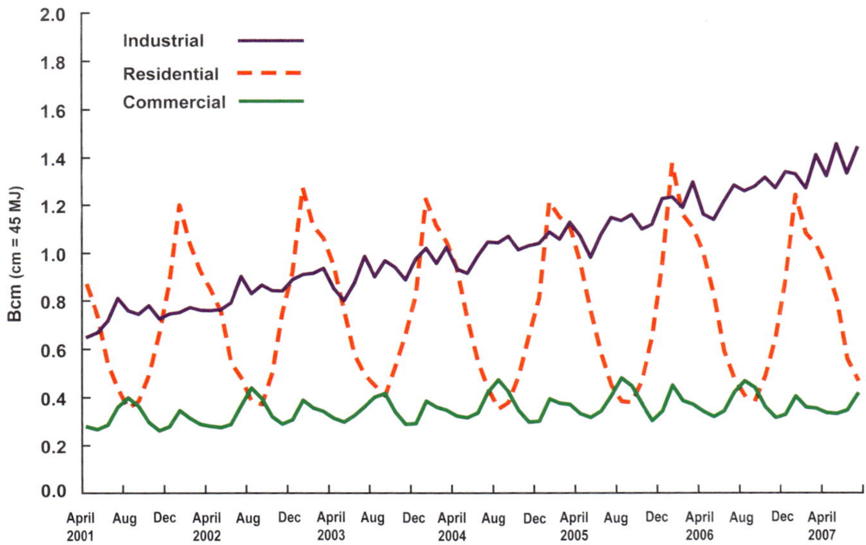

Figure 4.1: City Gas Sales by Sector 2001–07

Source: The Japan Gas Association

Table 4.9, spot purchases from Middle East countries such as Oman, Qatar, and the UAE, and from LNG exporters such as Algeria, Nigeria, Trinidad & Tobago and Egypt, which predominantly serve the Atlantic market, have been increasing. An interesting trend of these spot purchases is that despite a 500,000 tonne year-on-year decline in FY 2005, the prices paid (Figure 4.2) are substantially higher than long-term contract prices, and there were even cases of deals being done at the extremely high price of 19$/MMBtu.[14] The figures for spot transactions demonstrate that the LNG supply and demand balance was particularly tight in the Asia-Pacific region in the winter of FY 2005, when high prices were paid to divert cargoes from the Atlantic Basin. Converted to oil prices on the basis of calorific value, the spot price of 19$/MMBtu is equivalent to more than $110/bbl. It would certainly not be economically viable to use the LNG purchased at such a high price on a regular basis.

Why then, did LNG spot transactions reach such a high level? Basically, public service utilities such as electric and gas utilities operating in regulated markets have an obligation to maintain supplies. For this reason, they are unable to interrupt or cut back supplies to customers when feedstock prices are high. As market liberalisation progressed, utilities attempted to secure market share by locking customers into

Table 4.8: LNG Imports by Source 2000–06. Thousand tonnes

FY	*2000*	*2001*	*2002*	*2003*	*2004*	*2005*	*2006*
USA	1,260	1,266	1,253	1,242	1,210	1,250	1,127
Brunei	5,715	6,004	6,011	6,367	6,357	6,165	6,393
Indonesia	18,123	16444	17,522	17,490	15,545	13,813	13,951
Malaysia	10,923	11,,296	10,881	12,219	13,154	13,136	12,220
Australia	7,211	7,489	7,212	7,644	8,612	10,456	12,606
UAE	4,802	4,853	4,633	5,256	5,107	5,251	5,262
Qatar	6,000	6,386	6,640	6,608	6,762	6,396	7,707
Oman	123	681	867	1,656	1,104	1,101	2,864
Algeria						56	184
Trinidad & Tobago				56	55	56	276
Nigeria					112		165
Egypt							556
Total	54,157	54,421	55,018	58,538	58,018	57,797	63,309

Source: Trade Statistics of Japan, Ministry of Finance.

Table 4.9: Short-term and Spot LNG Trades in the Pacific Basin 2000–06. Thousand tonnes

	2000	*2001*	*2002*	*2003*	*2004*	*2005*	*2006*
UAE				109	295	373	530
Algeria						55	185
Nigeria					113		511
Oman		80	180	911	547	173	1314
Qatar		322		388	344	213	1025
Trinidad & Tobago				53	59		474
Australia		55		107	120	351	450
Brunei						105	182
Indonesia	243	1157		56			
Malaysia		109	57	466	294		240
Total	243	1702	238	2090	1773	1269	5076
Share in total imports (%)	0.4	3.1	0.4	3.6	3.1	2.2	8.0
ASIA TOTAL	1802	3185	1603	4221	4600	4503	13750
Japan's share in Asia	13.5	53.5	14.8	49.5	38.5	28.2	36.9

Source: The LNG industry, GIIGNL

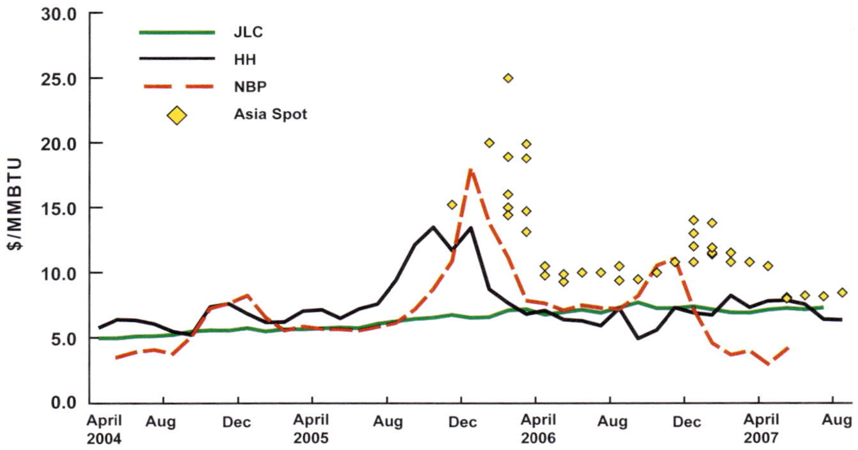

Figure 4.2: LNG Prices for Spot Transactions in Asia Pacific Market

Note: JLC (Japan LNG Cocktail) is the average import price of LNG

Source: JLC and Spot prices; Trade Statistics of Japan, Ministry of Finance. HH and NBP prices; Platts

long-term supply contracts, but many of these contracts arranged prior to the post-2003 oil price period had not envisaged such price levels or the tightness of the LNG supply/demand balance. Even if contracts permitted suppliers to reduce volumes, the idea that temporary problems could threaten security of supply to customers would break the trust relationship between supplier and customer and risk the loss of market share in the future. From the utilities' point of view, even if the price is one that a utility would normally be reluctant to pay, buying a few cargoes on the spot market is only a small part of the total volume that needs to be procured. It is circumstances like these that produced the high spot transactions for delivery in Asia in the winter of 2005.[15] If a similar situation recurs, there is the potential for prices to rise to such levels again.

Having experienced this situation, however, utilities have adopted new approaches to LNG procurement, and are actively working to avoid a recurrence of these conditions. This is an important consideration for the future. For example, one utility used to have long-term purchase contracts for slightly less than expected demand but incorporating upward flexibility and purchased spot LNG cargos to meet temporary demand. But, with the LNG supply/demand balance expected to be tight in the medium term, the utility has introduced a new approach using long-term contracts for larger volumes than envisaged demand, but with the ability to take up less than the full amount if demand turns out to be smaller than expected, for instance by diverting cargoes to other markets. Because the risks of high prices for spot cargoes increase in winter, many utilities are taking action to use their infrastructure to the full by buying LNG on the spot market at lower prices during off-peak seasons and storing it for winter use.

Long-term Contract Prices

As already noted, post-2003 oil price levels have led to substantial changes in relative energy price which has had a major impact on the outlook for natural gas demand. LNG price formulae in the Asia-Pacific market are linked to the Japan Crude Cocktail (JCC) price (Figure 4.3), but many contracts employ an S-curve which moderates the effect of extreme oil price movements (Figure 4.3).[16] Because of this mechanism, the competitiveness of natural gas relative to oil products has been greatly enhanced. Hence retail prices of city gas using natural gas as feedstock are starting to be competitive against fuel oil.

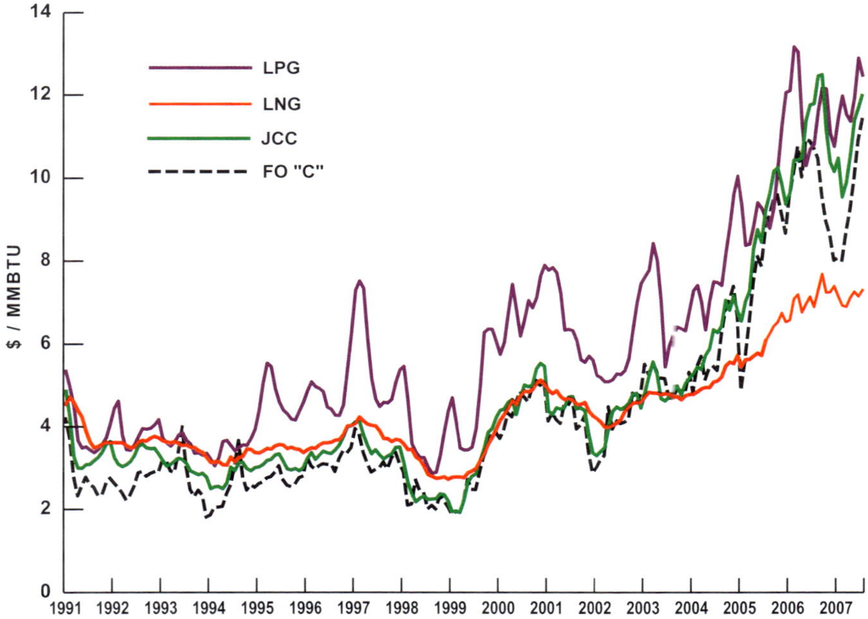

Figure 4.3: Import Prices for LNG and Oil

Source: Data from Trade Statistics of Japan, Ministry of Finance

In 2006, in terms of final energy consumption, the share of natural gas for industrial use was only about 7 percent.[17] Until recently, natural gas was uncompetitive against fuel oil, which has the largest share in the sector, but if part of industrial fuel oil consumption switches to natural gas, a substantial increase in natural gas demand would be expected to result.

Natural Gas Supply Infrastructure

LNG Receiving Terminals[18]

Natural gas demand is growing in both the metropolitan areas and in the local regions, and the construction of new LNG receiving terminals is planned as shown in Map 4.1. Japan currently has LNG receiving terminals at 27 locations. Those which started operating in 2003 or later are listed in Table 4.10, together with new terminals at the planning or concept stage. There are also reports (not listed in the table) of new terminals under consideration in Hokuriku, Hokkaido, and Kyushu.

In addition to constructing new terminals, many utilities are planning to upgrade existing plants by adding further LNG storage tanks to expand receiving capacity.[19] Upgrading port infrastructure and facilities in order to handle the larger carriers scheduled for use (particularly by

Table 4.10: New LNG Terminals*

Terminal	*Owner*	*Date of operation*
Nagasaki LNG Terminal	Saibu Gas	Sep. 2003
Sakai LNG Centre	Sakai LNG**	Jan. 2006
Mizushima LNG Terminal	Mizushima LNG***	Apr. 2006
Sakaide LNG Terminal	Sakaide LNG****	March 2010
Joetsu Thermal Power Plant	Chubu Electric Power/ Tohoku Electric Power	2012 or later
Joetsu-shi Naoetsu	Teikoku Oil	End of 2013
Yoshinoura Thermal Power Plant	Okinawa Electric Power	FY 2010 or later
Wakayama LNG Terminal	Kansai Electric Power	2016 or later
Omaezaki LNG Terminal (Under consideration)	Chubu Gas, Tokai Gas, Suzuyo Group	Not determined

* Starting operation in 2003 or later (including planned terminals)
** Sakai LNG: Kansai Electric, Iwatani International Corporation, Cosmo Oil Co. Ltd., and Ube Industries, Ltd.
*** Mizushima LNG: Chugoku Electric Power, Nippon Oil Corporation
**** Sakaide LNG: Shikoku Electric Power, Cosmo Oil, and Shikoku Gas

Source: Author, compiled from a variety of sources

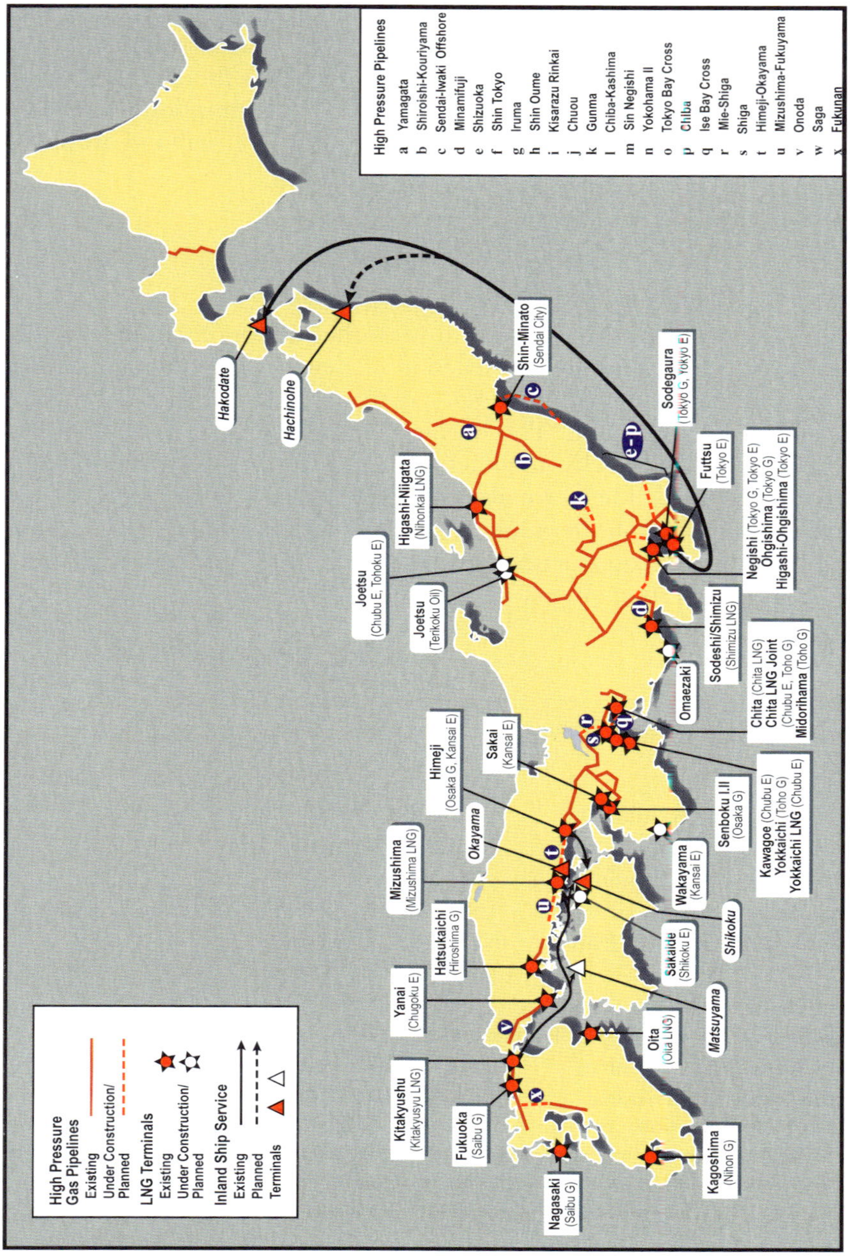

Map 4.1: LNG Terminals and Gas Pipelines

Source: Author, compiled from a variety of sources

Qatar)[20] and to upgrade smaller receiving terminals to be able to handle standard carriers is also under consideration.[21] These developments will all increase LNG receiving capacity and enhance economic viability.

Domestic Supply Infrastructure

Japan's natural gas supply system began with the construction of LNG receiving terminals on the outskirts of the major conurbations of Tokyo, Chubu, and Kansai, where the demand for energy is concentrated. This approach resulted in having nearly thirty receiving terminals dotted around a small country, a situation that has not arisen elsewhere in the world. A national high pressure pipeline network, found in most OECD countries, has not developed in Japan. The background to this situation includes the fact that no international pipelines have been laid, and because Japan has such a large population squeezed into a small amount of flat land, land prices and cost of pipeline construction are extraordinarily high. The complex combination of these and other factors has resulted in the current unique situation.[22] However, it has to be said that having LNG receiving terminals close to areas of demand permits a very efficient natural gas supply system. In Japan, trunk pipelines run from LNG terminals close to the big cities towards other natural gas demand centres, and these pipelines are gradually extended when there is demonstrable economic viability.

The current situation is that demand for natural gas is increasing nationwide, and so the means to supply natural gas to different regions have to be developed. This infrastructure includes pipeline extensions, LNG transportation by truck, railroad or coastal tanker systems, together with plant infrastructure such as small LNG receiving terminals (for coastal tankers) and satellite terminals.

This recent expansion of infrastructure is largely attributable to the liberalisation of gas and electricity markets. Taking the example of pipeline development, when the market for large users was liberalised, there were incentives to invest in pipelines by new entrants to the city gas market hoping to secure market share, and by the existing gas utilities newly permitted to supply gas outside the boundaries of their traditional service area. Similarly, investment in the construction of transportation facilities for coastal tankers (or tank trucks) has been accelerated as new entrants to the LNG business and the existing utilities compete to expand their supplies.

One important characteristic of the development of the supply infrastructure is an increasing shift from independent to interlinked supply systems. Where supply networks used to be closed, with a supplier's pipeline network linked only to its own receiving terminals, or domestic gas fields, pipelines are now being linked together to form a more widespread supply network with more supply sources (receiving terminals or gas fields).

A concrete example in the Tokyo metropolitan area at the end of 2006 was the linking of the supply networks of Teikoku Oil and Shizuoka Gas by the Minami Fuji Pipeline, a project by those two companies and Tokyo Gas. By 2012 Tokyo Gas plans to link its own network to those of the other two companies, completing an area-wide supply network that links a number of LNG terminals in the Tokyo metropolitan area, Niigata, and Shizuoka. Another notable example of progress is that in August 2007, INPEX Holdings announced a plan to construct a LNG receiving terminal at Naoetsu Port in Niigata. INPEX Holdings is a holding company formed by a merger between INPEX, which is involved in projects such as the Ichthys LNG in Australia (see Chapter 7), and Teikoku Oil, which develops domestic gas and operates natural gas pipelines running from Niigata to the Kanto area. This merger is of particular interest because it is an example of a new vertically integrated business model. Plans call for the new terminal to be operational by the end of 2013, starting with a handling capacity of between 300,000 and 400,000 tonnes, and gradually increasing to around 2 million tonnes.[23]

Infrastructure to link terminals is also being put in place in and around Japan's major metropolitan areas, both to increase the reliability of LNG supplies and to provide more supply flexibility. Projects like these include the Trans-Bay Gas Pipeline in the Tokyo metropolitan area and the Cross-Isewan Gas Pipeline in the Chubu area. The former would link TEPCO's Futtsu thermal power plant (in Chiba prefecture) with its Higashi-Ohgishima power plant (in Kanagawa prefecture) and is scheduled for completion in 2008. The latter is a project by Chubu Electric and Toho Gas to link the jointly operated Chita LNG terminal, with the Kawagoe thermal power plant on the other side of Ise Bay, continuing to Toho Gas' Yokkaichi plant which is due to be completed in 2013. Meanwhile, Chubu Electric is constructing the Mie-Shiga pipeline in association with Osaka Gas, to be completed in 2010, linking the Chubu infrastructure with that of the Kansai area. Natural gas supply networks are also being expanded in the Tokai area, the Seto Inland Sea area, and the northern part of Kyushu, adding to the accelerating development of Japan's supply infrastructure.

Domestic coastal shipping is a new method of transporting LNG within Japan. Supplies to Shikoku Gas and Okayama Gas from Kitakyushu LNG[24] started in 2003, followed in East Japan by LNG transported from Tokyo Gas' Sodegaura Terminal to Hakodate and Hachinohe

(Table 4.11). Coastal shipping resolves the problem of transporting gas to areas which cannot be reached economically by pipeline. The tankers used are in the $2500m^3$ class (carrying about 1000 tonnes of LNG) and, depending on the transportation distance, are capable of meeting annual demand in the range of 10,000–100,000 tonnes.

Table 4.11: Secondary LNG Terminals supplied by Coastal Shipping

Terminal	*Owner*	*Date of operation*
Hakodate Minato Plant	Hokkaido Gas	Feb. 2006
Shikoku Gas Takamatsu Plant	Shikoku Gas	Aug. 2003
Okayama Gas Chikko Plant	Okayama Gas	Aug. 2003
Hachinohe LNG Terminal	Nippon Oil Corporation	Mar. 2007
Shikoku Gas Matsuyama Plant	Shikoku Gas	Nov. 2008

Source: Author: compiled from a variety of sources

International Pipeline Prospects

The possibility of constructing international pipelines, specifically a pipeline from Sakhalin, has receded since the previous edition of this book was published.[25] Two factors behind this change are the greater likelihood of Sakhalin 1 exports to China, and the fact that Gazprom, which is already involved in Sakhalin 2, is seeking an opportunity to intervene in the project (see Chapter 6). However, the crucial obstacle is the reluctance of Japanese buyers to import natural gas by pipeline. Japan's many LNG buyers prefer importing gas in liquefied form rather than by pipeline. Given these attitudes, the project is unlikely to go ahead.

However, when this issue is considered in the context of Japan's energy policy and particularly in terms of securing energy resources, as emphasised in Chapter 6 of the New National Energy Strategy, importing natural gas by pipeline can be perceived as intrinsically beneficial. Moreover, in the vicinity of Sakhalin, there are many more undeveloped projects, and there is potential for the development of vast amounts of natural gas. Considering that securing overseas energy resources is likely to become increasingly difficult, it would be reasonable to expect some sort of consensus between the government and the private sector on completing an international natural gas pipeline from Sakhalin. Nevertheless, in 2007 there were no signs of such a consensus.

One positive factor regarding the potential completion of a pipeline from Sakhalin is that the Iwaki-Sendai pipeline[26] is going ahead. This project is being promoted by JAPEX, which was the principal investor in Sakhalin 1, and is planned with a view to resolving the technical and other issues that would be necessary for construction of a seabed pipeline from Sakhalin. It will be interesting to see how this project develops.

Liberalisation of Gas and Electricity Markets

The 'Japanese Approach'

Looking back at the energy markets at the time the first edition of this book was written, the Koizumi administration was pushing forward with regulatory reform in order to move the country out of its long-standing economic recession, and there was strong pressure to liberalise the electricity and gas markets. In that context, it is useful to examine how the electric and gas market systems were actually reformed, and to note briefly the extent of new entrants in both markets as a result of reforms.

Liberalisation of the electric power market proceeded in a series of steps, as shown in Figure 4.4, and since April 2005, approximately 63 percent of the electric market has been opened to competition.

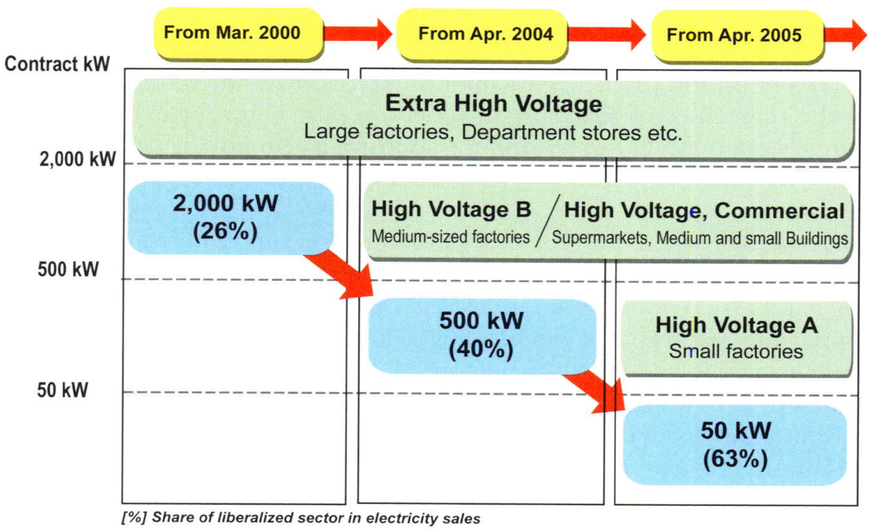

Figure 4.4: Opening of the Electricity Market to Competition

Note: (%) Percentage of the market open to competition

Source: METI

Electricity reforms in 2003 put in place a market system that included measures to retain integrated power generation and transmission.[27]

The outcome of liberalisation is that the number of new entrants selling electricity as power producers and suppliers (PPSs) increased gradually from just one company in August 2000, to eleven at the end of FY 2004, and just fourteen companies by February 2007 (Table 4.12).[28] The share of the deregulated segment actually captured by the PPS is still only about 2.3 percent. The power generated by PPSs in March 2006 was 878.9 MW, about 0.4 percent of the total power generated by the ten incumbent utilities.[29]

Table 4.12: PPS Entrants in the Electric Power Market 2001–06

	FY2001	*FY2004*	*FY2006*
PPS: power producer and supplier	6 companies	11 companies	14 companies*
Share of demand in deregulated segment	0.4 %	2.1 %	2.3 %

* A total of 21 companies have given notification, but only 14 actually made sales in 2006

Source: Provisional reports on total electric power demand from the METI website

As of May 2007, the government's Electricity Industry Committee was discussing the future of the electricity markets with a view to complete market opening including the residential sector, but the Committee was expected to conclude that 'it would be best to maintain a vertically integrated supply system without implementing full liberalization, and other dynamic reforms, and on that basis, to pursue "market liberalization with a Japanese approach", where the principle of competition functions to a certain extent.'[30]

In summary, in the context of Japan's new energy policy, it can be said that rather than deepening market liberalisation, the priorities are to strengthen supply security and carbon reduction through the promotion of nuclear power. Consequently, the effect of new market entrants on the structure of the industry is likely to be small, and in the medium to long term, it is unlikely that there will be a dramatic restructuring of the industry. Thus, in predicting natural gas demand in the power generation sector, the appropriate approach is to base predictions on the supply plans of the incumbent electric utilities.

As shown in Figure 4.5, liberalisation in the gas industry also proceeded in stages, and by April 2007 approximately 59 percent of the gas market had been opened to competition. Gas industry market reforms in 2003 enlarged the scope of liberalisation in retail markets, creating gas pipeline businesses, reinforcing and enhancing access to networks, promoting effective use of LNG terminals, and reviewing the regulatory system for wholesale supply.[31] The result, as shown in Table 4.13, is that over three years there has been a steady increase in activity by new entrants. Examining the new activity in detail reveals that electric utilities, oil companies and others have become gas pipeline and supply companies,[32] with electric utilities gaining more than half of the 119 customer contracts signed as of September 2007.[33] The larger new entrants are trading companies, LPG suppliers, steel and chemical companies. The share of supply volumes accounted for by these new market entrants grew from 5.0 percent at the end of FY 2003 to 11.7 percent at the end of FY 2006.

In terms of numbers of new entrants and their market shares, these results suggest that liberalisation has had a greater effect on the gas market than on the electricity market. Although, as with the electric industry, the likelihood of full liberalisation being extended to residential gas customers is low, competition in the gas market is likely to be

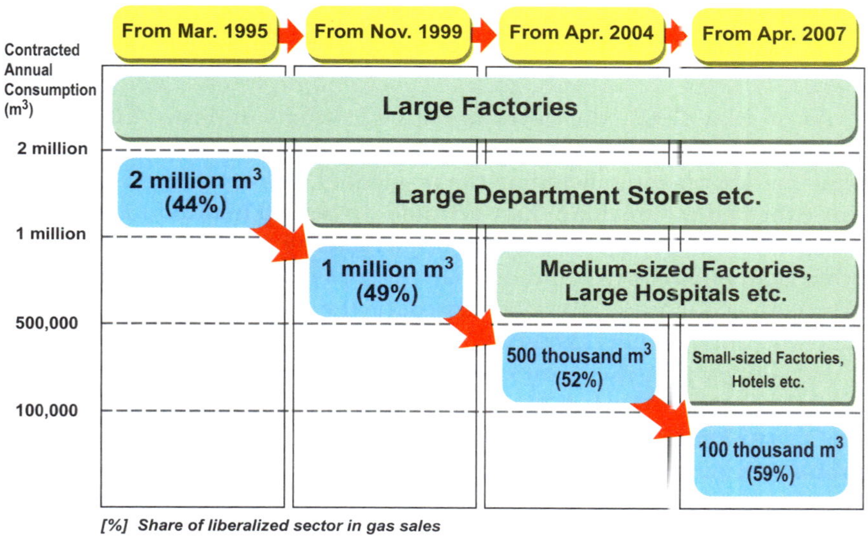

Figure 4.5: Opening of the City Gas Market to Competition

Note: (%) Percentage of the market open to competition

Source: METI

Table 4.13: Gas Market Penetration by New Entrants

Category		*April 2004*	*Sep. 2007*
Supply by general gas utilities to large customers outside their service area		27 contracts (0 new entrants)	143 contracts (39 new entrants)
Supply to large customers by companies other than general gas utilities	Gas pipeline companies	54 contracts (15 new entrants)	119 contracts (11 new entrants)
	Large-scale gas suppliers		67 contracts (17 new entrants)

Source: METI

promoted when supply infrastructure is further developed in both urban and local areas, along with a steady growth of gas consumption.

Responses of Gas and Electricity Utilities

How are electric and gas utilities responding to these changes in the market systems? The incumbent electric utilities have strong positions in the electric power market, and new entrants have struggled since liberalisation started. Furthermore, the post-2003 rise in crude oil prices had a detrimental effect on the viability of new entrants using oil-fired thermal power generation, and these companies are now in a difficult position. Idemitsu Kosan has already pulled out of electricity retailing completely, and Ishikawajima-Harima Heavy Industries and JGC Corporation have withdrawn from the GTF Research Institute, leaving it a 100 percent subsidiary of Mitsui & Co. Ltd., which is attempting to rebuild the business by concentrating on a biomass power generation project. Eneserve has pulled out of selling private power generation systems, which used to be its main business. As well as its effect on PPS businesses, the impact of high crude oil prices can be seen in the lack of progress by onsite generating businesses run by various energy companies including electric utilities.[34] Manufacturers that used to utilise oil-based fuels are increasingly switching to bought-in electricity to replace private generation. In FY 2005, there were 972 of these switches, accounting for over 950MW.[35]

In contrast, the impact of high oil prices on PPS businesses using natural gas as the energy source for thermal power generation has been smaller, but these businesses are not without their own problems. For example, eREX, which uses natural gas as its main source of power for thermal generation, had to withdraw from supplying to large customers at night when electricity tariffs are cheaper.[36] But there are also businesses like First Energy Service Company, which is constructing a 100 MW state-of-the-art CCGT power plant in Chiba prefecture[37] and planning to expand its business. Major city gas companies are also planning the construction of large-scale natural gas-fired thermal power plants. Tokyo Gas is to construct an 847.4 MW plant with Nippon Oil in Kawasaki,[38] a 1200 MW natural gas-fired power plant with Showa Shell in Yokohama[39] which, when added to its existing power plants, will give a total capacity in four locations of 2340 MW.[40] Osaka Gas' 1100 MW power plant at its Senboku LNG terminal is scheduled to come on line in April 2009, bringing its domestic generating capacity to 1800 MW by around 2010.[41] As examined in the next section, however, the economics of PPSs operated by natural gas will be strongly dependent on future LNG prices.

Electric utilities are steadily expanding their business as new gas market entrants. The basic business strategy of almost all incumbent electric utilities has been to announce themselves as total energy service companies, and they are therefore pursuing expanded sales of natural gas and other energies in addition to their electricity sales. They are particularly focusing on natural gas because of its growth prospects, and have been increasing their investments throughout the gas supply chain.[42] Their sales business includes both supply through pipelines from their own LNG receiving terminals, and supply via access to pipeline networks mainly owned by gas utilities. On a nationwide base, sales by electric utilities increased from around 500,000 tonnes in FY 2003, to more than 2.4 million tonnes in FY 2006.[43]

Electric utilities are involved in the gas business in a number of different ways. Tohoku Electric Power is involved in both pipelines and large-scale supplies through Tohoku Natural Gas, which it established together with JAPEX, and it is also selling LNG through the terminal operator, Nihonkai LNG.[44] This is a typical example of how electric utilities are collaborating with upstream companies to enter the gas business. Gas sales by Tokyo Electric are mainly direct sales through its own pipelines,[45] with a relatively small volume requiring access through city gas company pipelines or sales in the form of LNG; the company also has plans to upgrade its loading facilities for LNG trucks. Chubu Electric Power sold 360,000 tonnes of gas in FY 2006, of which 290,000 tonnes was accounted for by direct sales through its pipelines. The remainder was sold via trucks operated by its subsidiary, LNG Chubu. Chubu Electric is also collaborating with nearby gas utilities such as Toho Gas and Osaka Gas on pipeline construction. Kansai is the electric utility that has been the most active in promoting gas sales, and it has attained the largest number of customers. It is proactively growing its gas business through 100 percent subsidiary Kanden GASCO, and through Sakai LNG.[46] Chugoku Electric is involved in the gas business through Mizushima LNG Sales, established by Chugoku's 100 percent subsidiary, ESS (Energia Solution & Service), and eight other companies including Nippon Oil and city gas utilities such as Hiroshima Gas. Chugoku's terminals at Yanai and Mizushima each have 200,000/y tonnes of spare capacity for gas sales, allowing total city gas sales of 400,000 tonnes/y.[47] When the storage tank upgrade planned for Mizushima is completed in 2011, spare capacity for city gas sales may increase to around 1.2 million tonnes/y.[48] The company's active approach to the gas business can also be seen in its investment in the Setouchi Pipeline. Kyushu Electric sells gas and LNG through terminal operators, Kitakyushu LNG and Oita LNG, achieving sales volumes of about 210,000 tonnes in FY 2006[49] and was the first Japanese company to sell LNG by means of coastal shipping.

As can be seen from these examples, the electric utilities that are the main new entrants into the gas market, in many cases operate their gas business through increasingly close links with oil and city gas companies. These alliances represent major corporate innovation and will be crucial to the future development of the gas market.

Long-term Outlook for Natural Gas Demand

Japan's natural gas demand can be divided broadly into two sectors, power generation and city gas, but both of these sectors share the one factor that has a significant effect on future demand – the price competitiveness of natural gas. For example, industrial use of natural gas still accounts for only a small share of the total, but future improvements in natural gas price competitiveness could lead to substantial growth in demand. Conversely, reduced price competitiveness could cause the current level of demand to fall. Consequently, the approach taken here is to analyse LNG import prices in order to evaluate future price competitiveness of natural gas, and to consider its effect on demand.

Another important focus is the role of nuclear in power generation. With power generation accounting for 58 percent of total gas demand, and little likelihood of significant changes in Japan's electricity market or industrial structure, demand for natural gas can be inferred from the incumbent electric utilities' long-term supply plans (which are revised and published every year). However, since our experience of the past few years indicates that nuclear power can have a significant impact on future natural gas demand, these issues are worthy of specific consideration.

Accordingly, this section starts with a brief explanation of general trends in Japan's long-term energy demand, which form the basis for long-term natural gas demand. It then describes the latent demand for natural gas in Japan on a region-by-region basis. After covering those background situations, it considers the two important factors of price competitiveness and nuclear power before finally considering the potential for long-term natural gas demand.

Trends in Japan's Long-term Energy Outlook

The essential points concerning Japan's long-term energy supply/demand that are necessary to understand the long-term natural gas trends are summarised briefly here.[50]

First of all, the basic overall trend predicted for energy in Japan is for final energy consumption in the reference case to maintain an extremely low annual growth of about 0.1 percent by 2030,[51] reaching a peak soon after FY 2020 before starting to decline.[52] The background to this is declining population and long-term economic growth expectations of 1.4–1.5 percent per annum. Consequently, the point of discussion is whether gas can grow as a proportion of total energy demand which is barely expected to increase.

The second important point is that the structure of energy supply and demand is severely restricted by carbon reduction targets, and it is virtually certain that demand-side policies will continue to pursue further conservation and efficiency, and the supply side will increasingly shift to energy sources with lower CO_2 emissions. Because natural gas is the fossil fuel with the lowest carbon emissions, there are expectations that the market will adopt cogeneration and other technologies for efficient use of the fuel. Hence, natural gas will continue to be promoted. At the same time, in order to increase energy security as well as to respond to climate change, nuclear power is the main focus of electric power policy.

As described above, the government's latest long-term energy supply/demand outlook projects that by 2030, the share of gas in primary energy consumption will increase from the current level of 15 percent to 17.8 percent in the reference case and 19.1 percent in the policy target case.

Potential Demand by Region

The Electric Power Development Plan, which is based on long-term plans for individual power plant construction, provides a relatively accurate estimate of natural gas demand in the power generation sector. As far as the distribution sector is concerned, it is very difficult to predict the level of demand over the long term because the choice of fuels is usually determined principally on the basis of factors such as economics, environmental impact, convenience, and restrictions on choice of appliances. In particular, LNG prices, which are the decisive factor affecting demand in the sector, together with the prices of energies competing with natural gas, are potentially subject to sharp fluctuations. Therefore, there is little sense in making quantitative demand projections based on assumptions that are neither predictable nor transparent.

Accordingly, we first examine the extent of potential demand for natural gas, and then analyse the market circumstances that would lead to a shift to natural gas. In 2006, gas had a 14.4 percent share of primary energy consumption, of which power generation accounted for 58 percent. In the industrial sector, which represents 46 percent of final energy consumption, the share of natural gas is only around 7 percent,[53] so there is a significant opportunity for natural gas to gain market share.[54] Thus, we focus on the sectors where current potential demand for natural gas can be roughly estimated.

Market Competition with Oil Products. There is potential for switching to natural gas from:

- oil products (kerosene, diesel oil, fuel oil A, fuel oil C)[55] and steam coal in the industrial sector;
- fuel oil A in the commercial sector;
- naphtha, fuel oil A, fuel oil C and steam coal supplied for private power generation in the energy conversion sector.

The results of these calculations are shown in Table 4.14. The biggest area of potential demand is manufacturing industry where oil products are currently used (fuel oil A: 35 percent, fuel oil C: 27 percent, kerosene: 26 percent, diesel fuel: 11 percent).

To gain a deeper understanding of the competition between oil products and natural gas in this area, it is useful to refer to the result of 'the supporting program for promoting the conversion to natural gas of facilities with large energy consumption', which has been operating since FY 2002 under a government policy to facilitate switching to natural gas. Over the five years from FY 2002 to FY 2006, this project has subsidised the switch from oil products of 2.2 million tonnes of LNG equivalent.[56] The fuels replaced were fuel oil A 43 percent, fuel oil C 38 percent, kerosene 19 percent, and diesel fuel 0.2 percent. Boilers

Table 4.14: Potential Energy Demand by Sector which could Switch to LNG

Sector		Competing fuel	Potential to switch to LNG (mt LNG)
Final consumption sector	Industrial sector – manufacturing	Steam coal	5.5
		Oil-based fuels (kerosene, diesel fuel, fuel oil A, C))	15.5
	Consumer sector – commercial	Oil-based fuels (fuel oil A)	7.0
Energy conversion sector – private generation		Steam coal	7.5
		Oil-based fuels (fuel oil A, C, naphtha)	8.5
Totals		Steam coal	13.0
		Oil-based fuels	31.0
		Total	44.0

Source: author's estimate based on the energy balance table for FY 2005 by IEEJ

accounted for 73 percent of the facilities converted, with industrial furnaces accounting for 15 percent, water coolers/heaters 6 percent, and drying kilns 4 percent. The amount of fuel converted as a result of the subsidy was largest when oil prices rose in FY 2005 (approx. 570,000 tonnes), and FY 2006 (approx. 1,100,000 tonnes), with these two years accounting for 75 percent of the total. Financial assistance definitely encouraged conversion to natural gas from other fuels, but the simultaneous rise in oil prices is also thought to have provided a significant boost.

Table 4.15 shows the potential demand, from Table 4.14 above, categorised into regions by inference from statistics having a high correlation with energy demand for each sector, using indicators such as floor areas of commercial buildings, values of manufactured goods shipped, and consumption of self-generated electricity.

The resulting figures show that potential demand of 28.1 Mt, accounting for over 60 percent of the total, is concentrated in the three main metropolitan areas – Kanto, Chubu, and Kansai. These areas already have pipeline networks and other infrastructure for the supply of gas which, as described above, are still actively expanding. It is

Table 4.15: Potential Gas Demand by Region

*Region**	*Potential Demand (Mt)***	*Share (%)*
Hokkaido	1.6	4
Tohoku	3.7	8
Kanto	12.9	29
Chubu	8.9	20
Hokuriku	1.0	2
Kansai	6.3	14
Chugoku	5.3	12
Shikoku	1.2	3
Kyushu	3.1	7
Japan total	44.0	100

* Regions are defined as follows;

Tohoku: Aomori, Iwate, Miyagi, Akita, Yamagata, Fukushima, Niigata
Kanto: Ibaraki, Tochigi, Gunma, Saitama, Chiba, Tokyo, Kanagawa, Yamanashi
Chubu: Nagano, Gifu, Aichi, Mie, Shizuoka
Hokuriku: Toyama, Ishikawa, Fukui
Kansai: Shiga, Kyoto, Nara, Osaka, Hyogo, Wakayama
Chugoku: Tottori, Shimane, Okayama, Hiroshima, Yamaguchi
Shikoku: Tokushima, Kagawa, Kochi, Ehime
Kyushu: Fukuoka, Saga, Nagasaki, Kumamoto, Ohita, Miyazaki, Kagoshima

** The approach estimating potential demand by region is as follows:

Commercial use: Potential demand for commercial use is inferred from proportions of commercial building floor area by region
Industrial use: Potential demand for industrial use inferred from proportions of value of manufactured goods shipments (from Manufacturing Industry Statistics) by region
Private generation: Potential demand for private generation use inferred from proportions of power generated and consumed by private generators by region

Sources: Table 4.14 and FY 2005 Census of Manufacturers: Ministry of Economy, Trade and Industry, Energy Report Statistics: Ministry of Economy, Trade and Industry website, etc.

assumed that switching to gas is relatively unproblematic in these areas. The Toukaido corridor that links Chubu and Kanto has a large potential demand, and there are many plans for construction of new gas supply infrastructure in this region. Also, the Chugoku region has a high level of potential demand relative to the size of its economy. Demand is particularly concentrated in the industrial belt around the Seto Inland Sea, where there are currently many potential customers generating their own power using coal or oil.[57] There is also 9 Mt of potential demand in other regions, Tohoku, Hokuriku, Kyushu and Shikoku. These are also expected to benefit from a market environment that encourages switching to natural gas due to stricter air-pollution regulations[58] and the promotion of carbon-related energy-conservation policies.[59]

Competition with Electricity. These figures give a relatively clear picture of the potential demand relative to oil products, but for natural gas relative to electric power the situation is not so transparent, and the factors relating to electricity include many negative elements for future natural gas demand. In FY 2005, the share of electric power was 22 percent in industry, 51 percent in the commercial sector, and 44 percent in the residential sector. Electricity has already achieved a high share in each sector and it is expected that the trend of electrification will continue. So the focus will be on the relationship between natural gas – which can be used as both a primary and a final energy source – and electricity, which is a secondary source.

For example, in the industrial and commercial sectors, there has been active adoption of cogeneration systems using natural gas. Installed cogeneration capacity using natural gas in these sectors reached 3590 MW in FY2005 and is expected to reach 4980 MW in FY2010,[60] which will have a substantial effect on natural gas demand. When consumers consider installing a natural gas cogeneration system, the basis for the decision will be relative costs, in which electricity tariffs have a large weighting. Electricity price projections reflect the overall generation costs of three different power sources: nuclear, hydro, and thermal.[61] Therefore, even when oil prices are soaring, as they have post-2003, the overall cost still substantially reflects relatively inexpensive nuclear, hydro, and coal-fired power generation.[62] In addition, incumbent electric utilities have repeatedly lowered electricity tariffs to protect their market share as a response to market liberalisation. These factors are undoubtedly worsening the competitiveness of natural gas against electricity. Figure 4.6 shows electricity tariffs (industrial and commercial) and JCC prices (which are the basis of LNG prices) for the period 2002–07, demonstrating a trend that benefits electricity.

In the residential sector, there is also a battle for market share between natural gas and electricity. Recently, there has been a sharp increase in all-electric housing, actively promoted by the electric utilities. At the end of FY 2006, about 2.06 million households[63] were all-electric, out of a nationwide total of about 51 million households.[64] Although this only represents about 4 percent of the housing stock, the figure grew by

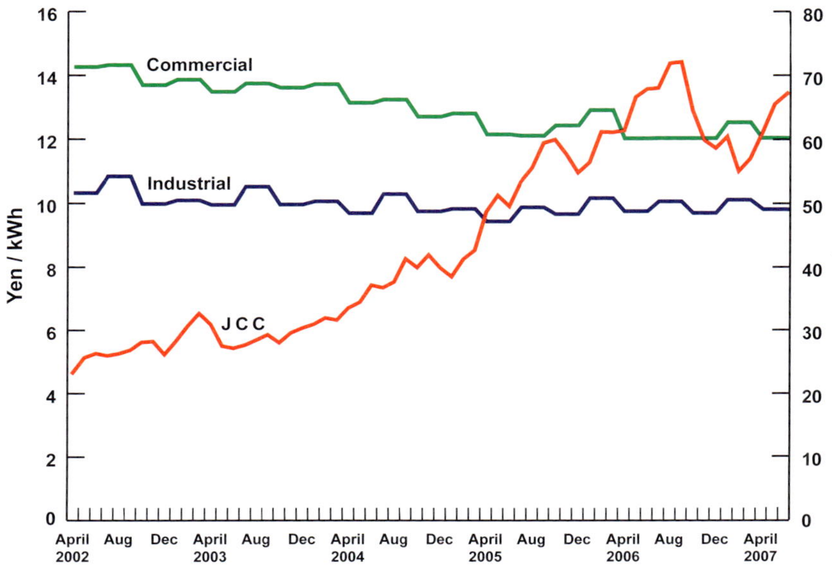

Figure 4.6: Electricity Prices for Large Customers 2002–07

Source: JCC prices; data from Trade Statistics of Japan, Ministry of Finance. Electricity prices; METI

396,000 households in FY 2005 and by 490,000 in FY 2006.[65] Since the total number of permits for housing starts in FY 2006 was 1,285,000,[66] the penetration of all-electric housing appears to be accelerating. To make a rough assessment of the effect on natural gas demand, about half the number of all-electric houses[67] was multiplied by the average gas consumption per house and converted to the amount of city gas that would otherwise have been consumed. This equals an annual LNG consumption in the order of 300,000 tonnes. Note, however, that the penetration of all-electric houses is thought to be higher in rural areas where city gas supplies are not available, and that within city gas service areas the all-electric trend particularly affects communal housing with only small levels of energy consumption per housing unit. Consequently, the actual effect of all-electric houses is likely to be smaller than the provisional calculation above. Nevertheless, the growth is extremely vigorous, and if this trend continues, it certainly has the potential to affect residential gas demand.

Price Competitiveness of Natural Gas

As outlined above in the section on recent market trends, the main factors determining future gas demand in the long term are the price competitiveness of natural gas against other fuels, particularly in the city gas sector, and the level of nuclear power for the power generation sector. These two factors are examined here to consider the influence they can be expected to have on natural gas demand.

Japan's LNG contracts are shown in Figure 4.7. Up to about 2010, existing contracts and new contracts that have already been signed (i.e. with sales and purchase agreements) will provide sufficient LNG procurement to meet anticipated demand. Projects that have started up since 2000 include Tiga (Malaysia), North West Shelf expansion (Australia) in 2004, Bayu-Undan (Australia) and Qalhat (Oman) in 2006. New contracts expected to come on stream by 2010 include Sakhalin II (Russia), Tangguh (Indonesia), and also contracts with Malaysia and Qatar.

During the period 2010–20, several existing contracts will terminate, including the Indonesia 73/81 contracts, and contracts in Brunei and Malaysia (Satu, Dua). There will be negotiations to extend them but, in the case of Indonesia, it is uncertain whether they will be successful (see Chapter 7). Grassroots projects due to start operating in or after 2010 that are currently at a stage where supplies are largely fixed include Pluto, Gorgon (Australia) and others.

A substantial share of Japan's LNG demand after 2010 is not yet covered by contracts. This is likely to require new long-term contracts, negotiations for some of which are already under way. However, as stated above there is strong upward pressure on new contract prices.

At the same time, in the domestic Japanese market, gas sales to large customers are subject to fierce competition between incumbent city gas utilities and new entrants, to the extent that prices appear to have fallen to levels which are only borderline viable for these companies.[68] Therefore, reducing LNG procurement costs will be a key factor in retaining market share and staying in business.

Considering these circumstances in the international and domestic markets, the sort of LNG price levels eventually agreed in negotiations between buyers and sellers will be an extremely important factor determining the scale of long-term natural gas market growth up to about 2020, particularly in the city gas sector.

A simple illustration of the effect is shown in Figure 4.8.

The Figure reveals the following points:

- If the LNG price is raised to the extent currently advocated by sellers (with LNG covered by existing contracts still at the current price), gas prices clearly lose competitiveness (price band E). LNG prices

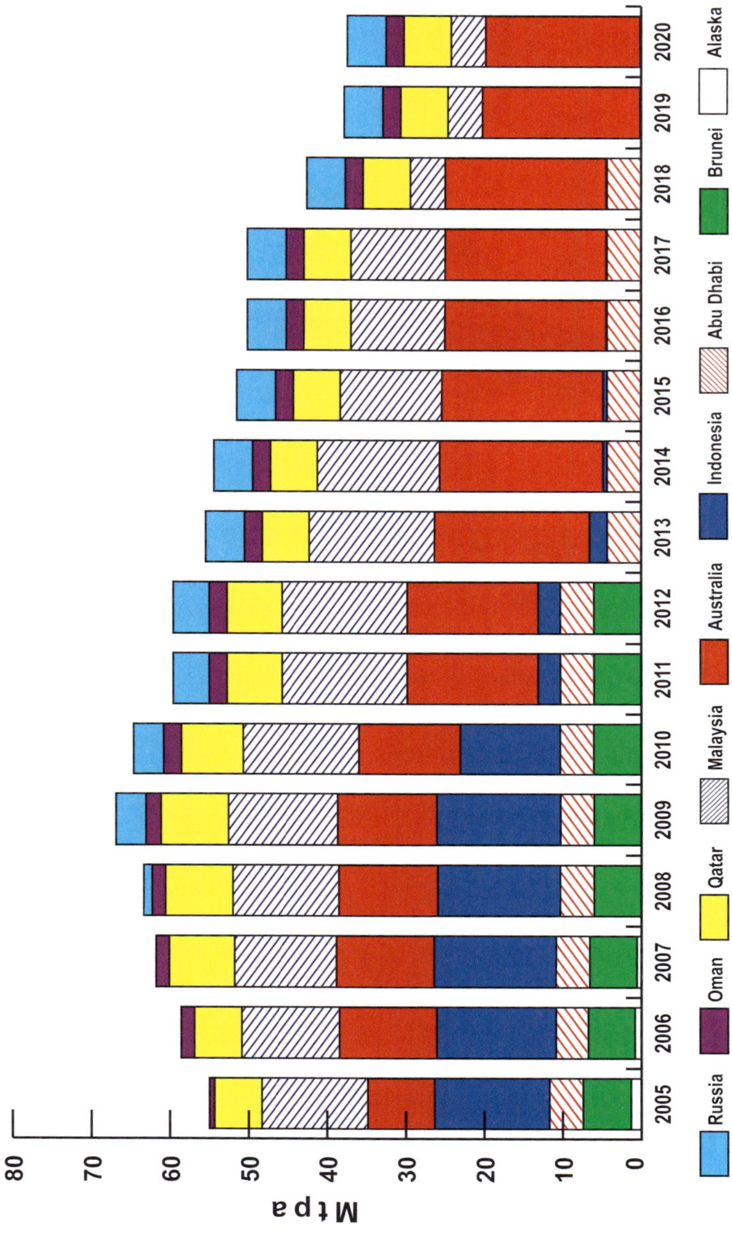

Figure 4.7: Contracted LNG volumes 2005–20

Source: Data from Osaka Gas

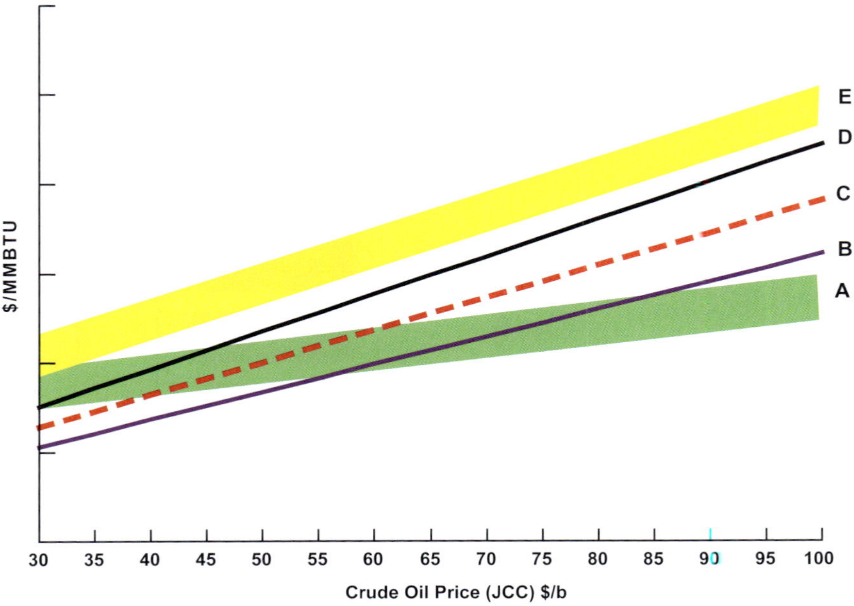

Figure 4.8: Price Competitiveness of Natural Gas against Oil Products*

* A: City gas price for large customers based on current JLC price (average LNG import price for Japan)
B: HSFO C (High Sulphur Fuel Oil C) price
C: LSFO C (Low Sulphur Fuel Oil C) price
D: LSFO A (Low Sulphur Fuel Oil A) price
E: Future city gas price as of 2015 for large customers based on raising LNG prices of all future LNG contracts to the so-called 'oil equivalent' price currently advocated by sellers in LNG contract negotiations

See footnote 55 for definitions of oil product groups.

Source: 'Outlook for Japan's natural gas prices', presented by C.Ishiguro, General Manager of Osaka Gas International Transport, 18 October 2007

based at 'crude oil equivalent' price levels in new contracts will be significantly higher than the price levels shown by the price band E in Figure 4.8,[69] becoming completely uncompetitive against fuel oil. This price level will not only make it impossible to secure potential demand, but will definitely lead to losses of existing demand as competitiveness against electricity will deteriorate substantially.

- If LNG were to maintain level A, and if crude oil prices were to remain at current levels (the JCC price for crude oil was about $67/barrel at June 2007) or higher, LNG would be price competitive against low sulphur fuel oil C (LSFOC), and would likely be able to secure a considerable amount of additional demand.

- Conversely, if crude oil prices fall to $40 or lower, price competitiveness against LSFOC would be lost, even at current JLC price levels, producing a situation in which it would not be easy to secure the potential demand; on the other hand, price competitiveness against electricity is likely to be recovered gradually.

In short, the fact that demand for natural gas has grown significantly as the price competitiveness of natural gas improved with crude oil price increases post-2003, is a confirmation of the potential for gas demand increases. Nevertheless, if LNG import prices are raised and price competitiveness deteriorates, it will become difficult for natural gas to realise this potential, and the share of gas in primary energy will stagnate.

Impact of Nuclear Power Generation

The next point to consider is the impact of nuclear power policy on natural gas demand. Details of Japan's current policy (relevant to this chapter) are set out in the Nuclear Energy National Plan.[70] The objective is to raise the proportion of electricity generated by nuclear power from the current level of about 30 percent up to 40 percent or higher around 2030. The policies to achieve this objective include practical initiatives on issues such as raising the current capacity utilisation rate, extending the working life of older facilities, deciding on ultimate disposal sites for high level radioactive wastes, and promoting the construction of new plants. Currently, the highest priority of these policies is raising capacity utilisation, and the target is to reach a load factor of 85–90 percent which is typically achieved by countries with the most successful nuclear sectors.[71]

Various nuclear power scenarios are at present being debated in Japan,[72] but for the purpose of this discussion a scenario involving construction of relatively few new nuclear plants is assumed. For the period to FY 2010, all the outlooks assume that only one new plant will be constructed, and it is fairly certain that the total of nuclear power generating capacity will be 50,140 MW.[73] Consequently, the effect of nuclear power on natural gas demand during this period depends on the load factor achieved by the nuclear plants. Approximate calculations suggest that with this total generating capacity, a 10 percent rise in capacity utilisation would result in an annual increase of 44 TWh in electricity generated, equivalent to the output of approximately four new nuclear plants.[74] In other words, if capacity utilisation of existing nuclear plants can be raised from about 80 percent to about 90 percent, the need to construct new plants could be restrained. While, if having all the nuclear plants in Japan working at 80 percent utilisation is considered to be the base level for FY 2010, then raising utilisation to 90 percent would result in generating additional electricity equivalent to cutting consumption of LNG for thermal power generation by 6 million tonnes. Conversely, if capacity utilisation declined to 70 percent, and if the reduction in electricity generated was all covered by LNG-fired stations, LNG consumption would rise by about 6 million tonnes.

Next, the long-term outlook for the period to FY 2020 is examined using the following assumptions. Total production by electric utilities in FY 2020 is assumed to be 1120 TWh,[75] but nuclear power generating capacity is assumed to have only a low level of new plant construction after FY 2010 (about five new plants), resulting in a total capacity of 57,000 MW.[76] This assumption is based on the energy industry consensus that the construction of new nuclear plants is unlikely to proceed in line with the individual electric power companies' long-term supply plans (published annually). It is generally believed that the main reason for this is difficulty in obtaining agreements and/or permission from the communities and local governments where the sites are located. However, particularly where plans involve construction of additional units at existing sites, local acceptance has, in most cases, already been obtained. A large proportion of new plant construction therefore has no significant public acceptance issues. The main reason for the discrepancy between the plans and the reality is uncertainty over levels of future demand.[77] The basic concept behind the Nuclear Energy National Plan is to encourage the construction of nuclear plants and wide-ranging power exchange between electric utilities. However, it is still unclear that such electric power exchanges will actually function as desired and, facing demand uncertainty, it will be difficult for power companies to go ahead with huge investments in nuclear plants.

In addition to these assumptions regarding plans for construction of new nuclear power plants, in an attempt to ensure that realistic figures (in terms of capacity) are used for the amount of power generated from sources other than LNG-fired stations, the FY 2007 Supply Plan figures have been used virtually unchanged.[78]

Estimates for LNG consumption at different load factors for nuclear power are shown in Table 4.16, using the assumptions outlined above, and assuming that the variations in capacity are fully substituted by varying the amount of LNG-fired thermal power generation. These calculations show that if nuclear power capacity utilisation can successfully

Table 4.16: LNG Consumption in the Power Sector in 2020, at different load factors for nuclear stations

Load factor for Nuclear power plants	*90*	*80*	*70*	*60*
LNG consumption (Million tonnes)	40.7	47.7	55.2	61.7

Source: Author's estimates

be raised to the 85–90 percent target, LNG demand in the power sector would increase by less than 5 million tonnes from its current level of about 38 million tonnes.

Because of the requirement to reduce CO_2 emissions, coal-fired power generation will be restricted, and if oil prices remain high, there is a strong likelihood that LNG (with the largest installed capacity) will be used as the main incremental fuel for thermal generation. Although current government policies for promotion of nuclear power basically assume no substantial increase in LNG demand for power generation, if nuclear plants fail to live up to expections, there will be significant additional LNG demand.

The description above outlines the basic approach when considering LNG demand in the generating sector. However, it must be noted here that medium- and long-term nuclear policies may be affected by the issue of the earthquake resistance of plants recently brought to our attention by the Chuetsu Offshore Earthquake in Niigata. This issue needs to be taken into account, but the extent to which it will affect policy is still unclear (see Appendix 4.4).

Having described two of the main factors affecting natural gas demand, the next step is to examine the outlook with these factors taken into account. Medium-term and long-term outlooks are examined separately.

Outlook for Natural Gas Demand until around 2010

There is general confidence that the economy will continue to perform well, and therefore that total energy consumption will increase, although continuing high oil prices are a cause of concern. Since this period includes the Kyoto Protocol target dates, the government is likely to exert strong pressure particularly on manufacturing industry to achieve CO_2 reduction targets, and this is certain to have the effect of raising natural gas demand.

In terms of competition and market structure, although the shift of policy towards security has brought market liberalisation to a halt, within the areas already open to competition, this trend will definitely remain. However, in the medium term, the likelihood of further liberalisation and competition is insignificant.

In the electric power sector, plans for construction of new power sources have been settled, and operational plans for all plants are basically already incorporated into the official 'supply plans', including estimates of load factors. The likelihood of large changes in demand for LNG and other fuels for thermal power generation is small.

However, looking at the international markets and taking into account the ability to supply LNG, including factors such as the start-up of new projects for the Asia-Pacific market and Indonesian supply shortages, LNG supply tightness looks set to continue. The significance of this for new long-term LNG contracts is that sellers are likely to be in a strong commercial position, and that prices of new LNG purchases will be higher than in existing long-term contracts.

In this situation, the factor with the greatest effect on natural gas demand over this period will be the price of natural gas, with oil price levels and their impact on international LNG markets holding the key. If oil prices remain at post-2003 levels, the effect of the S-curve built into existing contracts will ensure that average LNG import prices will have a relative advantage over oil products. However, if it becomes necessary to increase purchases beyond existing long-term contracts, it will then become necessary to rely on shorter-term or spot transactions. In such cases, if spot prices rise to the point at which competitiveness is lost, to increase the level of demand will be difficult, particularly in the city gas sector. As a result, the business prospects for new entrants in both the city gas and electricity markets would suffer, because with uncompetitive feedstock prices it is extremely difficult to capture new customers.

The situation described above seems to imply that it is relatively unlikely that the demand for natural gas will rise to substantially higher levels than the base case (shown in Table 4.17). The only conditions under which demand would rise above the base case would be:

- if the Pacific LNG supply/demand becomes less tight because of an unexpected decline in global demand;
- a situation where Japan's buyers had to quickly increase their purchases of LNG without regard for conventional economic rationality: for example, if buyers decided that city gas market share needed to increase even if this required purchasing LNG at relatively high prices, or if dependence on LNG-fired generation needed to increase because other generating capacity was not available.

Outlook for Natural Gas Demand until 2020

We have seen that primary energy demand is assumed to grow slowly up to 2020. Viewing energy supply from a long-term perspective, the market environment is likely to be one where the use of natural gas is promoted (as indicated above in the section on energy policy).

Japan's natural gas industry has frequently been criticised for failing to develop a pipeline infrastructure. Investment is being made in both large metropolitan areas and surrounding regions using a variety of transportation methods based on the scale of demand, transportation distance and other factors. While this will not take the form of a national, large diameter pipeline, the network is being gradually extended into existing and new demand areas. This is being facilitated by the progress towards linking the networks of different companies, which will strengthen and expand supply potential. But as far as an international pipeline from Sakhalin is concerned, there is currently an insufficient domestic consensus for this to be developed before 2020.

In terms of liberalisation, as of October 2007 the likelihood of full residential market opening for electricity and gas is low, even from a long-term perspective, and it seems likely that there will be no unbundling or similar drastic reform of industry structures. Nevertheless, markets for the electric and gas industries are now at least 60 percent open to competition, and will further develop, with the potential to substantially change the structure of the industry. In the small customer sector, electric utilities and oil companies are already starting to build soft corporate alliances, in some cases with city gas utilities, for sales, and pipeline (or other) transportation. The moves to interconnect different companies' networks (described above) are likely to further encourage this type of alliance. As soon as large numbers of supply sources (LNG terminals and domestic gas fields) are linked by pipelines, development of a variety of new LNG purchasing and supply and demand balancing arrangments can be developed. Such moves may further develop competition, as well as encouraging moves towards mergers or takeovers.

Global warming countermeasures will become a major policy objective. Future frameworks beyond the Kyoto Protocol are still uncertain, but it should be assumed that action on CO_2 emissions will become more vigorous. In power generation, commercial development of CO_2 sequestration and highly efficient IGCCs[79] may result in restrictions on coal use being removed or relaxed leading to a rise in its use. Japan's electric utilities and others are already increasing their efforts in these areas of R&D. Yet, the probability of this new coal utilisation technology coming into widespread use by 2020 is low. Development of renewable energy sources such as solar or wind power is also likely to be limited, so the main issues relate to existing power sources, in particular, LNG-fired and nuclear power. In the city gas industry, although a macro perspective reveals that energy conservation and other actions to cut CO_2 emissions are going to be strengthened in all areas, particularly manufacturing industry, the relatively high share of oil products in this sector suggests that switching to natural gas, or adopting efficient energy appliances such as cogeneration systems using natural gas, will be effective measures. Thus, from an environmental perspective, circumstances are expected to promote natural gas demand in both power generation and city gas.

This chapter has concluded that the factor with the most decisive influence on natural gas demand will be price competitiveness; but the load factors achieved by nuclear power stations will also be an important determinant. Moving on to consider natural gas demand up to 2020, the first step is to examine demand for power generation by incumbent electric power utilities. The supply plans by the electric utilities for the period up to FY 2016 include construction of nine new nuclear facilities (producing 61.5 million kW in total) at an 83 percent load factor. Based on these plans, even after 2010, LNG demand is likely to be about 39–40 million tonnes, a little higher than the 2006 level. Taking into account the LNG-fired thermal generating capacity shown in the plans, it is difficult to imagine a fall in LNG consumption from 2006 levels.[80] Consequently, this level can be perceived as the floor in demand for the electricity sector. In practice, however, there is little likelihood of successfully achieving the new construction schedules set out in the nuclear plans.

The most realistic (reference) case is for an additional six facilities (producing 57 million kW/year in total) to be constructed by 2020. In this case, if the load factor for overall nuclear plants is taken to be 80 percent, then the demand for LNG in the power sector can be predicted at around 48 million tonnes which probably represents the upper limit for the reference case. It is quite possible that the attempt to increase nuclear capacity utilisation will be successful and the load factor will climb to about 85 percent, in which case LNG demand would be about 44 million tonnes. Consequently, it is appropriate to consider the likely LNG demand in the electric power sector in around 2020 to be in a range between 44 and 48 million tonnes. If, as shown above, the load factor falls below 80 percent, there would be additional temporary demand for LNG.

The next step is to consider other demand, predominantly for city gas. In addition to sales made by city gas utilities in distribution, this category includes sales by electric utilities in the city gas market, demand for power generation by PPS companies, and demand for private power generation by manufacturers.

The demand outlook in this sector is based on the medium-term supply plans produced each year by the city gas utilities, and also takes into account the long-term outlook by the government and other research institutes. The demand is expected to be around 33 million tonnes in 2020, an increase over the present level of only about 5 million tonnes. Note, however, that this volume includes fuel for large new gas-fired power plants run by PPS companies (including the two major city gas utilities providing capacity of about 3000MW), as well as increases in the commercial and residential sectors. The demand figure appears to be projected by city gas utilities on the basis of assumptions that do not fully take into account points such as the price advantage gained by natural gas over oil products (due to oil price rises). Therefore, the expected switch to natural gas in the industrial and energy conversion sectors appears much smaller than the potential demand described above. Consequently, if the S-curve effect that provides price benefits over oil products, is maintained over the longer term, and at the same time if the absolute price level is competitive with electricity prices, then natural gas may be able to capture more potential demand than envisaged by these calculations. In reality, infrastructure projects are already under way to construct or upgrade facilities in the metropolitan areas, in the area along the Pacific coast of the Tokaido region and in the area along the Seto Inland Sea coast of the Chugoku region. So, it can be assumed that there is a reasonable possibility of securing several million tonnes of additional demand by 2020.

By contrast, if upcoming new LNG contracts and contract extensions incorporate terms which cause gas to lose its price advantage relative to oil products and electricity, then there is a strong possibility that

Table 4.17: Natural Gas Demand Outlook. Million tonnes

Sector FY	*2006*	*2010*	*2015*	*2020*
Current outlook				
Base case				
For power generation	38.2	41.0	41.0–44.0	44.0–48.0
For other purposes	27.6	29.0	31.0–33.0	33.0–38.0
Total	65.8	70.0	72.0–75.0	75.0–86.0
20% case				80.0–85.0

Source: author's estimates

demand will not meet the projection set out above. This would result in stagnating demand with new industrial customers and difficulties in retaining existing (including cogeneration) customers.

Having examined these projections, we can consider the share of Japan's primary energy supply that may be captured by gas. With reference to long-term supply and demand outlooks by government and by think tanks such as the Institute of Energy Economics, Japanese primary energy demand in 2020 is expected to be in the range of 520–550 Mtoe.[81] If it is assumed that the natural gas supply share is about 20 percent of the total primary demand, as it is in many OECD countries, this would require around 80–85 million tonnes of LNG equivalent. Comparing these figures with the natural gas demand outlook shown in Table 4.17 reveals the following points.

In the case where natural gas price competitiveness against oil and electricity is maintained, there is a substantially higher probability that the natural gas share will reach 20 percent, even with low demand for power generation. In contrast, if the price of natural gas rises to the extent that it loses market share to oil and electricity, the probability of its share of primary energy exceeding 20 percent is greatly reduced.

Conclusions

Six years ago, when the previous edition of this book was published, the Koizumi administration was implementing substantial structural reform policies, and the forces for liberalisation of power and gas markets were at their strongest. Even the electricity industry, which has a decisive influence on Japan's energy supply and demand structure, was uncertain about its future. The major issue for this chapter in the first edition was how to evaluate the outlook for liberalisation. At the time, Japan's economy was still in recession, energy demand was stagnant, and utilities discussing LNG procurement were concerned to avoid triggering take or pay penalties in their contracts. The previous discussion was generally pessimistic about future demand. However, the energy market was then hit by the shock of unexpectedly high oil prices, leading to the changes in the natural gas market described in this chapter.

The most important conclusion from the changes over the last five years is that all the major assumptions needed to project future gas demand – including energy policy, prices, infrastructure, and market systems – have become easier to interpret, and this has narrowed the range of future outcomes. Put another way, it is clearer that Japanese gas demand will grow substantially.

In the first edition, there was debate about the necessity of increased demand in the power generating sector as part of a 'gas shift' scenario. Since then, changes in government energy policy and the new emphasis on strengthening nuclear capacity overturned previous assumptions and reduced expectations for substantial demand growth in the sector. Despite this, one of the factors behind the expectations of growing demand in the power sector was foreseeable limits to the growth of demand in the city gas sector. That chapter was written during a period of low oil prices (and contained no inkling of post-2003 price levels), consequently, the price competitiveness of natural gas relative to oil products was weak, and it was considered too difficult to achieve the level of potential demand envisaged in this chapter. The 'gas shift' scenario in the previous edition was based on the key assumption that increased demand could be achieved only when the price of natural gas was substantially reduced by the rapid introduction of competition. Ironically, however, as the prospects for liberalisation receded and oil prices rose, this totally changed the dynamic for competition between energy sources, raising the price competitiveness of natural gas against the oil products that have the largest share of Japan's energy market, thereby substantially raising gas demand prospects.

The earlier chapter also proposed construction of an international pipeline from Sakhalin as a pre-condition for the 'gas shift' scenario, but this development looks unlikely to proceed in the foreseeable future. Nevertheless, from the perspective of securing energy resources, as emphasised in Japan's new energy policy, there could be a change in approach to the Sakhalin region with its promise of additional (oil and) natural gas supplies, and this has the potential to create a consensus between the public and private sectors regarding an international pipeline. Since introducing supply of natural gas by international pipelines is an essential means of securing large-scale energy supplies, from the perspective of energy security, debate on this issue should continue.

Growth in gas demand and market liberalisation are accelerating investment, and expanding supply infrastructure centred around LNG import terminals and linkages between local and regional networks.[82] At the same time the receding prospects for radical market restructuring revealed the future structure of the industry more clearly. It is very likely that electric utilities will be able to maintain their present structure over the long term, and this reduces uncertainties. Electric utilities were already in a very strong position with regard to LNG imports, and they are also expected to gain increasing influence in downstream gas markets. Even though major city gas companies will retain an influential market position, competition and collaboration between regional electric utilities, nearby city gas utilities, LPG companies and oil majors, are becoming important characteristics of the gas business.

Examining changes in LNG procurement reveals the collapse of the old convoy-style consortia.[83] This was a natural result of downstream market liberalisation and attempts by companies to establish their own identities. But with a few exceptions, it has resulted in Japan's utilities losing some of the bargaining power that came from being able to command large volume purchases. New arrangements for LNG procurement are evolving with electric utilities not only pursuing lower prices, but also greater flexibility to cope with future dips in nuclear power availability. Price competitiveness in the city gas sector is an absolute requirement. The traditional consortia are being replaced by alliances with clear objectives that meet the needs of individual participants.[84]

Finally, trends in the international LNG markets will have the biggest influence on Japanese utility procurement. Extremely high spot prices are a concern for the late 2000s and early 2010s, but in the long term it will be the price formulae agreed in new and renewed LNG contracts that have the greatest effect on natural gas demand. Agreements between LNG sellers and buyers need to be built on an awareness of the needs of Japan's end-consumers. Prices that ensure LNG remains competitive with other fuels will result in the potential for a substantial increase in natural gas demand, particularly in the city gas sector.

Appendix 4.1: New Japanese Gas Pipelines

Table A4.1.1: New Japanese Gas Pipelines

No.	*Pipeline*	*Owner*	*km*	*Operation*
a	Yamagata	Tohoku Natural Gas Co*	30	May 2006
b	Shiroishi-Kouriyama	JAPEX 80%, Tohoku Electric 20%	96	March 2007
c	Sendai-Iwaki Offshore	JAPEX	160	2011
d	Minamifuji	Teikoku Oil Shizuoka Gas Tokyo Gas	27	Dec. 2006
e	Shizuoka	Teikoku Oil	83	Dec. 2006
f	Shin Tokyo	Teikoku Oil	50	Dec. 2007
g	Iruma	Teikoku Oil	25	July 2005
h	Shin Oume	Teikoku Oil	49	2009
i	Kisarazu Rinkai	Tokyo Gas	na	Oct. 2008
j	Chuou	Tokyo Gas	na	Oct. 2009
k	Gunma	Tokyo Gas	100	March 2010
l	Chiba-Kashima	Tokyo Gas	80	Dec. 2010
m	Shin Negishi	Tokyo Gas	na	Oct. 2013
n	Yokohama	Tokyo Gas	na	Oct. 2013
o	Tokyo Bay Cross	Tokyo Electric	18	2010
p	Chiba	Tokyo Electric	na	na
q	Ise Bay Cross	Chubu Electric Toho Gas	19	2013
r	Mie-Shiga	Chubu Electric Osaka Gas	60	2010
s	Shiga	Osaka Gas	46	Oct. 2006
t	Himeji-Okayama	Osaka Gas	90	Under planning
u	Mizushima-Fukuyama	Setouchi PL	40	Dec. 2006
v	Onoda	Yamaguchi Gohdoh Gas	11	May 2006
w	Saga	Sanai Oil	28	2005
x	Fukunan	Saibu Gas	93	Dec. 2010

* Tohoku Natural Gas: Tohoku E 55%, JAPEX45%
** Setouchi PL: Hiroshina Gas 67%, Fukuyama Gas 20%, Chugoku Electric 12%, Mizushima Gas 1%

Source: company websites

Appendix 4.2: Long-term Energy Outlooks

Table A4.2.1: Long-term Energy Outlooks. 10^{15} kcal

METI Base Case

FY	*1990*	*%*	*2005*	*%*	*2010*	*%*	*2030*	*%*
Oil	290	57	257	47	230	44	234	42
Coal	86	17	117	22	92	18	97	18
Natural Gas	53	10	78	14	74	14	99	18
Nuclear	49	10	66	12	79	15	82	15
Hydro	22	4	17	3	19	4	18	3
New Energy	12	2	7	1	23	4	26	5
Total	512		542		517		554	
Final Consumption	344		374		364		388	
CO_2 emission	286		332		288		311	

Note: FY2010 Additional policy case See 'Energy in Japan 2006', pp.27–28 on METI website for details

Source: Advisory Committee for Natural Resources and Energy

METI Policy Implementation Case

FY	*1990*	*%*	*2005*	*%*	*2010*	*%*	*2030*	*%*
Oil	290	57	257	47	230	44	206	40
Coal	86	17	117	22	92	18	87	17
Natural Gas	53	10	78	14	74	14	99	19
Nuclear	49	10	66	12	79	15	80	15
Hydro	22	4	17	3	19	4	18	3
New Energy	12	2	7	1	23	4	26	5

Source: same as above

IEEJ Long-term Energy Outlook (April 2006)

FY	*1990*	*%*	*2005*	*%*	*2010*	*%*	*2030*	*%*
Oil	257		241	44	220	40	198	37
Coal	117		104	19	104	19	98	19
Natural Gas	78		84	16	92	17	95	18
Nuclear	66		83	15	97	18	104	20
Hydro and New Energy	24		31	5	34	6	37	7
Total	542		543		547		533	
Final Consumption	374		372		369		359	
CO_2 emission	332		312		303		284	

Source: Japan Long-term Energy Outlook, IEEJ, April 2006

Appendix 4.3: Nuclear Power Generation in Japan

1. Nuclear power plant incidents since 2001

Unscheduled Stoppages

2001	Nov. 7:	Chubu Electric Hamaoka No.1: Manual shutdown due to break in residual heat removal piping
	Nov. 14:	Chubu Electric Hamaoka No.2: Planned shutdown for the same reason above
2002	Shutdown of all TEPCO plants commences in July (Fukushima No.1 Reactor No.1 was the first to be restarted on July 13, 2005, and eventually all 17 plants shut down due to improprieties were restarted)	
	Apr. 26:	Chubu Electric Hamaoka No.1: Shutdown extended to March 2011
2003	Apr. 15:	All TEPCO plants shut down
2004	Aug. 9:	Kansai Electric Mihama No.3: Burst in secondary piping
	Feb. 21:	Chubu Electric Hamaoka No.2: Shutdown extended to March 2011
2005	Aug. 16:	Tohoku Electric Onagawa No.1, No.2, No.3: Stopped automatically on detecting earthquake
	(No.2 restarted in January 2006; No.3 restarted in March 2006)	
2006	March:	Hokuriku Electric Shika No.2 served with injunction to stop operating due to earthquake resistance design issues
	Jun. 15:	Chubu Electric Hamaoka No.5: Stopped automatically on low pressure turbine failure
	Jul. 5:	Hokuriku Electric Shiga No.2 stopped to take action on damage to low pressure turbine blades
	Aug. 18:	Kansai Electric Takahama No.3: Stopped automatically due to abnormal SG level
2007	Jul. 16:	TEPCO: Chuetsu Offshore Earthquake stopped operation at Kashiwazaki-Kariwa No.1–No.7

Source: company websites and other public data

Peak Power Demand and Nuclear Power Plans

Table A4.3.1 shows recent peak power demand statistics. These reveal that the average annual growth in peak demand from FY 2000 to FY 2006 was a tiny 0.2 percent, representing only an increase of 1915 MW. This is less than the capacity of two large nuclear power plants.

Outlooks by electric utilities and by the Ministry of Economy, Trade and Industry in the Nuclear Energy National Plan point out that although the increase in demand for electric power over the next ten years is expected to be about 17,400 MW as a total for the nine electric utilities, examining the situation at the individual utility level reveals that some regions do not expect sufficient growth of demand to make investment in a large nuclear plant viable (Table A4.3.2), and that the electric utilities need to work together and operate their businesses from an area-wide perspective.[85]

Table A4.3.1: Trends in Peak Power Demand for the 10 Incumbent Electric Utilities

FY	*2000*	*2001*	*2002*	*2003*	*2004*	*2005*	*2006*	*Average annual growth*
MW	173069	182378	179544	167267	174295	177697	174984	0.2%
Year-on-year growth	-	5.4%	-1.6%	-6.8%	4.2%	2.0%	-1.5%	

Table A4.3.2: Projected Growth in Peak Power Demand over the next 10 Years

	Growth in peak electric power from FY 2005 to FY 2015 (MW)	*Average annual growth (%)*
Hokkaido Electric Power	670	1.2
Tohoku Electric Power	1,350	0.9
Tokyo Electric Power	8,480	1.4
Chubu Electric Power	2,270	0.9
Hokuriku Electric Power	360	0.7
Kansai Electric Power	1,240	0.4
Chugoku Electric Power	1,170	1.0
Shikoku Electric Power	410	0.7
Kyushu Electric Power	1,520	0.9
Total of the nine above	17,400	-

Note: 2005 actuals are after compensation for air temperature; Peaks are in summer for all utilities except Hokkaido Electric; Winter peak is used for Hokkaido Electric.

Source: Nuclear Energy National Plan [*Genshi Rikkoku Keikaku - Nihon no Sentaku*], METI Agency for Natural Resources and Energy, November 13, 2006, p.97

Nuclear Capacity Utilisation in Japan and other Advanced Economies

Table A4.3.3 shows the utilisation rates in major countries with nuclear power plants. Note that in the USA, which has the largest capacity in the world, load factors in the 2000s have been close to 90 percent. Japan's nuclear power policy aims to catch up with the high utilisation rates achieved in those countries, and calls for more advanced operation

Table A4.3.3: Nuclear Power Plant Utilisation in Major Countries 2000–05*

Country	*MW*	*Reactors*	*2000*	*2001*	*2002*	*2003*	*2004*	*2005*
Japan	47700	55	80.9	81.0	78.4	57.4	69.9	68.7
USA	98254	103	87.1	88.3	89.1	88.7	89.5	89.4
France	63473	59	73.8	74.6	75.9	74.9	74.9	77.6
Germany	20303	17	86.9	87.4	84.1	84.2	87.2	86.1
UK	10982	19	63.9	71.1	73.7	82.2	65.4	57.0
Canada	12595	18	53.2	56.0	51.1	53.5	64.2	65.8
Sweden	8975	10	66.3	83.8	79.5	77.1	89.1	86.7
Spain	7442	8	90.8	93.1	91.7	88.9	91.9	83.9
South Korea	17533	20	90.1	92.9	92.4	93.7	91.8	95.0
Belgium	5728	7	91.4	88.3	89.4	89.6	87.7	89.1

* Capacity utilisation is based on the number of nuclear reactors at commercial power plants with capacities of at least 135 MW.

Source: *White Paper on Nuclear Energy* 2006, Atomic Energy Commission

Table A4.3.4: Nuclear Power Plant Utilisation by Japanese Electric Utilities 2000–05

Electric Company	*MW*	*Reactors*	*2000*	*2001*	*2002*	*2003*	*2004*	*2005*
Hokkaido	1158	2	85.8	84.8	92.9	80.2	80.4	87.5
Tohoku	3274	4	90.3	75.4	81.6	71.1	73.3	47.3
Tokyo	17308	17	79.4	80.1	60.7	26.3	61.7	66.4
Chubu	4997	5	87	69.5	33.7	53.2	51.9	63.1
Hokuriku	540	1	84.9	83.5	96.7	35.3	79.8	88.7
Kansai	9768	11	81.8	84.5	90.5	89.1	70.2	75.4
Chugoku	1280	2	60.3	91.6	95.7	68.5	65.4	82.9
Shikoku	2022	3	83.6	79.1	78.9	84.9	77.4	85.9
Kyushu	5258	6	85.8	79.7	85.9	88.9	86.2	86.8
JAPC*	2617	3	82.3	80	81	84.8	87.1	77.5

* The Japan Atomic Power Company

Source: FY 2006 Electric Power Supply/Demand in Japan 56, METI Agency for Natural Resources and Energy

and maintenance. The CO_2 emissions totals in the carbon reduction measures agreed by Japan's cabinet in 2005 were premised on achieving nuclear capacity utilisation in the range of 87–88 percent. In practice, its advisory council is already discussing changes to the approach taken for regular inspections of nuclear plants. These are currently conducted every 13 months, but a new system is being considered where plants would be able to operate continuously for up to 24 months.

Appendix 4.4: The Chuetsu Offshore Earthquake in Niigata and its Impact on Nuclear Power Generation

Outline

The extent to which earthquake resistance will affect policy is still unclear. The following is an outline of the most important issues raised by this incident. The Niigata Prefecture Chuetsu Offshore Earthquake[1] hit on 16 July 2007, stopping operation at all seven plants at TEPCO's Kashiwazaki-Kariwa Nuclear Power.[2] Damage to the generating facilities reported to date includes a transformer fire, very small leakage of radioactive material, and damage to a reactor crane;[3] more detailed surveys are planned. This is the first time that a nuclear power plant anywhere in the world has been hit by an earthquake of this strength, so the results of the surveys are keenly awaited to see what effects it had on safety measures.

The current earthquake resistance standards for nuclear power plants in Japan are determined by guidelines produced in September 2006 by the Atomic Energy Commission,[4] the first revision in 25 years of the old guidelines on seismic design for nuclear reactor facilities that were standardised in 1981. The main thrust of the revision to the guidelines was to enhance safety through ensuring earthquake resistance by implementing more detailed policies based on the sizes of earthquakes envisaged. Instead of being uniform throughout Japan, the resistance requirements are now set separately for each site with a nuclear plant, and vary according to surveyed geological conditions and records of earthquakes observed in the past. Based on these new guidelines, the electric power utilities are now planning to conduct geological surveys, determine basic earthquake ground motions, analyse the seismic responses of equipment, buildings, and so on, and evaluate seismic safety. The utilities have just started the work of conducting the first geological surveys.

Short-term Effects (as of August 2007)

In parallel with inspecting the structures and equipment for each of the seven reactors at the Kashiwazaki power station to confirm that there are no problems with safety systems, there will also be surveys to reassess active faults. Depending on the results, any repairs or upgrades to the facilities found to be necessary will be implemented. Since it is still very difficult to see the extent of the damage, it is also unclear when these reactors will be able to start up again. Media reports on 26 July 2007[5] suggested that it may be a year before it becomes possible to restart operation at even some of the reactors in this seven-reactor plant. Later reports are beginning to suggest that the plants may be shut down for even longer.

TEPCO announced that the temporary shortage of electricity resulting from shutting down this plant would be covered by means such as increasing the output of thermal and other power plants, arranging for additional power supply from other power utilities, delaying the scheduled inspection of its Fukushima nuclear power plants, and buying surplus electricity from private generators.

Long-term Effects

It is not possible to predict the long-term effects before examining the results of the detailed surveys into the damage, but it can be assumed that potential effects of this earthquake include the following.

- Extended shutdown of the Kashiwazaki Nuclear Plant (seven reactors)
- Earthquake resistance concerns revealed at other nuclear power plants, requiring countermeasures
- Intervention of political issues, including accommodation of local communities, holding back the policy of improving load factor of nuclear power plants
- Effects on the policy for extending the working lives of older nuclear power plants
- Effects on investment in construction of new nuclear power plants

Although there is a high possibility of the earthquake having a negative impact on the latest energy policies for increasing dependence on nuclear power for electricity supplies, it is not possible to predict the extent of such an impact at the current stage. For this reason, this report assumes that the intentions of the current energy policy remain unchanged, that there is no major impact on future nuclear power supply capacity, including plans for investment in new plants and plans to extend the lives of older plants, and assumes that the improvement in capacity utilisation, and so on is only impacted by being delayed a little.

Appendix Notes

1 Magnitude 6.8, centred out to sea, about 60 km southwest of Niigata City.
2 Total generating capacity of 8212 MW (Reactors No.1–5: 1100 MW/unit, Reactors No.6–7: 1356 MW/unit).
3 See TEPCO website for details.
4 In 1978, through revisions of the Atomic Energy Basic Law and related laws, the Nuclear Safety Commission was established by spinning off functions related to safety regulations from the former Atomic Energy Commission in order to strengthen the system for ensuring nuclear power safety. Organisationally, it is positioned directly under the Cabinet Office, and has much greater authority than ordinary advisory committees in that the Commission can make recommendations to relevant agencies in the name of the Prime Minister if it is necessary.
5 Comments by a professor of Tokyo University involved in the surveys related to the incident.

Notes

1 The Izanagi boom, lasting 57 months from 1965 to 1970, was the longest period of continuous economic expansion in Japan between the end of the Second World War and the end of the bubble economy in 1991.
2 Medium-term Forecast of the Japanese Economy: FY2006–2020, Japan Center for Economic Research, 29 Nov. 2006.

Also, the Japanese government's New Economic Growth Strategy envisages achieving real annual GDP growth of around 2.2 percent even with a declining population. For details, please refer to the Research Institute of Economy, Trade and Industry's '*Jinko Genshoka Deno Atarashii Seicho Wo Mezasu* [*Aiming for New Growth under Population Decline*]' published 15 August 2006.
3 Published by Agency for Natural Resources and Energy, METI in May 2006.
4 The Basic Act on Energy Policy was enacted in June 2002, and accordingly the previous Basic Energy Plan was formulated in October 2003. The latest Plan published in March 2007 was the revision of the previous one.
5 For example, see material dated 11 Oct. 2007 of the Environmental Committee (under the Industrial Structure Council organized by METI).

6 Indicates the volume of consumption as fuel for power generation by electric power utilities. Consequently, the volume of sales to large industrial customers and city gas utilities by electric power utilities, and the volume of sales for private power generation plants of manufacturers are not included.
7 Details are shown in Appendix 4.3
8 Yet, it is difficult to gain an accurate picture of how alternative power sources were used to provide the additional capacity to make up for the shortfall, since there was also an increase in demand, and some new or upgraded generating facilities came on line during the same period.
9 Note that as there was also a shortage of LNG in the international LNG markets and high spot prices, particularly in the winter of 2005–2006, utilities took action to respond to the issues. Thus there was no simple explanation.
10 Source: *Nikkan Kogyo Shimbun* 16 April 2007. According to the Business Plan for FY2007 of Tokyo Electric, the sales target of 42 billon yen for FY 2008 was revised upwards to 55 billion yen, and a sales target of 65 billion yen was set for FY 2009 for the gas sales business segment.
11 Source: *Nikkan Kogyo Shimbun*, 16 November 2006. Kansai Electric's FY 2003 business plan projected gas sales for FY 2007 of around 500,000 tonnes LNG equivalent (including a certain volume of oil-based fuels), but early performance resulted in this figure being revised upwards.
12 Source: FY 2007 '*Keiei no Mezasu Mono* [*Management Objectives*],' Chubu Electric's website.
13 Akira Miyamoto, 'Gas in Japan' in Ian Wybrew-Bond and Jonathan Stern (eds), *Natural Gas in Asia, The Challenges of Growth in China, India, Japan and Korea*, Oxford Institute for Energy Studies, 2002 (henceforth Wybrew-Bond and Stern), especially pp.118–20.
14 Estimated by data from Trade Statistics of Japan. In the winter of FY 2005, the highest priced transaction on the spot market was a purchase made by KOGAS from Oman at around 25$/MMBtu.
15 In Korea, insufficient long-term contracts for gas to meet demand (see Chapter 5) were a significant factor which caused the high spot prices in Asia. In that country, the government bears the obligation to provide supplies, and we can easily imagine that there was a strong need to secure supplies even if prices were extraordinarily high, which in turn caused the rise in the LNG price.
16 For a discussion of the S curve see Appendix 2 of this volume.
17 See data in Energy Balance Table FY2005; *Handbook of Energy & Economic Statistics in Japan 2007*, The Energy Data and Modeling Center (EDMC) for details.
18 This does not include coastal LNG shipping which is discussed below.
19 For example, Chita Midorigahama (Toho G), Mizushima (Mizushima LNG), Sodeshi (Shimizu LNG) and others.
20 Technical studies and consultation with port administrators and other parties have already begun at receiving terminals in the areas around major urban cities such as Tokyo, Chubu, and Kansai.

21 For example, Kagoshima Terminal (Nihon G).
22 See Wybrew-Bond and Stern, *op cit*, pp.132–5, for details.
23 *Nihon Keizai Shimbun*, 9 August 2007.
24 Later, LNG supply also commenced from Osaka Gas' Himeji terminal.
25 Wybrew-Bond and Stern, *op cit*, pp.146–51.
26 See Map 4.1.
27 See 'The Framework of Desirable Future Electricity Industry System' Advisory Committee for Natural Resources and Energy, Electricity Industry Committee Feb.2003 for details.
28 Notifications to METI amounted to 23 companies by the end of March 2006.
29 Total for the 10 companies was 201.8 GW. Source: *Electric Utility Handbook, FY 2006*, ed. Statistics Committee, Federation of Electric Power Companies of Japan.
30 Interviews with various experts on market liberalisation.
31 See Agency for Natural Resources and Energy, METI website for details.
32 A category defined through a series of market reforms, consisting of companies other than general gas utilities that own pipelines above a certain scale and operate gas businesses.
33 Including Kansai Electric: 41 customers, Tokyo Electric: 14 customers, Chubu Electric Power: 9 customers, and Tohoku Natural Gas (with equity provided by Tohoku Electric Power): 3 customers.
34 For example, Tokyo Electric has closed down its myEnergy business, and Kyushu Electric has cut back the scale of its Nishinippon Environmental Energy (NEECO) operations.
35 Source: *Nihon Keizai Shimbun*, 1 September 2006.
36 Source: Nikkei Business Daily, 19 March 2007.
37 Due to come on line in October 2007. Fuel will be city gas purchased from Tokyo Gas. It is reported that the CCGT plant provides a high generating efficiency of 50%.
38 Due to start operating in April 2008. Kawasaki Natural Gas Generation is 51% owned by Nippon Oil Corporation, and 49% by Tokyo Gas Source: Tokyo Gas website.
39 Ohgishima Power is 75% owned by Tokyo Gas, and 25% by Showa Shell. Construction plans for Yokohama were originally for three 400 MW plants to start operating in 2010, but media reports suggest that 800 MW of capacity will be constructed first, with the remaining capacity postponed. Source: *Nihon Keizai Shimbun*, 25 February 2007.
40 Tokyo Gas website.
41 Osaka Gas website.
42 This includes investments in oil and gas upstream business, LNG liquefaction and transportation business, LNG receiving terminal infrastructure and domestic pipelines.
43 Accurate figures are difficult to obtain because volumes, etc. sold by JVs with other companies are unclear. Estimates based on figures published by each company and Trade Statistics suggest a total of about 2.4 million tonnes.

For example, projected gas sales for FY 2008 by TEPCO are reported to be about 1 million tonnes, which is 1.4 times the total volume sold in FY 2006 by Saibu Gas, Japan's fourth largest city gas utility.

44 Tohoku E 42.33%, Development Bank of Japan 25%, Niigata Prefecture 16.67%, JAPEX 3.85%, etc.

45 Out of contracts for 740,000 tonnes at the end of 2006, 720,000 tonnes were direct sales by its pipelines. Source: *Denki Shinbun*, 26 November 2006.

46 Sakai LNG is owned by Kansai Electricity, Iwatani International Corporation, Cosmo Oil, and Ube Industries.

47 Source: Chugoku Electric website.

48 Estimated from various sources including a Nikkei Business Daily article on 14 December 2006.

49 Source: Kyushu Electric website.

50 For reference: 'Outlook for Energy Supply and Demand in 2030', Advisory Committee for Natural Resources and Energy, March 2005, See Appendix 4.2.

51 In the case that energy conservation policy will be promoted, the growth rate is minus 0.3%/year.

52 According to the reference case, domestic supply of primary energy will increase from a crude-oil equivalent of 588 million kl in FY 2000 to 607 million kl in FY 2030. In the policy target case, with greater energy conservation, higher economic growth, and low nuclear power usage, the supply will decline to 566 million kl.

53 Estimate based on Energy Balance Table for FY 2005 by EDMC.

54 As for the definition of the potential demand, there are currently two aspects to the growth of natural gas demand – geographical expansion and growth of market share within a specific geographical area. In the case of geographical expansion, in particular, it would be difficult to promote early conversion to natural gas if the areas lack pipelines or other supply infrastructure. The following assumption is adopted here; that from a long-term perspective, the demand will be covered by infrastructure which will be gradually established, starting from locations with the greatest economic benefit.

55 Fuel oil A consists of around 90% of diesel oil and around 10% of residual oil and is mainly used for small- and medium-size boilers and burners in commercial buildings, hospitals, hotels, schools and small factories. Also, it is consumed by small- and medium-size diesel engines in the agricultural and fishing industries. Fuel oil A is divided into two categories by sulphur content (though this is not specified by Japanese Industrial Standards but commercially defined by Petroleum Association of Japan); maximum sulphur contents of 0.1wt% for low sulphur fuel oil C (LSFO A) and sulphur contents more than 0.1wt% for high sulphur fuel oil C (HSFO A). Fuel oil C, which is the most inexpensive fuel oil, consists of more than 90 % of residual oil and mainly consumed by large-size boilers, burners and diesel engines in factories, power stations and so on. Fuel oil C is divided into two categories by sulphur content as well. Maximum sulphur contents of

0.5wt% for low sulphur fuel oil C (LSFO C) and sulphur contents more than 0.5wt% for high sulphur fuel oil C (HSFO C).

56 A government policy applied to businesses using recent model industrial furnaces, boilers, etc. that consume at least 50 kl/year (crude oil equivalent) of fuel in the form of coal, oil, and so on. Part of the cost of facility modification is subsidised when such businesses switch to a gas fuel of which natural gas is the chief component. Figures quoted here were published on Japan Gas Association website.

57 The Chugoku region accounts for a high 19% share of the country's total potential demand for self-generation.

58 In rural areas, regulations regarding SOx and/or NOx targeting air pollution will presumably be tightened to match levels in urban areas, and this is likely to be one factor in promoting a switch to natural gas from other fuels.

59 Energy conservation policy is the main pillar of global warming countermeasures, and there is increasing promotion of these policies, such as setting energy saving targets for each industry and maintaining greater control over such targets. There is also expected to be more widespread use of cogeneration systems and other means of using natural gas efficiently.

60 Source: Japan Gas Association.

61 Provisions for calculation of electricity tariffs are regulated by the Electricity Enterprises Law.

62 Regulated tariffs are computed by first calculating the overall cost and then distributing those costs according to specific contract types.

63 Source: Ministry of Internal Affairs and Communications: Figures at end of March 2006.

64 Source: *Sankei* website.

65 These figures include reform of existing houses as well as new houses. Source: as for Footnote 57.

66 Source: Building Construction Started statistics, MLITJ.

67 The assumption is based on the fact that the total number of residential city gas customers, approximately 26.1 million, represents about half the total number of households.

68 Judging from financial reports, some operators are making a loss in the large-customer gas sector.

69 This is because the prices are not averaged with those for current contracts.

70 '*Genshiryoku Rikkoku Keikaku - Nihon no Sentaku* [*The Challenges and Directions for Nuclear Energy Policy in Japan – Japan's Nuclear Energy National Plan*],' METI, 13 November 2006.

71 See Appendix 4.3.

72 In addition to the case studies in the government outlook, reference was made to the IEEJ's long-term energy supply/demand outlook and other sources. See Appendix 4.2 for reference.

73 Tomari No.3 in Hokkaido is due to come on line, providing 912 MW, but Tsuruga No.1, providing 357 MW, is due to be shut down.

74 Calculated on the assumption that one newly constructed nuclear plant would provide 1350 MW of additional capacity.

75 For example, the reference case in the long-term energy supply/demand outlook (April 2006) published by IEEJ assumes 1136 TWh for FY 2020, and 1189 TWh for FY 2030. The government's long-term energy outlook assumes 1187 TWh for FY 2030 in the reference case and 910 TWh in the energy-saving case.

76 According to the 'Outline of Supply Plan' (March 2007) published by the Central Electric Power Council, there are plans for construction of eight new plants by 2016, producing 11,350 MW in total, and there are further plans to add an extra four plants providing 4960 MW. For the purposes of this discussion it is assumed that some actions are taken to extend the life of the existing, ageing plants.

77 See Appendix 4.3 for details.

78 Coal fired: 220 TWh, oil-fired: 60 TWh, Hydro/others: 98 TWh. For details, see the Supply Plan referred to above footnote 76.

79 Integrated Coal Gasification Combined Cycle

80 LNG-fired generating capacity at the end of FY 2016 will be 64,520 MW, a 7.4% increase on the end of FY 2006. Even if some gains in thermal efficiency are expected, it is difficult to imagine LNG consumption falling below present levels.

81 Calculated by the conversion factor of 1 Mtoe=10^{13} kcal.

82 In addition to new LNG terminals, if terminals can handle the new, super large carriers, Japan will have sufficient receiving capability to easily import about 100 Mt of LNG per year.

83 Before market liberalisation, LNG was usually procured by a consortium formed by several electric and gas utilities partly because utilities had to aggregate demand and partly because utilities wanted to diversify supply sources. Creating relative advantages to other utilities in terms of LNG procurement was not so important for each utility.

84 For example, the establishment of CELT Inc. by Tokyo Electric and Mitsubishi Corporation.

85 Nuclear Energy National Plan [*Genshi Rikkoku Keikaku - Nihon no Sentaku*], METI Agency for Natural Resources and Energy, 13 November 2006.

CHAPTER 5

NATURAL GAS IN KOREA

Keun-Wook Paik

Introduction

Korea first imported LNG in 1986 and, aided by the early strategic decision to develop a nationwide pipeline network, natural gas has become a major industry. The market is maturing but faces the challenges of the introduction of competition in both the wholesale supply and retail distribution sectors as a result of the government's privatisation and deregulation policies for the gas and electricity industries. The potential introduction of pipeline gas from Russia will be another major factor for the industry.

This chapter will briefly review the country's energy demand situation. It will then examine the expansion of the natural gas industry based on the government's latest long-term gas supply plan including the role of power. Finally it will touch on the possibility of importing pipeline gas and its implications for the growth of the natural gas industry.

Energy Demand and Supply

Energy demand in Korea has experienced unprecedented growth driven primarily by the general level of economic activity. Total primary energy consumption increased 4.2 times from 45.7 million tonnes of oil equivalent (Mtoe) in 1981 to 192.9 Mtoe in 2000, greater than GDP growth over the same period. Energy consumption increased annually by 7.5 percent on average from 1990 to 2000, greatly exceeding the GDP growth rate of 6.1 percent.[1]

The energy intensity per unit of GDP slightly improved from 0.37 toe/million won in 1981 to 0.35 toe/million won in 1990. However, it increased annually by 1.4 percent on average in the 1990s and reached 0.40 toe/million won in 2000. Energy consumption per capita also increased from 1.2 toe in 1981 to 4.1 toe in 2000, similar to that of Japan and most countries in Europe where GDP per capita is 2 to 3 times larger than Korea.[2]

The ever-increasing growth in energy consumption came to a halt in 1998 in the wake of the 1997 Asian crisis. In 1998 South Korea suffered a 6.7 percent decline in GDP and the unemployment rate rose to 7.9 percent. The social effects of the recession included, in addition to unemployment and accompanying labour unrest, a doubling in the price of foreign goods and energy imports, a rise in the rate of personal bankruptcy and a rapid decline in salaries. However GDP returned to growth in 1999 (10.7 percent) and this trend was maintained through the first half of 2000.[3]

Table 5.1 provides historical data on Korea's energy balance up to 2005. The most striking change (not actually shown in the table) was the growth in oil consumption from a 12 percent share in 1965 to 63 percent in 1994. Two major new energy sources, nuclear power and LNG, were introduced in 1977 and 1986 respectively. The combined share of both energy sources increased from 17.4 percent in 1990 to 23.9 percent in 2000 and reached 29.4 percent in 2005.

Energy demand in Korea has returned to the same level as before the 1997 financial crisis as a result of high economic growth in 1999 and 2000, but the growth rate during the period 2001–05 is much lower than that of the 1990s. Korea has virtually no major indigenous energy sources and imported 73.5 percent of its energy needs in 1980

Table 5.1: Primary Energy Consumption by Source

	1970	*1980*	*1986*	*1990*	*2000*	*2005*
Oil (Mbbl)	62.70 (47.2%)	182.1 (61.1%)	200.6 (46.4%)	356.35 (53.8%)	742.6 (52.0%)	761.1 (44.4%)
LNG (Mt)	(-) (-)	(-) (-)	0.05 (0.1%)	2.33 (3.2%)	14.56 (9.8%)	23.35 (13.3%)
Coal (Mt)	11.6 (29.6%)	25.9 (30.1%)	42.9 (38.0%)	43.4 (26.2%)	66.5 (22.3%)	84.8 (24.0%)
Hydro (GWH)	1.2 (1.6%)	2.0 (1.1%)	4.0 (1.6%)	6.4 (1.7%)	5.6 (0.7%)	5.2 (0.6%)
Nuclear (GWH)	(-) (16.1%)	3.5 (2.0%)	28.3 (11.5%)	52.9 (14.2%)	109.0 (14.1%)	146.8 (16.1%)
Others (Mt)	16.7 (21.6%)	9.0 (5.7%)	5.3 (2.4%)	2.8 (0.9%)	2.1 (1.1%)	4.0 (1.7%)
Total (Mtoe)	19.7 (100%)	43.9 (100%)	61.5 (100%)	93.2 (100%)	192.9 (100%)	228.6 (100%)

Source: MOCIE & KEEI, *Yearbook of Energy Statistics 2006*, pp. 14–17.

rising to more than 97 percent in 2005. Even though dependence on oil fell to 44 percent in 2005 from 61 percent in 1980, the financial burden of oil imports is proportionately higher. Korea spent US$66.7 billion on energy imports in 2005, of which US$52.3 billion was spent on crude oil and oil products, and US$8.6 billion on LNG.[4] This is a very significant increase compared with US$37.6 billion recorded in 2000.

The rate of growth in energy demand through to 2010 will be much lower than that of the 1990s. In September 1998, the Asia Pacific Energy Research Centre (APERC) produced a comprehensive energy supply and demand outlook for the Asia Pacific Economic Cooperation (APEC) member countries. According to this, Korea's primary energy supply and final energy consumption in 2010 will be 231.3 Mtoe (of which natural gas accounts for 27.4 Mtoe) and 178.0 Mtoe respectively, based on a 'business as usual' scenario.[5]

The APERC projection figures are considerably lower than those prepared by the Korea Energy Economics Institute (KEEI) in 1998, and Korea's National Energy Plan in 2002. Bearing in mind that the 1994 IEA forecast of Korea's primary energy supply in 2010 was 325.5 Mtoe, both APERC and KEEI's modest figures confirm that Korea's energy demand in this decade is very unlikely to repeat the explosive growth witnessed in the 1990s.

As shown in Table 5.2, Korea's National Energy Plan announced in late 2002 – there was no 3rd NEP – envisaged that oil would remain the main energy source because of the demand in the transportation and industrial sectors, though the share of oil consumption will continuously decrease from 51 percent in 2001 to 49 percent in 2011 and 47 percent in 2020. The demand for natural gas will grow the

Table 5.2: Energy Demand Projections by Source. Mtoe

	2001	*2011*	*2020*	*Growth Rate (%)*	
				2001–11	*2001–20*
Petroleum	100.4	136.8	157.7	3.1	2.4
LNG	20.8	31.1	45.8	4.1	4.2
Anthracite	3.7	2.6	2.3	-3.5	-2.5
Bituminous	42.0	62.9	69.7	4.1	2.7
Hydro	1.0	1.2	1.2	1.6	0.5
Nuclear	28.0	41.2	50.6	3.9	3.2
New & Renewable	2.3	5.2	6.9	8.2	5.9
Total	198.3	280.9	334.2	3.5	2.8

Source: National Energy Plan (2002)

fastest and will account for 10 percent of total energy consumption in 2001, 11 percent in 2011 and 14 percent in 2020. A common element in the APERC and KEEI studies and Korea's National Energy Plan is the projection that the fall in the share of oil will be made up by a corresponding increase in the role of LNG and nuclear power. It remains to be seen how gas supplies will be split between LNG and pipeline gas.

The Natural Gas Industry

The Government's Long-term Supply and Demand Projections

According to Korea's city gas business regulations, the Ministry of Commerce, Industry and Energy (MOCIE) is required to produce a Long-term Natural Gas Supply and Demand Plan (LTNGSDP) every two years.[6] City gas demand in the 6th LTNGSDP (2001–2015) is much higher than in the 5th. The growth rate for the city gas sector during the projection period is 5.3 percent. Gas demand for power generation during the 6th LTNGSDP period is projected to make a modest increase with a growth rate of 2.0 percent, which is lower than the 2.2 percent in the 5th LTNGSDP. In terms of volume, gas demand for power generation during the period 2001–2010 is projected to be around 4.2–7.3 Mt. Like the 6th, the 7th LTNGSDP made a similar projection on city gas sector demand increase. However, gas demand from the power sector in the 7th plan (0.32 percent) was much lower than that of 2.0 percent in the previous one.

According to the 8th plan, natural gas demand is projected to grow at an average annual rate of 3.5 percent through 2020, to 33.5 Mt in 2015 and 40.4 Mt in 2020. The city gas sector is set to grow at 5.4 percent annually, but the power sector demand is projected to remain at 2006 levels after the gradual increase until 2011. During the 5–7th plan periods, government projections were always lower than actual demand. The 6th and 7th plans projected gas demand in 2006 at 21.1 Mt and 22.6 Mt respectively, but the actual figure was 24.8 Mt. There is a striking increase in the demand projection for 2015 between the 7th and 8th plans (Tables 5.3 and 5.4) from 29.3 to 33.5 Mt.

In 2004, the power generation sector became the biggest gas user with 42 percent of demand, a small increase compared with 1998 when the sector's share was 40 percent, but substantially more than the 35–38 percent in the period 1999–2003. In 2006, power generators purchased 13.0 Bcm (or 9.54 Mt) from the Korea Gas Corporation (Kogas), an

Table 5.3: Korea's Long-term Natural Gas Demand Projection 7th Plan. Mt

	Residential	*City Gas/ Commercial*	*Industrial*	*Sub-Total*	*Power Generation**	*Total Demand*
2003 A	6.457	2.198	3.324	11.979	6.648	18.447
2004 E/A	6.971	2.387	3.487	12.845	7.496	20.341/ 21.809
2005 E/A	7.328	2.572	3.758	13.658	7.612	21.270/ 23.350
2006 E/A	7.686	2.768	4.052	14.506	8.081	22.587/ 24.840
2007	8.043	2.974	4.369	15.386	8.572	23.958
2008	8.400	3.191	4.709	16.300	7.689	23.989
2009	8.699	3.391	5.039	17.129	6.653	23.782
2010	8.997	3.598	5.388	17.983	7.348	25.331
2011	9.289	3.808	5.755	18.852	6.992	25.844
2012	9.585	4.027	6.147	19.759	5.871	25.630
2013	9.887	4.257	6.565	20.709	5.931	26.640
2014	10.194	4.496	7.010	21.700	6.014	27.714
2015	10.503	4.742	7.477	22.722	6.554	29.276
2016	10.815	5.000	7.972	23.787	6.575	30.362
2017	11.130	5.269	8.494	24.893	6.764	31.657
Annual Growth Rate (2003–17)	3.97%	6.44%	6.93%	5.36%	0.32%	3.93%

Note: E/A means estimate/actual.

* The figures of power generation include CES and Independent Power Producers

Source: MOCIE, 7th LTNGSDP, p. 27 & 8th LTNGSDP; KEEI, *Yearbook of Energy Statistics 2006.*

Table 5.4: Korea's Long-term Natural Gas Demand Projection 8th Plan

	Power Generation		*City Gas*		*Total*	
	Mt	*(%)*	*Mt*	*(%)*	*Mt*	*(%)*
2006	10.55	(42.5)	14.29	(57.5)	24.84	(100.0)
2007	12.22	(44.4)	15.29	(55.6)	27.51	(100.0)
2011	13.81	(42.0)	19.09	(58.0)	32.90	(100.0)
2015	10.19	(30.4)	23.30	(69.6)	33.49	(100.0)
2020	10.57	(26.2)	29.78	(73.8)	40.35	(100.0)
Annual Growth %	0.1%		5.4%		3.5%	

Note: for the first time in this plan, MOCIE did not release a detailed sectoral supply and demand projection.

Source: MOCIE, 8th LTNGSDP

increase of 8.2 percent compared with 2005.[7] However, the Korean government is determined to increase the use of nuclear generation in the power sector, which will be an important factor limiting future increases in gas consumption.

In the city gas sector, growth slowed during the period 2000–2004 to 5.5 percent per annum, a stark comparison with growth rates of up to 30 percent per annum in the decade 1990–2000. In 2006, sales to retail gas companies decreased for the first time, down 0.5 percent from 2005 to 19.0 Bcm (13.96 Mt), due to warmer weather and a slowdown in economic growth. Considering that South Korea's gas expansion has been driven by city gas, this is a very significant development.[8]

Market and Infrastructure Expansion

In 1981, the structure and development of the gas market was created with the establishment of Kogas in 1983 to handle imports and distribution. The first LNG imports arrived late in 1986, the same year that a feasibility study to provide gas supplies throughout the country was started; its conclusions were approved in 1990.

Based on the Korea Gas Corporation Act of December 1982 and the City Gas Business Act of December 1978, Korea's natural gas business was divided into a wholesale and a retail sector. Under the terms of the Korea Gas Corporation Act, the Ministry of Commerce, Industry and Energy (MOCIE) provides administrative guidance and supervision for Kogas; there is no independent gas industry regulator (see below). In turn, Kogas was granted a monopoly to import, store, and distribute gas through main trunk lines in Korea. City gas companies (see Table 5.6) are regulated at the municipal level by governors of cities and provinces.

Since the end of 2004, Kogas' ownership structure has consisted of four main groups: the biggest, institutional and individual investors, held 38.82 percent, followed by the Korean government with 26.86 percent, Korea Electric Power Corporation (Kepco) 24.46 percent and local governments (six cities and eight provinces) 9.86 percent. Until Kogas is fully privatised it remains under state control. As Kepco holds almost a quarter of Kogas' equity, the privatisation of Kepco itself will have a major impact on Kogas because of the possibility of direct LNG imports by power generators and competition between those companies and Kogas for LNG supplies (see below).

Kogas has some 2400 employees in five divisions (administration, production, operations, marketing and construction) and R&D, and its training centre. It has three subsidiaries: Korea Gas Maintenance & Engineering Co Ltd., Korea LNG Ltd and Korea Ras Laffan LNG Ltd.

As shown in Table 5.5, LNG demand overtook that of LPG in 1994. Although LPG demand growth looks modest relative to that of LNG, its growth has been steady and continuous. The national LPG penetration distribution rate was estimated at some 60 percent in 1997, but this is falling as city gas is introduced more widely – although as Table 5.5 shows, LPG demand in the residential and commercial sectors has increased as a result of the growing use of gas for cooking and heating in areas not yet connected to the pipeline network.

Table 5.5: Gas Demand Expansion in Korea: 1987–2005. Mt

	LNG	*LPG*
1987	1.62	1.82
1991	2.69	3.71
1993	4.40	5.11
1994	5.86	5.46
1995	7.12	5.72
1997	11.38	6.16
2000	14.55	7.27
2005	23.35	8.00

Source: KEEI, Monthly Energy Statistics.

The rapid growth in demand for natural gas has been greatly facilitated by the expansion of the national pipeline network (Map 5.1). Unlike Japan, where power generation rather than the city gas sector provides the main market for imported LNG, in Korea city gas is the main market.

While Kogas is responsible for the gas import and high pressure pipeline business, private city gas companies run the low pressure distribution business. As shown in Table 5.6, at the end of 2005 there were a total of 33 city gas companies, of which 29 are LNG based and four use LPG.[9] Each of the 33 city gas companies has an exclusive franchise for its region and is therefore a geographically defined monopoly. There are seven city gas companies operating in the Seoul metropolitan area with roughly 7.6 million customers and over 10 Bcm/year supply. In Pusan, the annual gas supply capacity has already surpassed 1 Bcm, and that of Daegu recorded 0.87 Bcm. According to the 8th Long-term Natural Gas Supply and Demand Plan, as of 2005 a total of 82 cities and towns use natural gas and the figure is expected to rise to 115 during the years 2011–2015, and 124 after 2016.

Map 5.1: Pipelines and LNG Regasification Plants

Table 5.6: City Gas Companies as at the end of 2005

	Source of gas	*Date of supply*	*Number of Households (million)*	*Supply Volume (Bcm)*	*Supply Area*
Daehan	LNG	1980.2	1.144	1.470	Seoul (southeast) & Kyunggi (southeast)
YESCO (Kukdong)	LNG	1983.3	1.008	1.434	Seoul (northeast) & Kyunggi (northeast)
Seoul	LNG	1983.3	1,747	2.127	Seoul (west) & Kyunggi (west)
Kangnam	LNG	1985.12	0.318	0.374	Seoul (southwest)
Hanjin	LNG	1986.5	0.694	0.942	Seoul (north) & Kyunggi (part)
Samchuli	LNG	1983.10	2.098	3.250	Incheon (east) & Kyunggi (part)
Incheon	LNG	1984.1	0.548	0.685	Incheon (west) & Kyunggi (part)
Seoul Metropolitan Area Total			**7.557**	**10.282**	
Pusan	LNG	1982.10	0.691	1.028	Pusan
Daegu	LNG	1984.10	0.594	0.869	Daegu, Kyungsan, Chilgok & Koryong
Haeyang	LNG	1983.7	0.353	0.446	Kwangju, Hwasoon (part)
Chungnam	LNG	1987.12	0.378	0.548	Daejeon & Geryong
Kangwon	LNG	1984.12	0.064	0.071	Chuncheon & Hongcheon (part)
Chamvit Wonju	LNG	1989.2	0.060	0.081	Wonju
Chamvit Yongdong	LPG + Air	1996.9	0.016	0.015	Kangreung
Chamvit	LPG + Air	1995.7	0.012	0.010	Sokcho, Koseong (part)
Cheongju	LNG	1989.7	0.186	0.280	Cheongju, Jaecheon, Jincheon, Eumseong (part), Cheongwon(part)
Chamvit Chungbuk	LNG	1995.3	0.020	0.020	Choongju
Chungboo	LNG	1993.11	0.164	0.449	Cheonan, Asan, Boryong, Kongju & Yonki, Nonsan, keumsan, Booyeo, Seocheon, Cheonyang

Table 5.6: *continued*

	Source of gas	*Date of supply*	*Number of Households (million)*	*Supply Volume (Bcm)*	*Supply Area*
Seohae (Hanbo)	LNG	1996.12	0.017	0.155	Seosan
Jeonbuk	LNG	1984.5	0.165	0.229	Jeonju, Kimjae, Wanju (part)
Koonsan	LNG	1991.8	0.045	0.086	Koonsan,
Iksan	LNG	1992.1	0.062	0.085	Iksan, Jeongeup
Mokpo	LNG	1984.9	0.048	0.043	Mokpo, Mooan, Youngam
Cheonnam	LNG	1985.3	0.081	0.157	Suncheon, Kwangyang, kokseong, Koorae
Daehua	LNG	1988.5	0.043	0.055	Yeosu
Kumi	LNG	1988.10	0.124	0.365	Kumi, Kimcheon & Chilgok (part)
Pohang	LNG	1990.12	0.092	0.281	Pohang, Youngdeok, Wooljin
Kyungbuk	LPG + LNG	1995.2	0.027	0.047	Andong (LPG+Air) Youngju (LNG)
Seorabol (Shilla)	LNG	1996.6	0.046	0.069	Kyungju & Youngcheon
Kyungnam	LNG	1984.3	0.346	0.580	Changwon, Masan, Jinhae, Kimhae, Keojae, Tongyoung
Kyungdong	LNG	1984.8	0.302	0.969	Ulsan & Yangsan
GSE (Shin-Ah)	LNG	1999	0.044	0.072	Jinju & Sacheon
Jeju-e	LPG + Air	05.01	0.001	0.0008	Jeju Islands
Regional Area Total			**3.981**	**7.013**	
National Total			**11.538**	**17.295**	

Note: i) The date of supply does not necessarily coincide with the date of LNG supply; ii) LPG + Air means the LPG (62.5%) is mixed with Air (37.5%). Since LPG's calorific value is higher than that of LNG, city-gas companies using LPG should reduce the value by adding air.

Source: Korea City Gas Association, Dosi Gas Saoup Pyonram 2006 (City Gas Business Yearbook). *http://www.citygas.or.kr/index.jsp*

Gas Infrastructure

To meet the rapidly increasing gas demand, the Korean government has prepared an ambitious scheme to expand the gas import related facilities. In November 2002, Kogas began to supply gas to Kangwon province and by 2003 a nationwide transmission pipeline network of 2441 km had been completed (Table 5.7).

Table 5.7: Trunk Pipeline Construction Plan

	New section	*Accumulated Total*
2003 actual	-	2441 km
2004	110 km	2551 km
2005–07	164 km	2715 km
2008–10	40 km	2755 km
2011–17	50 km	2805 km

Source: MOCIE, 7th LTNGSDP

According to Korea's City Gas Association, at the end of 2005 a total of 5 trillion won had been invested in the development of the 177,034 km city gas distribution network. Up to 2010, another 2 trillion won (400 billion won annually for five years) will be invested, adding another 6332 km of distribution networks.

In 2007 there were four LNG import terminals, of which three are owned by Kogas, at Incheon, Pyeongtaek and Tongyeong. Together there are 38 storage tanks at the three terminal sites with a total capacity of 4.9 million litres (Table 5.8). In 2005 Korea's first private sector LNG terminal was completed.[10] The Pohang Iron and Steel Corporation's (POSCO) Kwangyang terminal has a capacity to import up to 1.7 Mt/y of LNG underpinned by a long-term supply contract signed in 2004. According to the 8th LTNGSDP, the Korean government aims to complete a small-scale LNG terminal in the Jeju Islands

Table 5.8: Korea's Gas Industry Infrastructure: As of Dec 2006

	Commissioning Date	*Storage Capacity million litres / nunber*	*Re-gasification Capacity Capacity Tonne / hour*
Incheon	Oct 1996	3.480 / 18	3,690
Pyeongtaek	Nov 1986	1.000 / 10	3,016
Tongyeong	Sep 2002	1.400 / 10	1,350
Total		4.880 / 38	8,056

Source: Kogas (2007)

Table 5.9: Gas Industry Investment Plan. Trillion won

	2006–10	*2011–15*	*2016–20*
Production Bases	3.2196	3.0163	1.2100
Storage Tanks	2.6013	2.4773	1.0603
Port facilities	0.1933	0.2111	-
Re-gasification facilities	0.4250	0.3280	0.1497
High pressure pipelines	0.5998	0.2347	-
Total	3.8194	3.2510	1.2100

Source: MOCIE, 8th LTNGSDP

by 2012 and a 20 inch pipeline between Jeju city and Seoguipo. The anticipated costs of these developments are shown in Table 5.9.

Kogas intends to construct a fourth LNG receiving terminal by 2013, to increase the storage rate from 13.1 percent to 14.0 percent. At the end of 2005, Kogas' base storage rate was only 8.9 percent, much lower than that of Japan (16–20 percent) despite the fact that the Korean turndown rate (see below) was much higher.[11] MOCIE has said that Kogas must maintain LNG storage capacity equivalent to 10 percent of annual demand, and the ratio could increase to 15 percent by 2012.[12] MOCIE's 8th Long-term Natural Gas Supply and Demand Plan (LTNGSDP) envisages a ratio of 16 percent by 2020.

Table 5.10: Seasonal Demand Fluctuations. Mt

	Highest Month	*Lowest Month*	*Turn Down Ratio*
2003 actual	2.580	0.981	2.63
2005	2.899	1.107	2.62
2010	3.536	1.261	2.80
2015	4.215	1.390	3.03
2017	4.597	1.493	3.08

Source: MOCIE, 7th LTNGSDP

As can be seen in Table 5.10 seasonal demand fluctuations are a key factor requiring storage capacity expansion. A 2007 IEA study explained the seasonality issue as follows:

> Seasonality is often expressed by a turn-down ratio (TDR), which is calculated by taking the demand in the maximum demand month and dividing by the demand in the minimum month. The city gas sector has a TDR of 4, meaning that four times as much gas is consumed in the peak winter month than in the lowest summer month. Within the city gas sector,

> residential gas use has a TDR of almost 10. The corresponding profile for gas-fired power generation (which covers 42 percent of total gas demand) is relatively flatter throughout the year, which means that it cannot offset the very high city gas winter peak demand. Winter peak power consumption is generally higher than in the summer, and peak power is most often provided by gas-fired generation, so gas used for power generation has a TDR of 2. In 2004, Korea had a TDR for its total gas usage of 2.63, while Japan has a TDR of about 1.4.
>
> Difficulties in managing a high TDR are exacerbated by Korea's source of gas, LNG. The economics of LNG production dictate that it should be produced at a fixed rate throughout the year; the LNG market is not designed to respond to customers that need larger amounts of gas at certain times of the year. In addition, all LNG importing countries are in the northern hemisphere and therefore experience peak heating season at the same time, compounding the problem.
>
> Korea is at a disadvantage when compared to many other countries because it does not have any geological capacity to store gas for long periods of time. Whereas most IEA countries have access to depleted oil or gas fields, or even old converted coal mines, Korea only has access to LNG storage which is very expensive when used as long term storage. Storing gas as LNG requires keeping the gas at -162 C, which has a very large energy requirement and is thus very costly.
>
> Kogas has chosen to contract for long term capacity throughout the year and buy spot cargoes from the international market in the winter, when global LNG spot market prices are at their highest. While this might be the least expensive strategy for securing sufficient winter supplies, it has risks because it relies on there being sufficient LNG volumes available on the global market.[13]

It is therefore not surprising that the Korean government is paying special attention to the storage capacity build-up domestically and internationally. In fact, Kogas signed an agreement with Oman's state gas company to build and operate two 0.2 million cubic metre storage tanks in the Sultanate.[14]

Gas Supply Issues

The security of gas supplies is the top priority of the Korean government's policy. The shutdown of the Arun LNG plant for several months as a result of civil disturbances in Indonesia[15] was a cause of serious concern for Kogas and the Korean government, even though the interrupted LNG supply volumes were replaced without too much trouble by purchasing spot cargos. A disruption of gas supply was the last thing the government had expected, and these events justified the policy of diversifying its LNG import sources. As shown in Table 5.11,

Table 5.11: Kogas Long-term LNG Import Contracts

	Project Name	*Quantity (Mt)*	*Supply Period*	*Import Terms*
Indonesia	Arun III	2.0	1986–2007	Ex-ship
	Arun III	0.3	1992–2007	Ex-ship
	Korea II	2.0	1994–2014	FOB
	Badak V	1.0	1998–2017	FOB
Malaysia	MLNG II	2.0	1995–2015	FOB
	MLNG III	1.5 / (0.5)	2003–2010	Ex-ship
	MLNG III	0.71	2004–2008	Ex-ship
	MLNG II	0.4	2004–2008	Ex-ship
	MLNG III	1.5 / (0.5)	2008–2028	DES
Qatar	Ras Laffan	4.92	1999–2024	FOB
	Ras Laffan	0.96	2004–2008	Ex-ship
	Rasgas	2.1	2007–2026	Ex-ship
	Rasgas	2.1	2009–2028	-
Oman	OLNG	4.06	2000–2024	FOB
Brunei	BLNG	0.7	1997–2013	Ex-ship
Australia	NWS	0.5	2003–2010	Ex-ship
Yemen	YLNG	2.0	2008–2028	FOB
Russia	Sakhalin Energy	1.5	2008–2028	FOB
Domestic	Donghae gas	0.4	2004–2019	PNG

Note: (figure) means optional supply
Source: Byung-Ho Lee, 'Energy Security of Korea and Gas Procurement Strategy', presented at KEEI 2006 International Symposium on Northeast Asia Energy Cooperation titled 'Toward Regional Energy Cooperation in northeast Asia: Key Issues in the Development of Oil and Gas in Russia', September 15, 2006, Seoul; *Kogas Annual Report 2006*, p. 28 & *International Gas Report, 2006* issues

Korea's dependence on Middle East LNG rose significantly due to two 2.1 Mt/year contracts with Qatar.

Kogas has signed eleven major long-term gas supply contracts with Indonesia, Malaysia, Qatar, Oman, Yemen, Russia and Brunei. Qatar became a major LNG supplier to Korea in 1999 and Oman started its LNG deliveries from 2000. In 1992 Kogas signed a letter of intent with Ras Laffan LNG Co. (Rasgas) to import 2.4 t/y of LNG over 25 years from Qatar's North field commencing in 1998, with an option for an additional 1.6 Mt/y. Following that, in 1995 Kogas signed a letter of intent with Oman LNG for the supply of 3 Mt/y of LNG for 25 years commencing in 2000. After 1997, Kogas did not sign any long-term contracts because the impending privatisation (see below) would have been a major complication for new supply contracts.[16] In addition, Kogas was trying to reserve a significant part of the market for pipeline gas imports from Russia.

However, growing demand caused a lot of problems[17] and consequently Kogas was forced to adopt a relatively short-term contract strategy. This involved signing short-term contracts to cover 3–5 years from 2004, and then pursuing long-term contracts for the post-2007 period (Table 5.12). For this reason Kogas signed two seven-year supply contracts with Australia and Malaysia during the first half of 2003. It was intended that decisions on new long-term LNG contracts would be made after the 7th LTNGSDP was finalised, but steps needed to be taken before then. In August 2004, Kogas invited bids for 5 Mt/y of long-term LNG supply, signalling a policy change by MOCIE and Kogas. Seven bids – Sakhalin 2, Tangguh, Gorgon and NWS, Yemen LNG, Iran LNG, and MLNG – were submitted to Kogas (Rasgas and Oman LNG did not submit bids). Five bids were short listed and Kogas finalised contracts for 5.0 Mt/y of LNG supply from 2008 composed of 2 Mt/y supply from YLNG, and 1.5 Mt/y supply from MLNG and Sakhalin 2 respectively.[18] The price was US$203/tonne, [US$3.81/MMBtu] 38 percent less than previous contracts, based on a US$40/ barrel oil price. The contract was protected from oil price increases above US$40/barrel, and is not affected if the price rises significantly above that figure.[19]

Table 5.12: Korea's LNG Import Contract Structure. Percent

	Number of Producing Countries	*Number of Projects*	*Contracted Volume (Mt/year)*	*Ratio (%)*
Long-Term 15–25 years	8	8	22.88	83.4
Mid-Term 6–14 years	2	2	2.50	9.1
Short-Term <6 years	2	2	2.07	7.5
Total	9	12	27.45	100.0

Source: Kim and Cho.

In late 2006, Kogas CEO Soo-Ho Lee and the company's labour union warned that there could be a very serious gas supply shortage from 2008.[20] The main reason for this was MOCIE's underestimation of future demand, but also the Ministry's policy of introducing multiple gas importers. Kim and Cho, members of the Korean Parliament's Industry and Resources Committee, have argued that, as a result, Korean companies may have paid up to 17 trillion won (US$18.2 billion) more than necessary for LNG since 1999.[21] This sensational criticism was a reminder that wrong judgments by MOCIE could impose an astronomical financial burden on gas consumers.

Soon after this warning, Kogas secured two long-term supply deals from Qatar, totalling 4.2 Mt/y at a very high price (see below). Deliveries under the first contract started in 2007, and the second is expected to begin in 2009. In May 2007 GS Caltex signed Heads of Agreement with Chevron Australia Pty Ltd (CAPL) and Chevron International Gas Inc (CIGI) to import 0.5 Mt/y of LNG for twenty years after 2010. This supply contract is GS Caltex's first direct LNG import since 2004 when it secured approval for LNG imports. GS Caltex has already done the negotiation for the use of Kogas's LNG facilities and main trunk pipeline.[22]

Due to its very high dependence on imported gas, the Korean government placed security of supply at the top of its policy agenda. Most LNG supply is based on long-term take-or-pay contracts. All contracts are linked to international crude oil prices and approximately one-third have an S-curve in the price formula.[23] As noted earlier, in recent years Kogas changed its policy of signing long-term take-or-pay deals to a more flexible policy that depends heavily on shorter-term contracts and spot purchases. Korea is the most active player in the spot LNG market mainly because of its very seasonal demand for gas. Consequently, in the post-2003 price environment Korea's price burden has been much higher than for any other buyer in Asia.[24]

Reductions in Indonesian LNG exports (see Chapter 7) have alarmed Koreans about the future security of supply. Indonesia accounts for 25 percent of supply to Korea and a similar proportion of supply to Japan, forcing the two largest LNG importers in the world to search for back-up supply sources This situation creates pressure on global LNG supplies with Korea's demand soaring and its old contracts soon requiring renewal.

Even though the prospect of gas supplies from abroad was very gloomy, there was good news domestically. The Korea National Oil Corporation (KNOC) made a small gas discovery in 1998 from the Gorae V structure, Block 6-1, located 60 km offshore from Ulsan, Kyungsang South province. Reserves were estimated at 7.0–8.4 Bcm, of which 4.76 Bcm are recoverable, which would provide the equivalent of around 3.4–4.0 Mt of LNG. In February 2000 the government announced that commercial production from 'Donghae 1' would begin in July 2002, supplying gas to Ulsan and Kyungsang provinces for fifteen years. However, the production was delayed and did not start until 2004. KNOC is also considering using this gas field for storage once reserves have been exhausted. Even though the scale of gas production is quite small, the development of the Donghae 1 project will have a positive impact on Korea's gas expansion plans.

It is worth mentioning MOCIE's announcement confirming gas hydrate-bearing rocks from Donghae (or Sea of Japan). The gas hydrate pool was found 100 km south of Ulleung Island and 135 km northeast of Pohang, and 6.5 metres below the sea bed. The Korean government estimates up to 600 Mt of hydrate reserves in the area, and plans to begin commercial production from 2015.[25]

Supply Shortages and Prices

On 7 September 2006, one of the main Korean newspapers reported that both MOCIE and Kogas had confirmed that only 13.5 Mt of LNG had been secured for the period October 2006–March 2007. Considering that the projected demand increase during the same period is 16.0 Mt, this left a supply shortfall of at least 2.5 Mt/y.[26] Increased LNG demand in power generation was regarded as the main cause of the shortage. Gas demand for power generation increased very rapidly due to delays in construction of nuclear- and coal-fired power plants since 2004. On top of this, the high oil price has accelerated the conversion of heavy oil plants to LNG for power generation. LNG demand in the power sector for 2006 was projected to be 8.0 Mt but that figure was subsequently raised to 10 Mt.

To cover this supply shortfall another gas import contract for either LNG or pipeline gas needed to be secured by Kogas and the government well before 2005. To avoid increasing its dependency on Middle

Table 5.13: Projected Gas Supply Shortage. Mt/y

	Domestic Incremental Demand	*Possible Volume under negotiation*	*Possible shortfall*
2007	3.72	2.60	- 1.12
2008	6.26	2.60	- 3.66
2009	5.43	2.72	- 2.71
2010	9.03	5.81	- 3.22
2011	13.12	6.50	- 6.22
2012	14.03	6.50	- 7.53
2013	14.98	13.00	- 1.98
2014	17.17	16.50	- 0.67
2015	19.26	16.50	- 2.76
2016	19.48	16.50	- 2.98
2017	21.85	16.50	- 5.35

Note: The domestic incremental demand volume is compared with the 7th and 8th LTNGSDP.

Source: Kim and Cho.

East gas, and in particular gas supplies needing to pass through the Straits of Hormuz, the Korean government was very anxious to explore all possible ways of diversifying its gas import sources. But due to the debates on the privatisation of Kogas and the introduction of pipeline gas, no new LNG was secured prior to 2005, by which time the LNG market situation had changed fundamentally. Strong global demand favoured sellers and, with rising crude oil prices, LNG prices had increased significantly (Table 5.14).

Table 5.14: Gas Price Comparison. $/MMBtu

	Korea *CIF*	*Japan* *CIF*	*EU* *CIF*	*UK* *Heren NBP index*	*USA* *Henry Hub*
1985		5.23	3.83		
1986	2.77	4.10	3.65		
1989	3.30	3.28	2.09		1.70
1991	3.56	3.99	3.18		1.49
1993	3.35	3.52	2.53		2.12
1995	3.37	3.46	2.37		1.69
1996	3.80	3.66	2.43	1.85	2.76
1998	2.83	3.05	2.26	1.92	2.08
2000	5.03	4.72	3.25	2.68	4.23
2001	4.84	4.64	4.15	3.22	4.07
2002	4.24	4.27	3.46	2.58	3.33
2003	4.98	4.77	4.40	3.26	5.63
2004	5.64	5.18	4.56	4.69	5.85
2005	6.86	6.05	6.28	6.69	8.80
2006	-	7.14	8.77	7.87	6.76

Source: *BP Statistical Review of World Energy* (June 2007) & KEEI

The cost of these delays was significant. On 16 November, MOCIE confirmed that Kogas had agreed to buy an additional 2.1 Mt/y of LNG from Qatar on a long-term basis. Under the deal with Qatar's Rasgas, Kogas will be taking the extra volumes for twenty years, starting in January 2007, at a very high price of more than US$10/MMBtu (assuming a Japan Crude Cocktail price of US$60/barrel). The price was significantly higher than the US$7.90/MMBtu agreed by Australian North West Shelf buyers to renew their contracts (assuming the same crude oil price).[27]

Given that the gas prices in the three LNG contracts with Sakhalin 2, MLNG and YLNG were around US$3.8–4.2/MMBtu (based on US$40/barrel[28]) and (as Table 5.15 shows) that the POSCO and K-Power contract prices were in the range of US$3.3–3.5/MMBtu (based on US$60/barrel), the price of over US$10/MMBtu paid for

Table 5.15: Price Advantage during the Buyer's Market period. US$/MMBtu

	NWS–Guangdong	*Tanggu–Fujian*	*Tanggu–POSCO*	*Tanggu–K Power*
US$ 10 / bbl	2.78	2.79	2.72	2.90
US$ 20 / bbl	3.04	3.04	3.02	3.20
US$ 30 / bbl	3.30	3.29	3.32	3.50
US$ 40 / bbl	3.30	3.29	3.32	3.50
US$ 50 / bbl	3.30	3.29	3.32	3.50
US$ 60 / bbl	3.30	3.29	3.32	3.50

Source: Kim and Cho, Repercussions of the Government's Inadequate Gas Policy.

Qatari LNG was excessive. This price burden provided huge incentives for Kogas and other Korean private companies to aggressively pursue equity gas purchases.

Kogas' existing long-term contract with Rasgas does not have the 'S' curve mechanism (which limits the impact from extremely high or low oil prices) in the LNG price formula. In other words, the company is very exposed to the impact from high oil prices. It also bought five spot LNG cargoes for delivery in January–March 2007 to cope with winter demand. Kogas paid record high prices of up to US$26/MMBtu for spot LNG cargoes during the winter of 2005–06 as its inventory sank below operational safety levels amid a global shortage of the fuel.[29]

Overseas Gas Development and Partnership

Until recently Kogas had never been active in upstream business as it lacked the manpower and expertise. However, in 2006 it set up a joint venture (JV) with Uzbekneftegaz to develop the Surgil gas field located close to the Aral Sea.[30] It also started a JV called KG to explore the joint development zone 06-102 between East Timor and Australia.[31] In May 2007, Kogas announced that it would farm in ENI's seven blocks in East Timor (A/S/C/E/H blocks), Mozambique's Area 4, and Indonesia's Krueng Mane;[32] this is likely to be only a foretaste of much more extensive upstream operations.

To a certain extent, Korea's private firms such as SK have been much more aggressive in seeking overseas gas supply sources than Kogas itself. In fact SK Corp is pursuing two major LNG projects in Yemen and Peru as part of consortia.[33] In addition, in early 2004, Daewoo International (a Korean trading house) made a major discovery offshore Myanmar. Block A-1 is estimated to have 128.8–184.8 Bcm of gas and is expected to begin production around 2009/2010. In 2006, Daewoo International made another discovery with estimated reserves of 50.4–95.2 Bcm in Block A-3 and is scheduled to start second-phase drilling on the block soon. Daewoo International has a 60 percent operating interest in blocks A-1 and A-3 located in the Rakhine basin offshore Myanmar. Its partners in the project are Kogas (10 percent), India's ONGC Videsh Ltd (20 percent), and Indian gas utility GAIL (10 percent). Myanmar which has the right to 60 percent of the blocks' output has been in negotiations with several parties over gas sales from the two blocks. Kogas and Japanese trading house Marubeni were named as preferred bidders to buy 3.7 Mt/y of LNG from Myanmar's offshore blocks A-1 and A-3 being developed by Daewoo International. The two firms offered a price of US$7.0/MMBtu, overwhelming other competitors from China, India, and Thailand, and the price played the key role in the selection of Kogas and Marubeni's proposal.[34] However, by late 2007, the Myanmar government appeared to have decided to export the gas to China by pipeline (see Chapter 7).

Daesung group, whose main business is city gas, is driving a very large offshore pipeline project, 'The Asian Gas Grid'. The core of this project is to link Indonesia's East Natuna gas resources (with 6.2 Tcm of gas, of which 1.29 Tcm are estimated to be recoverable) to markets in China. The project would require the laying of the world's longest sub-sea pipeline of almost 5000 km.[35] Thus far there has been no substantial progress in this super-giant project.[36]

What is really needed to accelerate the upstream business development is cooperation between Korea National Oil Corp and Kogas. If the two state firms join forces to bid for exploration blocks worldwide, the chances of winning will be much higher. But the short-sighted policies of labour unions of both companies are blocking this new development opportunity due to their concerns that such a merger would lead to massive redundancies.

Besides this upstream sector activity, there were a number of attempts to pursue overseas partnerships. In early 2005, Kogas signed a MOU with China Gas Holdings to set up a JV for gas business development in China, and Kogas would be able to purchase up to 10 percent of China Gas equity.[37] As of 2006, China Gas possesses natural gas franchises in more than fifty cities and districts in provinces and regions along the lines of West-to-East Natural Gas Transmission and Chuan-to-East Natural Gas Transmission. China Gas supplied natural gas to 280,000 households (as of March 2004) and more than 2000 industrial and commercial users (as of 2006).[38] The JV between Kogas and China

Gas did not happen and was replaced by a JV between SK Corp and China Gas signed in March 2007.[39]

In December 2005, the Daesung group signed a MOU with Rizhao city in China's Shandong province for a 0.5 Mt/y LNG receiving terminal but suspended the project in August 2006, firstly because the required investment rose from US$30 million to US$60 million, and also because the Chinese partner was not enthusiastic about the project.[40] Daesung underestimated the complexity of obtaining the necessary permits, and from the outset the likelihood of approval from the China National Development and Reform Commission was very low.

Kogas has also entered into a number of new partnerships. CNOOC announced on 29 November 2006 that it had signed a memorandum of understanding with Kogas for cooperation in the areas of LNG and gas production. The Shanghai Daily reported that the two companies had agreed to increase their ties in spot LNG trades, cargo swaps and overseas joint ventures.[41] Kogas also signed an accord with Mitsubishi Corporation (the Japanese trading house) to cooperate in LNG trading, gas swap deals and leasing of LNG carriers.[42] Kogas attempted to swap its term LNG cargoes scheduled for summer delivery for winter arrival cargoes to ensure adequate winter supply, with limited success.

In March 2007, Kogas and LG Corp formed a consortium with Houston based Gulf Coast LNG Partners (GCLP) and EMS Group to be operator of the Texas based Calhoun LNG terminal with two 160,000 cm storages.[43] In May 2007, Kogas and the Sultanate of Oman agreed to set up a joint venture for LNG trading, and construction and operation of storage. This agreement is a follow-up to the Korea–Oman government's Energy Resources Cooperation Agreement of November 2005, and Memorandum of Understanding on joint LNG business between Kogas and the Oman Government in June 2006.[44] The rapid expansion of Kogas' overseas business development is very likely to continue and strengthen in the coming years.

Gas Industry Privatisation, Restructuring and Liberalisation

Privatisation and Restructuring

In 1995, the government announced plans to privatise Kogas, and requested the Korea Energy Economics Institute (KEEI) to undertake a study of the different options including the structure of the industry, the form of competition, the extent of privatisation and the nature of regulation.

Several factors led the government to consider privatising its gas industry. In addition to there being a shift in energy policy, the government held the view that placing Kogas on a more commercial footing would increase the overall efficiency of the company in areas such as cost, negotiations with suppliers and customers, and diversification of LNG sources including contract terms. Kogas also has an ambitious expansion programme requiring substantial funding, and private sources of finance would reduce pressures on the government to provide these funds.[45]

The KEEI study recommended delaying the implementation of the privatisation until the main domestic pipeline network had been completed. However in 1998, the government announced a privatisation plan for eight major public enterprises, including Kepco and Kogas. The measure was partly a response to the requirements of the World Bank in providing loans to support the Korean economy at the time of the 1997 crisis. Late in 1999, the government finalised its plans for the restructuring of the gas industry.

The main objectives of the plan were (i) to achieve the better use of resources through the introduction of competition together with an increase in management efficiency, (ii) the promotion of a technologically advanced gas industry and (iii) to improve the quality of service by offering customers a choice of supplier.[46]

In July 1998, the Plan for Gradual Privatisation of Kogas was announced, and in November 1999 the Basic Plan for Restructuring the Gas Industry was introduced to enhance competition in the gas sector in Korea.[47] The plan was submitted to the National Assembly at the end of 2001 and envisaged the division of Kogas into three subsidiaries, two of which would be privatised.[48] To reduce its market share as a way of encouraging gas market competition, Kogas was not allowed to sign new long-term contracts so that other companies could enter the market. The plan stipulated that LNG receiving terminals and pipelines should be subject to open access.

The partial privatisation of Kogas before dividing it into different private entities led to an initial public offering of 43 percent of its equity in November 1999. By the end of 2004, the government had reduced its stake in Kogas to 26 percent. The other major owners were Kepco (24.5 percent), local government (9.9 percent), and private investors (38.8 percent).[49]

The original liberalisation plan aimed at the establishment of a regulatory commission specifically for the gas industry. Up to now there has been no specific gas industry regulator or regulation other than supervision by MOCIE with the Fair Trade Commission (FTC) responsible for competition regulation.[50] The ambitious gas industry restructuring plan established in November 1999 was buried in the wake of the closing of the 16th Parliament mandate period [2000–2004]. The Roh Moo-Hyun government gave up the idea of privatising two supply companies at the end of 2004 when international LNG market conditions changed. However, from 2005, as increasingly favoured sellers, MOCIE decided to introduce the option of direct LNG imports.

Kogas' supply and import monopoly was officially ended in 2005 but the government opened the market to competition much earlier and allowed POSCO (Pohang Iron and Steel Corp) to build the first independent LNG import terminal at Kwangyang. It was hoped that termination of Kogas' monopoly would introduce more competition based on the fact that it is the world's largest single LNG buyer and South Korea as a whole is the world's second largest LNG market. However, in negotiations for long-term LNG supply contracts, Kogas experienced the increase in the market power of buyers and consequently took a very different view from the government with regard to the introduction of competition. Despite losing its supply monopoly, Kogas appears determined both to expand LNG supplies and to accelerate Russian pipeline imports.

Kogas' new investment plans are more focused on consolidating its dominant position. At the end of May 2006, it announced an investment of 855.4 billion won (US$911.86 million) to expand its LNG storage and re-gasification capacity (see Table 5.9). Kogas storage capacity will increase by 27 percent with six new 200,000 kilolitre storage tanks, and three new re-gasification trains are also to be built. The company's three existing LNG terminals at Incheon, Pyeongtaek, and Tongyeong have a combined processing capacity of 56 Bcm (although Incheon and Pyeongtaek are already undergoing expansion programmes). In September 2006 another 313.3 billion won (US$334 million) was announced to provide one additional re-gasification train and three 200,000 kilolitre storage tanks.[51]

Liberalisation

A 2006 KEEI study,[52] suggested the introduction of supply competition in the gas industry through the elimination of entry barriers and by inviting new entrants into the market. Under this scheme, the incumbent monopoly Korea Gas Corp (Kogas) is set to maintain its current functions of facility operation and gas sales. New entrants would have to compete with Kogas, relying on access to the latter's essential facilities. The study pointed out that the most important task will be to ensure non-discriminatory access to these facilities. The study also identified the treatment of legacy import contracts as the most difficult and contentious problem to be addressed in restructuring the gas industry. The KEEI study confirmed that liberalisation will be a difficult task.

The IEA's 2007 study on Korea's gas industry encourages Korea to set out a clear plan – with milestones and dates – for reform of gas and electricity markets including the necessary ingredients for successful liberalisation. The IEA's policy advice on the natural gas sector is as follows:[53]

i) To build on earlier efforts to open the gas market by establishing a timetable for comprehensive gas market reform, beginning the process as soon as possible.
ii) To establish an independent regulator to oversee the gas market as the first step in the liberalisation process.
iii) To institute open access/regulated third-party access on pipelines and ensure that the regulator monitors open access implementation and has the power to penalise network operators who fail to comply.
iv) To ensure that the relevant regulator monitors the activities of the incumbent gas supplier so that these activities do not result in the abuse of market power once the market is liberalised.
v) To effectively unbundle and regulate transmission and distribution from other parts of the value chain to enable competition while ensuring fair access.
vi) To let price signals encourage customers to use gas in ways that reduce overall seasonal variations in natural gas consumption.
vii) In the absence of sufficient price signals, to implement cost-effective policy measures to smooth gas demand across winter and summer.
viii) To consider revising the existing gas emergency planning procedures to take account of structural changes in the power market that may render it less effective.

A Merger of KNOC and Kogas?

It is safe to say that the Kogas privatisation scheme has been abandoned. By contrast, some members of the parliament proposed a merger between Korea National Oil Corp (KNOC) and Kogas. During the Parliament's 2005 Annual Inspection Session, Kyo-Heung Kim MP argued that, rather than gas industry restructuring, the policy priorities should be medium- and long-term LNG supply and the creation of a major Korean energy company.[54]

The idea of a Korean oil and gas 'super-major' was floated during 2004 when it was reported that the ruling party and MOCIE were pursuing a merger of KNOC, Kogas and Korea's Mining Corp. There was no opposition to the concept of merging a part or whole of KNOC and Kogas, but MOCIE was focusing on resource development, while the ruling party was seeking comprehensive restructuring (Table 5.16).[55]

Table 5.16: Options of KNOC and Kogas Merge

Ruling Party		*MOCIE*
Option 1	KNOC and Kogas overseas development section merger	Development section : Merger between KNOC and Kogas
Option 2	Holding Company: 3-4 independent accounting firms - Overseas Development firm - Domestic Sales firm - Storage Business firm	Kogas focuses on energy trading
Option 3	Merger of KNOC and Kogas	

Source: Korea's *Financial News* (October 13, 2004)

The suggestion was repeated during the 2006 Parliament Annual Inspection Session. MP Cheol-Kook Choi argued that the merger of KNOC and Kogas would create the world's 34th largest energy firm with assets of 18.5 trillion won (US$19.8 billion), of which 7.5 trillion won (US$8 billion) from KNOC and 11.0 trillion won (US$11.8 billion) from Kogas.[56]

Given that Kogas is the biggest LNG buyer in the world but has virtually no experience in upstream business, the establishment of a Korea National Oil and Gas Corporation (KNOGC), could create synergies by maximising Kogas' buying power and KNOC's upstream expertise. But there is very strong opposition from the labour unions of KNOC and Kogas who are concerned about substantial job losses. A merger would see KNOGC become the holding company for KNOC and Kogas and expand particularly into overseas oil and gas asset development. If the possibility of KNOGC is raised during the 18th Parliament Session (2008–2012), there is a chance of making this concept a reality.

Gas for Power and Direct LNG Imports

Although electricity demand has increased by around 9 percent per year over the past two decades, the Korean government predicts this rate will fall to 4 percent per year over the next decade. Most of this new consumption will be met by gas-fired power plants, while Korea Electric Power Co (Kepco) is expected to convert all of its remaining oil-fired facilities to run on gas feedstock in the near future.[57]

In November 2004, MOCIE approved direct LNG imports by Kepco's four generation companies. Given that Kogas had only just shortlisted bids for short-term LNG supplies (see above), the timing of this decision was the worst and it delivered a very negative impact on Kogas negotiations.[58] However, since the government had already given the approval for LNG imports by POSCO and K-Power, it was very difficult for MOCIE to refuse the Kepco GENCOs.[59]

There is a consensus that, rather than promoting competition, the government's decision to allow direct LNG imports by the Kepco GENCOs damaged Korean gas prospects. The GENCOs contacted the Sakhalin 2 sellers and consequently the terms offered to Kogas by Sakhalin 2 were cancelled. In November 2004, MOCIE forced Kogas and Kepco's four power generation companies not to compete with each other for the LNG imports.[60] This move by MOCIE indirectly confirmed the ferocity of the competition between the companies and that the government had underestimated the GENCOs' determination to pursue LNG imports.

In the meantime POSCO completed its LNG receiving terminal at Kwangyang in the southern part of Korea in 2005. Both POSCO and K-Power (a JV between SK and BP) contracted to import 1.6 Bcm/y of LNG. Based on this supply contract, POSCO has negotiated third-party access rights to the Kogas high pressure network from Kwangyang to Pohang.

In summer 2006, MOCIE decided to give Kogas sole responsibility for LNG imports until 2012. This suspended for six years the government decision to introduce competition. MOCIE was forced into this policy reversal by the change to a sellers' market in LNG supply and the rapid increase in LNG prices. With government approval, Kogas will extend its medium-term supply contracts with Australia and Malaysia, and initiate new long-term supply contracts with Qatar.[61]

The twenty-year contract between Kogas and Kepco expired on 30 November 2006, and in December, Kogas and Kepco's four generating companies signed another twenty-year contract, with a condition that if the government approves the direct LNG imports for the GENCOs, these volumes will be excluded from the long-term contract.[62]

The ambitious plans to liberalise the Korean gas market have made little progress and the government has no plans to allow direct LNG imports for city gas use. So far, only POSCO, K-Power and GS Caltex have been allowed to import LNG independently. As discussed earlier, GC Caltex signed a gas supply contract in spring 2007 aiming at a 2012 start-up date for a planned terminal in Boryeong, near to one of the sites that Kogas is also considering. GS Caltex's project is at the permitting stage and has obtained about fifty permits (of more than seventy required), including the right to import LNG directly for its own consumption. The firm could start by purchasing about 2 Mt/year, half of which would be supplied to its 650,000 b/d Yeosu refinery for fuel conversion from oil to gas. The other half would be used as fuel at its three existing power plants with a total capacity of 1.45 GW. An affiliate, GS Power, has two units with a total capacity of 950 MW in the Anyang and Bucheon areas.[63]

As shown in Table 5.17, Kepco GENCOs, Daelim Industry, and Korea District Heating Corp have not given up pursuing their own

Table 5.17: LNG Import Plans of Power Generators

Project Developer	Volume Mt/y	Destination	Import timing
POSCO	0.55	Kwangyang & Pohang plant	2005.07
K-Power	0.60	Yeosu Power plant	2006.01
GS Caltex	1.90	Yeochun Complex (1.0), GS Power (0.5), LG Energy (0.4)	2008 Approved
Daelim Industry	0.70	Inchon Songdo power gen	2010 (plan)
KEPCO GENCOs	5.70	South (3.0), West (1.2), Central and East (1.5 each)	
KDHC*	0.80	Affiliate regional heating fuel	
POSCO	0.50	POSCO Power	
GS Caltex	1.00	Boogok Combine additional expansion	
Total	11.75	Intended direct LNG import volume in 2010 : 8.7 mt	

* Korea District Heating Corp

Source: Kim and Cho

LNG imports, and if these private companies receive approvals, then gas demand from Independent Power Producers (IPPs) will increase (Table 5.18). However, in 2007, there was an attempt to scrap the law that has allowed direct gas imports for self-use (i.e. like POSCO which imports LNG to generate its own power). Ten members of Korea's Parliament are aiming to block any further imports of gas for self-use, and this initiative will revive debate on the merits of competition in the gas industry.[64]

Table 5.18: Projection of LNG Demand by Sector. Mt

Year	*Town Gas*	*Power Generation*	*IPPs + Community Energy*	*Total*
2003 Actual	11.98	6.47	0.472+0.0	18.45
2004	12.85	7.50	0.592+0.0	20.34
2005	13.66	7.61	0.622+0.0	21.27
2007	15.39	8.57	0.624+0.46	23.96
2010	17.98	7.35	0.631+0.599	25.33
2015	22.72	6.55	0.631+0.751	29.28
2017	24.89	6.76	0.631+0.753	31.66

Source: MOCIE, 7th LTNGSDP, Dec 2004, p, 23.

Pipeline Gas Imports

Both MOCIE and Kogas have been pursuing pipeline gas imports from Russia since 1996. However, they have made no progress due to the Russian government's UGSS (Unified Gas Supply System) plan, unsuccessful price negotiations between Gazprom and CNPC, and Gazprom's objection to exports of Kovykta gas. This section will briefly review Gazprom's Asian policy to explain how strongly its Asian export policy would affect its exports to (China and) Korea, and will focus on Russian gas supply to the Korean peninsula.

Gazprom's Asian Policy and its Implications towards the Korean Peninsula
Gazprom's Asia policy dates back to the early 1997 speech by Gazprom CEO Rem Vyakirev.[65] Until the early 2000s, no major steps were taken towards significant development of Russia's gas exports to northeast Asia, but since 2002, when Gazprom became the coordinator of Russia's eastern natural gas development, the company has pursued cooperation with all of the countries concerned.

In early August 2004, the joint venture negotiations for the West–East pipeline between PetroChina and a consortium composed of Shell,

ExxonMobil and Gazprom collapsed (see Chapters 2 and 6). Gazprom subsequently informed PetroChina indirectly that the route of the Kovykta gas export pipeline (see Chapter 6) would be changed. Immediately following the collapse of the negotiations with the Chinese, Gazprom informed Kogas that the firm would not support a pipeline crossing the Yellow Sea.[66] This provided indirect confirmation that Gazprom had rethought its strategy for exports to northeast Asian countries and had decided to use the 'Korean card'. This is also a good illustration of how pipeline politics has affected Russia's gas export planning in Asia.

The first draft of the Eastern Gas Program, prepared jointly by Russia's Ministry of Industry and Energy and Gazprom, was published in September 2007 (see Chapter 6). Russian newspaper *Kommersant* reported 'The dates of the field's development and exports will depend on results of Gazprom's negotiations with buyers in China and Korea. China's refusal to accept Russia's terms would kill the export component of the programme.'[67]

By 2007, there was no breakthrough in Russian exports to Asia, partly due to Gazprom's decision not to prioritise Kovykta gas for export to China and Korea before 2015. Gazprom never initiated the trilateral price negotiations with CNPC and Kogas, and bilateral negotiations between Gazprom and CNPC left little room for compromise between the two parties. The price disagreement between Gazprom and CNPC is a serious obstacle to Russian gas exports to the Korean peninsula.

South Korean Initiatives for Russian Gas to the Korean Peninsula[68]

Korea's interest in a transnational pipeline dates back to the late 1980s when Hyundai Group founder Jung Ju-Young began to explore the possibility of Sakha gas development and a pipeline to the Korean Peninsula. In July 1992, a Korean consortium led by Korea Petroleum Development Corporation (PEDCO, now Korea National Oil Corp: KNOC) was established, and Daewoo Corp became the driving private sector in pursuing this project.

In November 1994, a preliminary feasibility study between Korea and Sakha Republic was signed. The 12-month study cost $20m, of which $10m was in the form of data provision from Russia, and a $10m cash payment from Korea. The route was as follows: from gas fields in the south-western part of Sakha Republic – Yakutsk – Tynda – Blagoveshensk – Khabarovsk – Vladivostok – DPRK – ROK, 5143 km in length, of which the Russian section was 4383km. At that time, there were only five gas fields with over 100 Bcm in the Sakha Republic;

Chayandinskoye's proven reserves were only 210 Bcm, although they increased to 755 Bcm in 1997 and further to 1240 Bcm in 2002.

In December 1995 the feasibility study was completed and the verdict was not positive due to poor commercial viability. No further steps were taken, and Korean interest switched from Sakha gas to Kovykta gas. As Irkutsk Oblast had asked Korean companies to develop its giant gas field in 1994, Kogas decided to take an initiative towards the Irkutsk region's Kovyktinskoye gas field development. In mid-1995, the Kogas consortium was established to pursue first the feasibility study followed by an investment decision.

Unlike the approach by Kogas consortium, the Hanbo Group took a decisive step by purchasing a 27.5 percent share in Rusia Petroleum which held an exploration and development licence in the Kovyktinskoye (and Verkhnechonskoye) field in the Irkutsk region. The total investment for this equity share was a mere US$44 million from the East Asia Gas Company (EAGC), a 100 percent subsidiary of the Hanbo Group. In November 1997, following the bankruptcy of the Hanbo Group in the wake of the Asian economic crisis, EAGC surrendered 20 percent of the equity to Sidanco which was eventually taken over by BP (see Chapter 6). However, the Korean government has never taken any step to protect the remaining 7.1 percent equity in Rusia Petroleum, and never appreciated the value of the equity.[69]

In 1998, a Japanese consortium led by JNOC (Japan National Petroleum Corp) and Sumitomo Corp proposed a five-country feasibility study on a pipeline connecting the Kovykta gas field with China, Korea and Japan via Mongolia, but the initiative was not supported by the Chinese government and BP. The collapse of these negotiations in 1998 meant that the region had little choice but to return to the formula of Sino-Russian co-operation. Without Korea's participation, it would have been a bilateral project. In May 1999, Korea expressed its interest in the Sino-Russian feasibility study on the Kovykta gas project, and in November 2000, Kogas, CNPC and Rusia Petroleum signed an agreement for a full feasibility study on Kovykta gas development; which was completed in November 2003.

However, the Moscow authority did not approve the conclusions of this study, arguing that the Kovykta pipeline project should be a part of the grandiose Unified Gas Supply System (UGSS) and that Kovykta could be planned or pursued independently. In August 2004, Gazprom's official response was that although Kovykta gas could not be supplied to China and Korea, the company would promote a pipeline from Sakha Republic gas fields (and possibly Sakhalin) to Nakhodka and then to South Korea via an offshore route.[70]

The possibility of Sakhalin 1 gas supply to South Korea via North Korea was explored at the beginning of the Roh Moo-Hyun government, but the option was eliminated due to poor management by the president's office of a highly political but multi-functional proposal. Sakhalin I project operator Exxon Neftegas Ltd, a subsidiary of ExxonMobil never wanted to take any political or commercial risks associated with pipeline gas to the Korean Peninsula[71] and the US government showed no official interest.

MOCIE and Kogas were both against the Sakhalin option due to political risks and a preference for the Kovykta project. The option of Sakhalin gas supply to the Korean Peninsula was sidelined, and the ExxonMobil group first proposed a pipeline to Japan (see Chapter 6) and then pursued the option to supply 8 Bcm/year of gas to Heilongjiang province in north eastern China. Negotiations with CNPC began in 2004, and in October 2006, a preliminary agreement was signed on future gas supplies from its Sakhalin 1 project on which the ExxonMobil representative said: 'The intermediate deal ... allows us to reach final sales and purchase agreement.'[72] Unfortunately the project looks unlikely to be implemented as Gazprom has no intention of giving up its co-ordination rights for Russian gas export to Northeast Asian countries, in particular China and Korea,[73] and the price to the Chinese gas consumers in the Heilongjiang province is still too high.

In 2006, Kogas floated the idea of an equity swap with Gazprom.[74] This could be a very attractive proposal for Gazprom to secure a sizable equity stake in Kogas, but such a swap would leave Gazprom with a much higher share in Kogas than Kogas' share in Gazprom and this is not acceptable to the Korean government. As an alternative, the government could try to secure a minimum strategic level of Gazprom equity to influence the timing of pipeline gas, or Kogas' participation in Sakhalin 3 project.[75] But these ideas go far beyond the current aspirations of Kogas even though, since its May 2003 cooperation agreement with Gazprom, both firms have held regular meetings at vice president level.[76]

On 17 October 2006, Kogas CEO Soo-Ho Lee and Gazprom CEO Alexei Miller met in Seoul following the inter-governmental agreement during the summit between Presidents Roh and Putin in September 2004 in Moscow.[77] The two CEOs agreed to initiate the commercial negotiations for gas supply from 2012–13. They discussed the possibility of short-term LNG swaps and short-term trade, and also talked about building LNG capacity in eastern Russia, gas-to-liquids technology and joint gas exploration, production and marketing in third countries. However, Gazprom refused to specify Kovykta as the source of pipeline supply to Korea.

In short, the options for pipeline gas exports to the Korean peninsula before 2015 are very limited. The Eastern Programme of September 2007 prioritised exports based on Sakhalin gas. The Sakhalin–Vladivostok gas transportation system will be built to gasify the Primorye territory and for further gas exports to China and South Korea. Development of the Kovykta field will also depend on the resolution of helium recovery problems and the development of a petrochemical industry. Gazprom has prioritised Kovykta gas for local use or supply for domestic use in western Russia (see Chapter 6). Unless those plans are pursued, the development of Kovykta will not begin before 2017.[78]

The decision is a heavy blow to both Kogas and MOCIE's planners as the chance of pipeline gas from Sakhalin or Sakha (Chayandagas) to the Korean peninsula before 2015 is very slim. Without a breakthrough on border price negotiation between Gazprom and CNPC before the summer of 2008,[79] there is no chance for Kovykta gas exports to China and Korea before 2015.

The Korean government is desperate to import at least 10 Bcm/y of pipeline gas from Russia to compensate for anticipated LNG shortages in the 2010s (see above). Without pipeline gas, Korea will have no choice but to chase large long-term LNG supplies to start from 2013–14. Given the increase in LNG prices (see Table 5.14) and very high prices in recent contracts signed with Qatar, this is not an attractive prospect.[80] If pipeline gas supply cannot be implemented before 2015, Korea will have no choice but to wait for another eight to ten years to import Russian gas either from Sakhalin 3 or (Sakha's) Chayandagas. Ironically this situation will accelerate Kogas initiatives to expand upstream business development and considerably increase entry barriers for Russian pipeline gas to Korea after 2015.

The DPRK's (North Korea's) Views on Imports of Pipeline Gas

The DPRK authority has been reluctant to express any interest towards the idea of pipeline gas despite the fact that it has been studying the import option since the mid-1990s. To observe the progress of trans-national natural gas pipeline projects in Northeast Asia, the DPRK established a Natural Gas Research Society, DPR Korea (NGRS DPRK), under the leadership of DPRK's Asia Pacific Peace Committee in 1998. NGRS DPRK sent delegates to the 1998 Ulaan Baator (4th) and 1999 Yakutsk (5th) Northeast Asian Gas & Pipeline Forum conferences.[81]

In September 2001, South Korea announced that it had reached a preliminary agreement with the DPRK on a joint feasibility study for a pipeline passing through DPRK territory.[82] Later, however, Kogas argued its work on this section could not progress as the authorities had demanded Kogas should renounce any future pipeline plan that would not pass through DPRK territory.

2001 Dutch Agreement

On 6 April 2001, the NGRS DPRK signed an unpublished 18 point Memorandum of Understanding with a consortium of three Dutch trading companies (HS International Trading, Tamalone International, and Boscalis International).[83] The MOU gave the consortium exclusive rights to build the North Korean portion of the pipeline from the Russian border to the South Korean border. DPRK expected that the Dutch consortium would act as an intermediary in promoting the pipeline project with ExxonMobil, Japanese companies, and South Korean gas officials. The MOU envisaged the construction of three gas-fired power stations along the pipeline route with a total capacity of 500 MW.

2002 KoRus Agreement

On 3 August 2002, FSI Energy signed an agreement with Kyung-Bong Kim, chairman of NGRS DPRK and former head of the DPRK Academy of Science. The agreement gave FSI Energy exclusive transit rights with regard to the pipeline from Northern Sakhalin to the Korean Peninsula. Under this agreement, FSI Energy needed to identify the source of gas supply by 1 June 2003. FSI Energy sought the support of Congressman Curt Weldon (Republican, Pennsylvania), chairman of the House Armed Services Committee. The project was doomed to fail as it had no supply source and no interest from the US and South Korean governments. FSI Energy failed to secure US$10 million feasibility study funds from the US Department of Energy and neither ExxonMobil nor Kogas took the project seriously. Unfortunately the KoRus project did incalculable damage to NGRS DPRK as it clearly exposed the latter's political and business naiveté.[84]

Both the 2001 MOU and 2002 agreement indirectly confirmed that the DPRK authorities were interested in pipeline gas, but were not ready to officially discuss the issue with Washington and Seoul since this would potentially signal its willingness to compromise with respect to the settlement of the Korean Peninsula Energy Development organisation (KEDO)[85] project.

2003 NSC Advisor's Initiative

Unlike the Kim Dae-Jung government which focused on East Siberian gas supply and Sakhalin LNG supply to South Korea, the Roh Moo-Hyun government has shown interest in a Sakhalin pipeline to the

Korean peninsula. The so-called 'gas for peace' concept was that the US administration would guarantee the safety of the DPRK regime in return for its decision to renounce its entire nuclear programme, with South Korea as the main driving force behind the pipeline gas introduction via DPRK. In an interview National Security Advisor Ra said that:

> ...thermal power stations supplied by Russian gas would provide a peaceful alternative to Pyongyang's nuclear programme... This is one of the possibilities we are looking at... Gas could be supplied from either Irkutsk or Sakhalin. Advisor Ra added that Seoul's plans for a gas pipeline were at an early stage and had not been discussed in detail with its allies or North Korea.[86]

The views expressed in the interview were not coordinated with Korean government and business stakeholders[87] and the result was very strong opposition from both MOCIE and Kogas and consolidation of the status of Kovykta gas project. As far as MOCIE and Kogas were concerned, supply security was the main obstacle to the Sakhalin pipeline option and the feasibility study on Kovykta gas was in the process of finalisation. On 14 November 2003, the three-year feasibility study was finally completed, but did not lead to an inter-governmental agreement since, as noted above, Moscow refused to agree.[88]

It is worth clarifying the fundamental difference between the KoRus project and the Gas for Peace Formula. KoRus was the initiative of a small private entity driven by commercial interest, while the Gas for Peace formula was a state level initiative, driven by political ambition.

2004 UN Special Envoy's Initiative

The United Nations was more open-minded about pipeline gas passing through DPRK territory. In August 2003, a meeting on the DPRK nuclear crisis between US State Department assistant secretary James Kelly, and special envoy to the Secretary General of the United Nations Maurice Strong, was held as a follow-up to the meeting between Kofi Annan, Secretary General of the United Nations and Secretary of State Colin Powell.[89] These meetings created the United Nations study on energy scenarios for DPRK.

Maurice Strong's view was that

> Energy is a humanitarian issue. DPRK urgently needs energy and it is only possible with the international community's support. To tackle the DPRK's long term energy shortage problem, Russia's natural gas could be an option. In particular the Sakhalin gas project is worth taking note of as it can be

> done with the shortest pipeline route with a short working period. For this project, an energy specialist group is being established and feasibility work is being studied.[90]

At a meeting with the DPRK Military Commission chairman Jong-Il Kim, both sides agreed that the UN would study the options of long-term energy and economic aid to the DPRK.[91] Maurice Strong made clear that if the DPRK wanted to obtain UN aid, the authorities needed to make the necessary data available and Chairman Kim unexpectedly agreed. As a result, in August 2005, the UN's Working Group on Energy for the DPRK led by Maurice Strong produced a comprehensive report entitled 'Energy Scenarios for the DPRK-Report of the Working Group' by the University for Peace, under the United Nations.[92]

2005 Gazprom CEO's DPRK Visit

As discussed earlier, China's decision not to include foreign companies, including Gazprom, in its West–East Pipeline development in summer 2004, switched Gazprom's focus towards Sakhalin gas supply to the Korean Peninsula. In early 2005, China Xinhua News Agency reported that Gazprom CEO Alexei Miller had quietly visited Pyongyang and met DPRK premier Bong-Ju Park and deputy premier Du-Chul Ro to discuss energy cooperation between Russia and the DPRK.[93] No details were made available, but it was clear that Russia was taking the option of pipeline gas to the Korean Peninsula very seriously.

Despite these developments, the DPRK authority did not pursue the pipeline gas supply option. A senior diplomat at DPRK's London Embassy said that 'Pyongyang authority has virtually rejected the 2005 summer proposal by Minister Dong-Young Jung on his electricity supply from the South to North as the proposal would force DPRK to depend on an external energy supply source.'[94] The remarks gave a strong hint that the same logic would be applied to pipeline gas supply. The DPRK authority will be reluctant to show any interest in pipeline gas until the benefits and costs are clearly defined.

Despite its negative stance, Kovykta is the only chance of Russian gas supply to DPRK before 2015. If Gazprom decides to start Kovykta development in 2008 to ensure exports to China and South Korea (via the Yellow Sea) from 2014, this could include the option of a small-scale – say 1–2 Bcm/year – branch pipeline from Dandong to Kaesung via Pyongyang. As the total distance is well below 1000 km, this branch line could be completed within a year. However, this option will only be possible following a political breakthrough in the six-party talks, in particular in the negotiations on nuclear weapons between Pyongyang and Washington.

Conclusions

Korea's gas industry is set to expand, even though future growth rates will be much lower than during the 1990s. In the wake of the completion of a 2500 km nationwide trunk pipeline, Korea's gas industry is entering a more mature phase of development. MOCIE's 8th LTNGSDP projected that gas demand would be 56 Bcm, based on a growth rate of 3.5 percent per annum during the period 2007–2020, which is somewhat higher than the 51–53 Bcm projected by the 2002 National Energy Plan. In late 2007, the Korean government envisaged gas demand at 50–55 Bcm in 2020, accounting for 15–16 percent of primary energy consumption, slightly higher than the current 13–14 percent rate. This target looks realistic, but is much lower than the average share of gas in primary energy demand in OECD countries.

Liberalisation and competition reforms in the gas industry lost momentum during Roh Moo-Hyun's government mainly due to lack of political will and strong opposition from labour unions.[95] It remains to be seen whether the new government, which will be responsible for the period 2008–12, will implement the IEA's 'text book' policy recommendations, and whether it will have the political will to review the introduction of competition in the industry. But the task of introducing competition by breaking-up the Kogas monopoly, which was tried by the Kim Dae-Jung government (1997–2002) and the Roh Moo-Hyun government (2003–2008) will be very difficult in current and anticipated market conditions.

The real challenge for gas market liberalisation in South Korea is not internal politics but the anticipated shortage of LNG in the face of increased Asian demand and South Korea's increasing need for seasonal supplies to manage its peak winter demand. Kogas has used swaps with Japanese utilities as a way to manage seasonality and its cargoes have been diverted to Japan to meet the summer air conditioning load whilst Japanese cargoes have been redirected to Korea in winter. In addition, Kogas has negotiated LNG supply deals with increasingly high seasonality, trading on its position as a reliable off-taker and major buyer in times of oversupply.[96]

But even these initiatives cannot compensate for the absolute shortage of LNG supply in the next few years. The worst period for Korea will be 2010–2015 as the gap between committed and required LNG supply will increase, unless a breakthrough is made in relation to pipeline gas imports (Table 5.19).

It is for this reason that Kogas and the Korean government are still very keen on pipeline imports from Russia before 2015. For most outside

Table 5.19: Korea's LNG Demand and Supply Gap. Mt/y

	2011	*2015*
Demand	32.9	33.5
Supply	24.3–24.8	21.6–22.1
Gap	8.1–8.6	11.4–11.9

Note: Kim and Cho calculated that the gap will be greatest around 2011–12 reaching 6.2–7.5 mt; by 2015, the gap will be only 2.8 mt assuming that Korea will secure 6.5–10.0 mt of new LNG supply starting from 2013-14.

Source: Calculated based on Tables 5.11 and 5.4.

observers, including this author, the Kovykta gas development is the only realistic chance to deliver Russian gas to China and Korea before 2015, but Russia's September 2007 Eastern Gas Programme ruled out this option. The Programme does not envisage the pipeline gas supply to China and Korea before 2015, with Sakhalin and Sakha gas as the favoured sources of these exports whenever they commence.

The long-delayed price negotiations between Gazprom and CNPC have been the major obstacle to Russian gas exports to China, but the Korean government was willing to consider the option of increasing import volumes from 10–15 Bcm/year to reduce the price burden. If Russian gas exports depend on the development of Sakhalin (offshore) and Sakha (Chayandagas) reserves, this will delay their timing to (China and) Korea until well after 2015. The conclusion is that it will be at least ten years before pipeline gas can be delivered and this is a very gloomy picture for Korea's gas supply since, without Russian pipeline gas, it will have no leverage in future LNG price negotiations.

Interestingly this delay to Russian pipeline exports could breathe new life into the prospects for a pipeline through North Korea. Since Russian gas exports to Korea, starting around 2017–18, will probably be based on Sakhalin 3 or Chayanda (Sakha) gas such a pipeline could pass through North Korean territory if the nuclear crisis is settled peacefully. The option of Russian gas supply to Korea via an offshore (Donghae or Sea of Japan) pipeline looks unrealistic as the price of gas will be too high.

Up to 2015, the Russian pipeline gas option will be very attractive for Korea, but after that the situation may change. Kogas, with government support, has begun to explore prospects for foreign equity gas ownership. Kogas planners are likely to aim for at least two trains, roughly 8 Mt/y of equity LNG imports for the period 2016–2020. The best way to maximise equity gas prospects would be to merge KNOC and Kogas, creating the Korea National Oil and Gas Corp (KNOGC) or substantially to expand Kogas' upstream business. It remains to be seen whether Korea will blindly wait for the Russian gas by pipeline or LNG, or aggressively pursue the option of equity gas supply. Should the latter option prove successful, then prospects for substantial pipeline gas supplies might be delayed beyond 2020.

References

IEA 2007a: International Energy Agency, Energy Policies of IEA Countries: *The Republic of Korea 2006 Review*, OECD/IEA: Paris, 2007.

IEA 2007b: International Energy Agency, *Natural Gas Market Review 2007: Security in a globalising market to 2015*, OECD/IEA: Paris, 2007.

International Gas Report (Platts) various issues

Kim and Cho: Hyung-Joo Kim and Jeong-Shik Cho, *Jeongbooyi Gassaneop Jeong-check Pahengkwa ge Younghyang* (*Repercussions of the Government's Inadequate Gas Policy*), prepared for Korea's Parliament Industry and Energy Committee Hearings (October, 2006). (The authors are Members of Parliament for the ruling party.)

MOCIE Long-term Natural Gas Supply and Demand Plans (LTNGSDPs): Ministry of Commerce, Industry and Energy (MOCIE):

4th Long-term Natural Gas Supply and Demand Plan (LTNGSDP: 1998–2015), March 1999.

5th LTNGSDP (1999–2015), March 2000.

6th LTNGSDP (2001–2015), November 2002.

7th LTNGSDP (2003–2017), November 2004.

8th LTNGSDP (2006–2020), December 2006.

Notes

1 Ministry of Commerce, Industry and Energy (MOCIE) and KEEI, *Yearbook of Energy Statistics 2006*.

2 *http://www.keei.re.kr/web_keei/en_news.nsf*

3 For the cause and implications of this 1997 financial crisis, see Stephen Haggard, *The Political Economy of the Asian Financial Crisis* (Washington, DC: Institute for International Economics, 2000) ; Gordon De Brouwer, *Financial Integration in East Asia* (Cambridge: Cambridge University Press, 1999); Ippei Yamazawa, *Strengthening Cooperation Among Asian Economies in Crisis* (Tokyo: Japan External Trade Organisation and Institute of Developing Economies, 1999) ; Seiichi Masuyama, Donna Vandenbrink and Chia Siow Yue, eds, *East Asia's Financial Systems: Evolution and Crisis* (Singapore: Institute of Southeast Asian Studies and Nomura Research Institute, 1999).

4 Ministry of Commerce, Industry and Energy, *Yearbook of Energy Statistics 2006*, pp. 22–3.

5 For the figures under the environmentally friendly scenario and under protracted case scenario, see Asia Pacific Energy Research Centre, APEC Energy Demand and Supply Outlook: Energy Balance Tables, updated September 1998 (Tokyo: APREC, 1998), pp. 62–5.

6 For details see MOCIE Long-term Natural Gas Supply and Demand Plans.

7 Sang-Ryul Lee, member of the Korean Parliament, pointed out during the Parliament's annual inspection session that LNG shortage was caused by Kepco and its generation companies. During the first half of 2006, the power generation sector's oil product demand (1.85 Mt) decreased 31 percent, while that of LNG (5.24 Mt) increased 25.9 percent. In fact, a survey confirmed that during Jan. 2005 and March 2006, B-C price was cheaper than LNG during Jan. 2005 and May 2005 but Bunker-C oil price became dearer than LNG after June 2005. This is the reason why power generation firms switched from B-C to LNG for their power generation. See, *The Gas Industry News*, 25 April 2006, and 18 October 2006.

8 International Energy Agency, *Natural Gas Market Review 2007: Security in a globalising market to 2015*, p.13.

9 Since the calorific value of LPG is higher than that of LNG, some city gas companies using LPG have suggested equalising the quality of the two products by adding air.

10 *The Gas Industry News*, 4 July 2005.

11 *The Gas Industry News*, 5 April 2006. In September 2007, Kogas announced that the firm had chosen Samcheok in Kangwon province as the priority negotiation site for the 4th LNG receiving terminal. Up to 2019, Kogas would invest 2.3 trillion won and build 200,000 kl scale storage tanks, 14 units and port facilities and re-gasification facilities. See *The Gas Industry News*, 14 Sept. 2007.

12 *International Gas Report*, 26 Feb. 2007, p. 13.

13 International Energy Agency, *Energy Policies of IEA Countries: The Republic of Korea 2006 Review*, pp. 100–101.

14 Based on the MOU signed between Kogas and the Oman government for LNG joint business in June 2006, both parties agreed to set up a JV for LNG business on 29 May 2007, and the initial JV capital will be US$ 10 million. *The Gas Industry News*, 29 May 2007.

15 *International Gas Report*, 30 March 2001, pp. 2–4.

16 In 2002, Kogas requested MOCIE to approve Shell's offer of 2.0 Mt/y of LNG from Sakhalin 2 since it offered a unique opportunity for Kogas to have a strategic partnership with Shell, but MOCIE rejected the request due to privatisation and competition considerations. Consequently Kogas missed the opportunity to secure a very special status in the Sakhalin 2 project, and the cost of this now appears to be very great.

17 In late April 2003, Kogas had to suspend the supply of LNG to power companies and consequently during 25 and 30 April, the six power gen-

eration companies under Kepco had to use heavy oil or Kerosene instead of LNG. During the 2002–3 winter period the LNG supply shortage was caused by excessive demand from the power companies, but this April 2003 supply shortage was caused by Kogas failure to provide the contracted volume to the companies. Power companies under Kepco argued that a total of 309.5 billion won (roughly US$258 million) loss was caused due to the replacement of LNG with heavy oil and kerosene from 15 November 2002 to 25 March 2003.

18 *The Gas Industry News*, 8 & 11 Oct. 2004.

19 *The Gas Industry News*, 18 July 2005.

20 Korea Economy Shinmoon, 16 Nov. 2006.

21 Kim and Cho, 2006.

22 *The Gas Industry News*, 11 May 2007. GS Caltex also signed the LNG supply contract with Qalhat Oman which started to produce LNG from 2004 with a capacity of 3.7 Mt/y. The volume of import has not yet been decided. See *The Gas Industry News*, 28 May 2007.

23 The S curve maintains gas prices at maximum or minimum levels when oil prices rise above or fall below certain defined thresholds, see Appendix 2 of this volume.

24 International Energy Agency, *Natural Gas Market Review 2007: Security in a globalising market to 2015*, p. 189.

25 600 million tons would equate to 23.5 trillion cubic metres of gas which, for comparison would give Korea gas resources similar to Qatar and Iran. MOCIE has said future prospects seem optimistic as the sediment extracted from Donghae (Sea of Japan) is top quality containing 99 percent methane. The Korean government has budgeted 226 billion won for a three stage project to run until 2014. The first phase that began in July 2005, costing 67 billion won, will end in 2007. Five additional areas will be drilled starting in September 2007. See *The Korea Times*, 25 June 2007.

26 Dong-Ah Ilbo, 7 Sept. 2006. The figure of 2.5 Mt is somewhat exaggerated compared with the possible shortfall of 1.1 Mt shown in Table 5.13.

27 *International Gas Report*, 26 Feb. 2007, p. 13.

28 *The Gas Industry News*, 19 Feb. 2005.

29 *International Gas Report,* 17 November 2006, pp. 17–18. India's state-run Petronet LNG which operates a 5 Mt/y of LNG import terminal at Dahej in western Gujarat state, plans to buy around 20 LNG cargoes from the spot market in 2007. A Petronet LNG official said 'The seven cargoes were bought at around US$ 8.0/MMBtu from several sources including Qatar and Algeria.' See, *FT International Gas Report*, 23 April 2007, p. 28.

30 The Surgil field is located in Ustyurt region in Uzbekistan, and the estimated gas reserves are 113 Bcm. Due to the high ethane contents, the Korean consortium is considering the pipeline gas production from the second half of 2007, and the gas chemical complex from 2011. When the project is on the right track, the production capacity of pipeline gas, high density polyethylene and polypylene is expected to reach to 2 Mt/y, 0.362 Mt/y and 0.083 Mt/y respectively. *The Gas Industry News*, 2 March 2007.

31 *The Gas Industry News*, 21 May 2005.
32 *The Gas Industry News*, 22 May 2005.
33 As for the Yemen LNG project, the proven reserves are 9.15 Tcf, and the annual production capacity will be 6.7 Mt/y. Yemen LNG's shareholders are Total 39.62 percent, Hunt 17.22 percent, Yemen Gas Co 16.73 percent, SK Corp 9.55 percent, Korea Gas Corp 6.0 percent, Hyundai Corp 5.88 percent, and The General Authority for Social Security & Pensions (5.0 percent). In the case of the Peru project, the annual production capacity is 4.45 Mt/y and the LNG plant shareholders are Hunt Oil 50 percent, SK Energy 20 percent, Repsol YPF 20 percent, and Marubeni Corp 10 percent (In the case of gas field ownership, Hunt Oil 50 percent, SK Energy 30 percent and Repsol YPF 20 percent). See *The Gas Industry News*, 29 August 2007, *www.yemenlng.com/ws/en/go.aspx?c=ylng_glance* & www.perulng.com
34 *International Gas Report*, 29 January 2007, p. 32; *The Gas Industry News*, 24 January 2007.
35 Robert Prichard, 'Legal and Regulatory Challenges of the Asian Gas Grid', presented at UNESCAP-PEG Workshop on the Asian Gas Grid, 16 Nov. 2005, Busan, Korea. For the background to the Asian Gas Grid, see APERC (Asia Pacific Energy Research Centre), *Energy Supply Infrastructure Development in the APEC Region*, 2001 March, Chapter Six (pp. 67–94).
36 In late August 2006, the Daesung Group chairman met China's powerful National Development and Reform Commission chairman Ma Kai to discuss the project.
37 *The Gas Industry News*, 1 Feb. 2005.
38 *http://www.chinagasholdings.com*
39 SK Corp said it expects to account for 10 percent of China's gas market by 2020 through cooperation with China Gas. Xinhua News Agency, 16 March 2007.
40 *The Gas Industry News*, 14 Aug. 2006.
41 *International Gas Report*, 4 Dec. 2006, p. 20; *The Gas Industry News*, 30 Nov. 2006.
42 *International Gas Report*, 17 Nov. 2006, p. 15.
43 *The Gas Industry News*, 23 March 2007. Kogas plans to spend up to US$905 million. This move is part of an effort to meet government plans to double storage by 2017 to 8.64 Mcm in an effort to increase security and meet seasonality requirements. See, 'Korea – the End of Liberalisation', *Gas Matters*, December 2006, pp. 13–16.
44 *The Gas Industry News*, 26 June 2006 and 29 May 2007.
45 International Energy Agency, *Asia Gas Study*, OECD/IEA: Paris 1996, pp. 142–3.
46 Ministry of Commerce, Industry and Energy, Basic Plan of Korea Gas Industry Restructuring, November 1999, p. 5 (in Korean).
47 IEA , *Natural Gas Market Review 2007: Security in a globalizing market to 2015*, p. 193.
48 Ibid.
49 Ibid.

50 Ibid.
51 *FT International Gas Report*, 20 Oct. 2006, pp. 10–12.
52 H. J. Doh and J. K. Seo, Lowering Entry Barriers and Competition Policy in Natural Gas Industry, KEEI, Dec. 2006 (published in Korean).
53 International Energy Agency, *Energy Policies of IEA Countries: The Republic of Korea 2006 Review*, p. 10.
54 *The Gas Industry News*, 11 Oct. 2005.
55 Korea's Painenshal (Financial) News, 13 Oct. 2004.
56 *The Gas Industry News*, 1 Nov. 2006.
57 *International Gas Report*, 20 Oct. 2006, pp. 10–12.
58 *The Gas Industry News*, 14 Nov. 2004.
59 Under Kepco, there are Korea Hydro and Nuclear Power (KHNP), and five non-nuclear GENCOs, Korea Southern Power (KOSP), Korea Midland Power (KOMIP), Korea Western Power Co (KOWP), KOSEP (Korea South-east Power), and Korea East-West Power Co (KOEWP). As of 2005, LNG based installed capacity of the five GENCOs are: EWP 2.1 GW, SPO 3.6 GW, WP 2.28 GW, SEP 0.9 GW, and MIP 3.841 GW. See IEA, *Natural Gas Market Review 2007*, p. 118.
60 *The Gas Industry News*, 4 Dec. 2004 & 19 Feb. 2005
61 *The Gas Industry News*, 5 July 2006.
62 *The Gas Industry News*, 3 Jan. 2007
63 Another private player SK E&S, an affiliate of SK Corp, plans to build an LNG import terminal by 2013 but the project is still at the feasibility study stage. SK E&S already has 25 percent of the city gas market share in Korea through its ownership of ten city gas companies. See, *FT International Gas Report*, 26 Feb. 2007, p. 1.
64 *The Gas Industry News*, 5 April 2007.
65 For a detailed review of Gazprom's Asian Policy, see Keun-Wook Paik, 'Sino-Russian Oil and Gas Relationship: Implications for Economic Development in Northeast Asia' presented at Northeast Asia Cooperation Dialogue XIII: Infrastructure and Economic Development Workshop organised by Institute for Far Eastern Affairs, Russian Academy of Sciences, and Institute of Global Conflict and Cooperation, University of California, Moscow, 4 Oct. 2002; 'Natural Gas Expansion in China and its Implications towards East Asia Market', presented at an International Workshop titled 'Changes in the World Natural Gas Market and Challenges for East Asia' organised by Japan Bank for International Cooperation, Tokyo, 20 January 2005; 'Sino-Russian Oil and Gas Pipelines: The Reality and Implications at the Second Colloquium on Eurasian Pipelines and East Asia: A Path to Integration or a Marriage of Convenience?: Eurasian Oil and Gas to China, Korea and Japan', co-sponsored by Harriman Institute and Weatherhead Institute, Columbia University, New York, 30 Nov./1 Dec. 2006 ; Issues from Interfax China Energy Report Weekly, Dow Jones China Energy Report, and Xinhua News Agency's China OGP.
66 Dong-Ah Ilbo, 5 August 2004.
67 Kommersant, 'China to OK Eastern Gas Program', 10 September 2007.

The Russian Industry and Energy Ministry believes that the most effective option for selling gas from Eastern Siberia and the Far East is the East option, which involves supplying all gas produced at the Kovykta field into the UGSS, while Sakhalin gas will be sent to the Asia-Pacific market. According to this scenario, gas production in Eastern Siberia and the Far East in 2030 will amount to 113.6 Bcm per year, with domestic consumption in the region of 30.2 Bcm, supplies to the unified gas supply system of 35 Bcm, 8 Bcm being used for technical needs, 15.3 Bcm used to produce liquefied natural gas and 25 Bcm being exported – 15 Bcm to China and 10 Bcm to Korea. Under this scenario, Russian gas in China will cost $26 per 1000 cubic metres, and in South Korea – $140–$160 per 1000 cubic metres. Total capital investment in geological exploration, production, processing and transportation of gas through Russia, China and Korea will amount to $31.6 billion. See, Gazprom's Discussion Package for the interagency Working Group to develop a Programme for creating a unified Gas Production, Transportation and Supply System in Eastern Siberia and the Far East with potential exports to markets in China and other countries in Asia and the Pacific (2004), and Oleg Sinyugin, 'Eastern Dimension of Russia's unified Gas Supply System', Northeast Asia Energy Focus (KEEI), vol. 2, no. 4, Nov. 2005.

68 Keun-Wook Paik, 'Pipeline Gas Introduction to the Korean Peninsula', prepared for Korea Foundation project on 'Energy and Environmental Co-operation in the Korean Peninsula', January 2005, *http://www.chathamhouse.org.uk/pdf/research/sdp/KPJan05.pdf* ; Chris Ajemian, 'Energy For Security: A Natural Gas Pipeline Solution to the North Korean Security Threat, *The Non-proliferation Review*, vol. 14, no. 2, July 2007, pp. 329–49 ; Selig S. Harrison, 'Towards Oil and Gas Cooperation in Northeast Asia: New Opportunities for Reducing Dependence on the Middle East', Woodrow Wilson International Center for Scholars, Asia Program Special Report, No. 106, December 2002.

69 Keun-Wook Paik, 'Natural Gas Expansion in Korea', in Ian Wybrew-Bond and Jonathan Stern, *Natural Gas in Asia: The Challenges of Growth in China, India, Japan and Korea* (Oxford: Oxford University Press, 2002), pp. 218–26.

70 This initiative was a part of Alexei Miller's Tokyo announcement in June 2003.

71 Author's interview with senior officers of Exxon Neftegas Project Services Inc (Tokyo based) and Exxon Neftegas Limited (Yuzhno-Sakhalinsk based) confirmed that the US major was considering the pipeline up to Vladivostok only. They had no interest in the extension of the pipeline to the Korean Peninsula at all.

72 On 2 Nov. 2006 deputy energy minister Andrei Dementyev said 'All the rights of Sakhalin 1 investors are outlined in the production sharing agreement. No one is going to challenge the investors' rights'. However, one day earlier Dementyev told deputies of the Duma, Russia's lower chamber of parliament that a decision on future gas supplies from Sakhalin 1 would be taken after Russia has approved a long-term programme of gas resources

development in East Siberia and the Far East. See, *FT International Gas Report*, 3 Nov. 2006, p. 17.

73 The Letter of Intent (LOI) signed by Russia, China and Korea on 14 Nov. 2003 for Irkutsk gas purchase expired on 13 May 2005 without any progress. *The Gas Industry News*, 4 June 2005.

74 Kogas CEO Soo-Ho Lee said Kogas was considering the option of swapping equity with Gazprom to reduce Korea's heavy dependence on gas supply from the Middle East. See, *The Gas Industry News*, 27 July 2006.

75 The author was told that the minimum strategic level of Gazprom equity would be 1 percent.

76 The 5th VP level meeting was held in the Jeju Islands. During the meeting with Deputy CEO Ananenkov, Kogas CEO Soo-Ho Li proposed to Gazprom a joint participation in Sakhalin block 3 bidding. *The Gas Industry News*, 6 April 2007.

77 *The Gas Industry News*, 17 Oct. 2006.

78 Interfax Russia & CIS Oil & Gas Weekly, 6–12 Sept. 2007.

79 In September 2007, Alexander Medvedev, Gazprom Deputy Chairman said agreements for up to A$ 55 billion or US$45 billion worth of Australian LNG signed by PetroChina had strengthened the Russian company's hand in talks on the sale of some of its own huge reserves of gas to China. See, Peter Smith, 'China's LNG deals good for Gazprom discussions', *Financial Times*, September 10, 2007. According to Xinhua News Agency, CNPC and Shell agreed that the ceiling CIF price will be US$ 10/MMBtu or 2.7 yuan/cubic metres, much higher than the present gas price in China. See, China OGP, 15 September 2007, p. 2. Interestingly, during the 10th Northeast Asian Gas and Pipeline Forum in Novosibirsk, CNPC gas specialists told the author that there is no chance that CNPC would accept such a high price for LNG supply. Time will tell what price PetroChina actually agreed to pay. The author believes it will be very difficult for PetroChina to accept the level of US$10/MMBtu as China's gas consumers cannot afford such a price burden.

80 For discussion of LNG prices see Dong-Ah Ilbo, 3 & 17 Nov.; Hankook Kyungje Shinmoon, 16 Nov. 2006.

81 The delegation was led by Prof. Kyung-Bong Kim, former head of the DPRK Academy of Science (deputy premier level academician). Since 2000, NGRS DPRK has not attended the pipeline conference.

82 During 6–7 Sept. 2001, a six-person South Korean delegation led by Mr. Jong-Sool Kim, senior vice president of Kogas visited Pyongyang. At that time, the DPRK argued that it would cooperate on the pipeline feasibility study only when South Korea guaranteed that the pipeline route would not bypass DPRK territory. However, the demand was not acceptable.

83 Selig S. Harrison, 'Towards Oil and Gas Cooperation in Northeast Asia: New Opportunities for Reducing Dependence on the Middle East', Woodrow Wilson International Center for Scholars' Asia Program Special Report, No. 106 (December 2002) and author's interview with Mr. Harrison.

84 Had the DPRK authority known the weakest point of the agreement, that

is that no gas supply source would be available without the commitment on gas market opening from Kogas, Pyongyang would not have signed. It was solid evidence of the DPRK's lack of professional understanding of the natural gas business.

85 In October 1994, the United States and the Democratic People's Republic of Korea (DPRK, or North Korea) signed the *Agreed Framework*, under which the DPRK agreed to freeze and ultimately dismantle its nuclear programme. The Agreed Framework sought to resolve comprehensively the issues arising from the programme. The facilities subject to the freeze included an operational 5 MW experimental graphite-moderated reactor, a partially complete reprocessing facility, and a 50 MW power reactor under construction, all at the Yongbyon Nuclear Research Center, as well as a 200 MW power reactor under construction at Taechon, DPRK. The Executive Board of KEDO decided on 31 May 2006 to terminate the LWR project. See, *http://www.kedo.org*

86 Andrew Ward, 'Deal for gas pipeline could solve Korean nuclear crisis', *Financial Times*, 31 March 2003. Korean newspapers and journals reported that the author had played a pivotal role in introducing the Gas for Peace formula to the presidential transit committee (during Dec. 2002– Feb. 2003) and this was the case. See, *Weekly Chosun*, 13 March 2003; *Sisa Journal*, 31 March 2003; *Meil Kyungje Shinmoon*, 8 April (author's interview with the newspaper).

87 The major stakeholders being: National Security Advisor's office, Ministry of Commerce, Industry and Energy (MOCIE) & Korea Gas Corp (Kogas), and Presidential Committee on Northeast Asian Business Hub.

88 Observers said that Gazprom was very unhappy that the detailed feasibility study exposed the cost breakdown of Kovykta gas production, as it was not acceptable to allow both China and Korea to dictate the price negotiation based on the study. The cost issue was one of the key factors that influenced Gazprom's stance towards the Kovykta gas project.

89 Author's interview with Mr. Maurice Strong, in Seoul, August 2003.

90 Dong-Ah Ilbo, 24 Nov. 2003. This remark was confirmed in his interview with NHK on 4 March 2004 when he stated that a natural gas pipeline to the Korean peninsula passing through DPRK can be studied assuming that the DPRK nuclear crisis would be resolved. See, Joong Ang Ilbo, 5 March 2004.

91 Joong-Ang Ilbo, 19 May 2004 & 20 June 2004.

92 The author was a member of this study group.

93 Kyung Hyang Shinmoon, 24 Jan. 2005. Russia is very anxious to restore its influence with the DPRK. Moscow has proposed exporting its crude oil to DPRK with no tax. The proposal was made in August 2005 when the Russian Federation's Far East region's presidential plenipotentiary Konstantin Fulikovsky met DPRK premier Bong-Ju Park. The main proposal was to refine Russian crude oil at DPRK's Seungli PetroChemical Plant in Rason City (Hamkyung North Province) with Russia exporting the crude oil tax-free. Even though the refining capacity has not been fully utilised since the

1990s when the Former Soviet Union suspended crude oil supply to the DPRK, the capacity is much bigger than that of Bonghua PetroChemical plant in Pyongan North Province, which depends on crude oil from China. This proposal suggested that Moscow has not given up the option of supplying oil and gas to the Korean Peninsula. See, *Yonhap News*, 3 Sept. 2005.

94 Author's interview with senior diplomat, DPRK London Embassy, In July 2005.

95 'Korea – the End of Liberalisation', *Gas Matters*, December 2006, pp. 13–16.

96 Ibid.

CHAPTER 6

RUSSIAN AND CENTRAL ASIAN GAS SUPPLY FOR ASIA

Jonathan Stern and Michael Bradshaw

Introduction

This chapter explores the potential for Russian and Central Asian gas supplies to Asia up to 2020. The aim is to provide a perspective against which readers can judge the progress of relationships and projects which have already stretched over many decades. Given the lead times required for gas export projects, 13 years is an entirely appropriate time-frame to consider likely developments. However, because of the pace and extent of change which has already occurred during the 2000s, any attempt to project developments over such a long period may be far too ambitious.

The Soviet Era 1965–91

An important starting point for any discussion of future Russian and Central Asian gas trade with Asia is a brief assessment of developments in the Soviet period. At the same time that Soviet gas exports to European markets were getting under way in the late 1960s, similar discussions were taking place about the possibility of developing pipeline and/or liquefied natural gas (LNG) trade with Japan. Discussions about oil and gas development around Sakhalin Island originate in 1965–66 and the first Japan–Soviet Economic Co-operation Committee. For most of the following 25 years, until the break-up of the Soviet Union, the possibility of exports of Sakhalin gas to Japan was considered in terms of LNG, although pipeline gas was considered as an option from the beginning.[1]

During the 1970s and 80s, the progressive discoveries of oil and gas fields offshore Sakhalin Island established a major hydrocarbon province, dominated by natural gas, but with significant quantities of oil.[2] During the Soviet period, a project – which eventually became Sakhalin-1 – was pursued in cooperation with a consortium of Japanese companies headed by the Japan National Oil Corporation and funded by Japanese government credits.[3]

The other major gas export project during the Soviet era involving the Japanese (but also the US) market was the Yakutia LNG project, whereby gas would be piped from fields in Yakutia (now Sakha Republic) to be liquefied at a coastal site (Nakhodka was most frequently mentioned but other locations such as Olga and Magadan were also considered). By the late 1970s, around 1000 Bcm of reserves which were required for the project had been proven.[4] During the 1970s, proposals were also made for a pipeline link between the Sakha fields and a possible Sakhalin export project. Such a link was crystallised at the end of the Soviet era with the publication of the 'Vostok' (East) plan to link the major oil and gas resources in Eastern Siberia and the Far East to both local markets and exports to Japan, China and Korea (North and South). The Vostok plan was a typically grandiose Soviet concept with a great many pipelines carrying large quantities of oil and gas, with little attention to how they might be financed and by whom.[5]

Only in the post-Soviet period did it become clear that, in addition to the Sakha and Sakhalin fields other significant gas deposits had been discovered in the Irkutsk region including Verkhnechonskoye and Kovytinskoye.[6] At the time of discovery, little attention or publicity was given to these discoveries, probably because of the lack of immediate market prospects.

During the Soviet era, gas export projects to Asian markets failed to make progress because of a mixture of political, commercial and institutional obstacles. Given that until 1990, Japan was seen as the only realistic Asian market for Soviet gas, the lack of a Peace Treaty between Japan and the Soviet Union after the Second World War was a serious problem. But the most high profile issue in bilateral relations centred on the territorial dispute over four small islands in the Kurile chain which were occupied by the Soviet Union after the Second World War.[7] The following 45 years saw periodic negotiations and initiatives to reach a settlement. At the close of the Soviet era, a commentary from Europe concluded:

> Soviet–Japanese relations must be considered among the most bizarre of the major bilateral relations among great powers…The experience of the past century only reinforces the sense of unease and tension. There is little reason to expect these two states ever to be close friends...The ending of the Cold War has meant that the Pacific region, which was always more multipolar than bipolar, is developing a new balance of power with vast options for change and instability…For those seeking the real implications of the coming Northeast Asian détente, one would be better advised to look further to [the] grander issues, than remain bogged down in the niggling details of an old and insignificant territorial dispute. [8]

But this 'old and insignificant territorial dispute' continued (and continues) to have substantial significance for both countries. From the Soviet side, the dispute and lack of a Peace Treaty with Japan were symbols of a failure to agree on the post-World War 2 status quo. Japan's tendency to follow the political lead of the United States reinforced the lack of trust between the two countries during the Cold War era. For these reasons, there never existed political conditions between the governments which would have provided the basis for multi-billion dollar trade involving substantial and long term economic, and thus considerable political and strategic, inter-dependence. Soviet governments did not wish to make any concessions which might open up post-War territorial disputes elsewhere, notably in Europe. The political relationship with the Soviet Union was mirrored in the lack of enthusiasm on the part of Japanese gas buyers to import Soviet supplies. The attitude of these buyers in respect of Soviet gas stood in sharp contrast to their much greater keenness to develop LNG trade with Southeast Asian (and Middle East) countries. Thus while Soviet–Japanese discussions on (oil and) gas projects periodically took place, no significant commercial progress was made and the Soviet side – increasingly preoccupied with European trade – gradually lost interest. That is not to say that Soviet–Japanese cooperation did not promote regional economic development in the Soviet Far East. During the second half of the 1970s a series of long-term compensation agreements promoted the development of the South Yakutian Coal Complex and the development of the region's forestry sector, both of which supported the construction of a port at Nakhodka, not far from Vladivostok. There was also an agreement to explore Sakhalin's continental shelf for oil and gas fields.[9]

What is particularly striking about the 1965–91 period is that gas trade commenced and expanded enormously in both the Soviet Union and Japan. In 1965, Soviet natural gas exports were minimal and had not yet commenced to Western Europe. By 1991, Soviet gas exports to European countries – several thousand kilometres from the principal gas fields in Siberia – had reached 105 Bcm, of which 63 Bcm were delivered to 7 West European countries.[10] Likewise in 1966, Japanese LNG imports had not yet commenced but 25 years later the country was importing around 50 Bcm of LNG from six countries which – in the cases of Australia, Abu Dhabi and Alaska – involved a shipping distance of several thousand kilometres.[11] For 25 years therefore, the two countries which were becoming the world's biggest natural gas exporter and LNG importer over huge distances were unable to agree terms for a trade which would have involved a pipeline or shipping distance of around a thousand kilometres.

Until a few years before the collapse of the Soviet Union, natural gas trade aspirations – and other large-scale commercial dialogues involving Asia – were focused entirely on Japan; and largely (aside from a brief period in the mid- to late-1970s) on Sakhalin resources. As noted above, there was an agreement between Japan and the Soviet Union to explore Sakhalin's continental shelf; however, this project was reliant on western technology and in 1979 was one of the specific targets of US sanctions following the Soviet invasion of Afghanistan.[12] For the majority of the period, the lack of any positive or substantial Sino-Soviet or Soviet–South Korean political relationships meant that neither could be considered as potential large-scale business partners. Likewise, there was no suggestion that Central Asian resources should be targeted at any markets other than the USSR and Europe.

By 1990, this had begun to change with the addition of new political and commercial initiatives encompassing both China and Korea. In particular, the visit of President Gorbachev to Beijing in 1989 marked the start of a new era in Sino-Soviet relations, including energy cooperation.[13] But these early beginnings were not to be crystallised until the post-Soviet era when individual Russian, Turkmen and Kazakh initiatives and project opportunities presented themselves. The foundations of gas exports to Asia which were laid in the Soviet era were resource foundations. Political and commercial issues had been sporadically and unsuccessfully addressed in the case of Japan, and were in a very early stage of unfolding with other Asian countries. But it was clear by 1991 that, if political and commercial issues could be resolved, the resource potential to support large-scale gas trade with Asian countries was considerable.

Russian Supplies in the post-Soviet Era: Projects and Policies

Western Siberia

Given the geographical and market orientation of this chapter, the West Siberian resource base needs to be considered from an eastern perspective since, for the past 25 years, this has been the main gas producing region of the Soviet Union and Russia. West Siberian resources can be most conveniently divided between the currently producing region of Yamal Nenets, and the established resource base of the Yamal Peninsula.[14] At the end of 2005, Gazprom's West Siberian gas reserves were estimated at 16.9 trillion cubic metres (Tcm) proven and probable, of which approximately 12 Tcm were in areas under development, and 4.9 Tcm were in fields on the Yamal Peninsula.[15]

Since 2000, concerns have been increasingly expressed about the ability of Gazprom – the dominant Russian gas company – to maintain production in Western Siberia. To summarise a complex situation as briefly as possible, Soviet and Russian gas production has been sustained for more than 20 years by three supergiant fields in Western Siberia (Medvezhe, Urengoy and Yamburg). These fields are all well into their decline phase with production in the 2000s declining each year by between 20 and 25 Bcm. In 2001, the supergiant Zapolyarnoye field was brought into production and this masked the decline in the older fields. Zapolyarnoye reached its plateau production level in 2004. From that year, Gazprom has been forced to compensate for declines at the older fields by a combination of small newfields and the purchase of producing assets from other companies.[16] The major question for Gazprom – and the entire Russian gas industry – is the timing of investments to open up the company's next generation of supergiant fields on the Yamal Peninsula. The company announced in October 2006 that Bovanekovskoye, the first of these Yamal Peninsula fields, would start production in 2011, increase to 115 Bcm/year and eventually to 140 Bcm/year.[17] But by late 2007, a 2011 start date was looking somewhat optimistic for Yamal gas and a consensus was forming that production in excess of 100 Bcm/year would be more likely towards the end of the 2010s.[18]

Significant delays in Yamal gas development have potentially adverse consequences for Gazprom's ability to meet its future commitments to different customer groups, particularly the Russian market. But for the purposes of this chapter, the important point is that for the next decade, until the Yamal Peninsula fields are fully on stream, West Siberian gas will be required for Russian and CIS/European markets where it can be delivered through established pipeline networks and be sold at attractive prices.[19] Therefore it is unlikely that significant volumes of West Siberian gas would be available for export to Asia prior to 2020 and this should be borne in mind in the discussion of the Altai Pipeline below.

Eastern Siberia and the Far East

Detailed characterisation of the Eastern Siberian and Far Eastern gas resource base is in some respects entirely straightforward. Multi-trillion cubic metre gas resources are largely concentrated in what Gazprom has declared the four principal centres: Krasnoyarsk, Irkutsk, Sakha and Sakhalin, which is substantially in excess of the potential domestic requirements of the regional population.[20] However, numerically and conceptually, resource estimations are misleading for various reasons: new and conflicting data constantly reveal higher potential resources; there is seldom any estimation of the proportion of these resources which may be economically and technically recoverable at different price levels; and there is rarely any estimate of whether such resources could be considered commercially viable when delivered to markets via transportation systems requiring investments of several billions of dollars.

With those caveats, Table 6.1 provides an estimate of the initial potential gas resource base for the Siberian and Far East Federal Districts.[21] Clearly the potential is huge: only 122 Bcm been produced in these regions out of proven reserves of nearly 5 Tcm (and possible total resources of more than 57 Tcm); and still less than 10 percent of the regions has been explored for hydrocarbons.[22] Table 6.2 provides estimates of East and West Siberian reserves from specific fields which are most likely to be delivered to these regions or exported to Asian gas markets. We shall see later that there is active consideration of delivering West Siberian gas from the existing producing fields (or possibly the Yamal Peninsula) to domestic and export markets. Also, in the following section the Sakhalin resource base and export projects are covered in detail. From this it is clear that the confirmed reserve base, although only a small fraction of the potential resource base, is still very substantial. Table 6.2 further refines the reserve base in terms of the major fields identified in all four regions.

In 2004, the Siberian and Far Eastern Federal Districts combined produced 16.2 percent of Russian GDP, the Siberian and the Far Eastern districts accounting for 11.5 percent and 4.7 percent respectively.[23] The 2003 Russian Energy Strategy placed significant emphasis on the development of Far Eastern gas resources, advancing the possibility that production could expand up to 106 Bcm by 2020 (Table 6.3) and projecting that in the same year the region would account for 15 percent of total Russian gas exports.[24] Yet, such is the scale of the resource base identified above (Tables 6.1 and 6.2), that even these very large increases appear well capable of being met without placing restrictions on any of the proposed projects, domestic or export.

The Sakhalin Projects[25]

As noted above, during the Soviet era interest in gas exports to Asia centred largely on the oil and gas fields offshore Sakhalin Island. In the immediate post-Soviet period, there was renewed interest from

Table 6.1: Natural Gas Reserves and Resources in Siberian and the Far Eastern Federal Districts (at January 1, 2005). Billion cubic metres

	Initial Resources	*Production and Losses*	A+B+C1	*Reserves* C2	C3+D	*Extent of Exploration %*
Onshore						
Siberian FD:	37,873	35.3	2597	5665	31,676	6.9
Tyumen	974	20.6	283	28	644	31.1
Novosibirsk	58.6	0	1	0	58	1
Evenkiya	9202	0	283	782	8137	3.1
Krasnoyarsk	5421	0	91	224	5105	1.7
Irkutsk	10,365	0.8	12569	2456	6338	15.1
Ust Ordinsk	662	0	0	0	662	0
Omsk	1	0.05	1	0	0	100
Taimyr	11,191	13.8	371	75	10,732	3.4
Far East FD:	13,795.1	86.7	1354	1133	11,985	9.9
Yakutia	12,792	39.8	0	1103	10 365	10.3
Magadan	4	0	7	0	4	0
Chukotka	351	0	16	3	432	1.9
Kamchatka	485	0		7	462	3.3
Koryakski	354	0	0	0	354	0
Sakhalin	379	46.9	47	128	267	24.7
Amur	82	0	0	0	82	0
Khabarovsk	91	0	0	2	89	0.40
Primorsk	14	0	0	0	13	0
Jewish AR	8	0	0	0	8	0
Total	52,430	122	3950	4698	43,661	7.8
Offshore						
Okhotsk	6225	0.4	872	320	5033	14.0
Japan	333	0	4	8	328	1.1
Laptev	2240	0	0	0	2240	0
East Siberian	3346	0	0	0	3346	0
Chukchi	2020	0	0	0	2020	0
Bering	715	0	0	0	715	0
Pacific Ocean	76	0	0	0	76	0
Total Seas:	14,955	0.4	876	321	13,758	5.9
Total On- and Offshore	67,385	122.4	4825	5019	57,418	8.3

Source: Program for the Unified System of Gas Production, Transportation and Supplies in East Siberia and the Far East Considering the Potential Opportunities for Exports to the Markets of China and other APR Countries, Gazprom: 2007, Draft; cited in *A Long Term Vision of Natural Gas Infrastructure in North East Asia – 2007 Version*, North East Asian Gas and Pipeline Forum, September 2007, Table 2.1.2, pp. 8–9

Table 6.2: Reserves in Eastern and Western Siberian Gas Fields Available for North East Asian Markets

Region	*Field (Owner)*	*Reserves (Gas in Bcm, Oil in Mt)*
Irkutsk	Kovyktinksoye (Rusia Petroleum)	1,932 Bcm + 90 Mt condensate+ 2.3 Bcm helium
	Verkhnechonskoye (Rusia Petroleum)	280 Mt + 96 Bcm
Republic of Sakha*	*Chayandinskoye**** (Sakha Republic Gov)	1240 Bcm + 50 Mt
	Sredne-Botuobinskoye (Sakhaneftegas)	171 Bcm
	Taas-Yuriakskoye (Sakhaneftegas)	114 Bcm
	Talakanskoye (Surgutneftegas)	124 Mt + 50 Bcm
*Krasnoyarsk***	Yurubchonskoye (Yukos)	282 Mt + 374 Bcm + 29 Mt condensate
	Kuyumbinskoye (Slavneft)	154 Mt
	Sobinskoye (Gazprom)	159 Bcm
West Siberia	Palkliahinskoye, Bolshehetskaya (Gazprom)	3,021 Bcm, of which C1 751 Bcm, C2 596 Bcm, and C3 1,203 Bcm

* As of 2002, the estimated recoverable oil and gas resources in Sakha Republic (Yakutia) were 2.39 billion tonnes of oil, 9,420 Bcm of gas, and 409 million tonnes of condensate respectively.

** The geological oil and gas resources of Yurubchen-Tokhomskaya area, composed of Yurubchen, Kuyumbinskoye, and Tersko-Kamovskoye fields in 2002, were estimated at 1.2 billion tonnes and 1,000–1,200 Bcm respectively.

*** licence to be auctioned.

Source: Keun-Wook Paik, *Pipeline Gas Introduction to the Korean Peninsula*, Project on Energy and Environmental Cooperation on the Korean Peninsula, Chatham House, September 2004, *Interfax Russian and CIS Oil and Gas Weekly*: May17–23, 2007, p.14

international companies in the region because, unlike many other potential development opportunities within Russia, long distance transit through Russia and other CIS countries was not necessary for the export of these resources. There are a number of different Sakhalin

Table 6.3: Projected Gas Production in Eastern Siberia and the Far East. Bcm

	2005	*2010*	*2015*	*2020*
Central	8	31	86	95
Optimistic	8	52	97	106

Source: *Russian Energy Strategy*, 2003, Figure 11, p.72

projects, known by their numbers 1–6 (see Table 6.4 and Map 6.1), but the most advanced are Sakhalin-1 and 2 which are the largest foreign investment projects under way in Russia.[26]

In analysing the prospects for Sakhalin, it is useful to distinguish three generations of projects, each with different time lines and prospects in terms of exports of natural gas. The 'first generation' projects are those based on the initial exploration activity funded by the Japanese Government in the 1970s and continued in the 1980s by Sakhalinmorneftegaz (SMNG), now a subsidiary of Rosneft. Development rights to these fields were awarded in 1991 and were subsequently the subject of production sharing agreements (PSAs) in 1994–95 and have become the basis of the Sakhalin-1 and Sakhalin-2 projects (current project membership is shown in Table 6.4).

The 'second generation' projects relate to the blocks tendered in 1993 that are collectively known as Sakhalin-3. Three blocks – Kirinskii, Ayakhskiye and East Odoptinskiye – were awarded to foreign companies, Exxon, Mobil and Texaco. Unlike the first generation projects, which were based on already known reserves, these projects were first and foremost exploration projects. Initial efforts focused on the Kirinskii block and, following various mergers, the 'Pegastar' consortium was created involving ExxonMobil and ChevronTexaco. Rosneft and SMNG were also brought into the consortium to provide Russian content and some initial exploration work was done. ExxonMobil said that a sum of $60 million had been spent. However, the foreign investors decided to wait for the project to be awarded a PSA before committing substantial investment. The PSA never materialised and in January 2005 the Russian Government annulled the licences of the Sakhalin-3 blocks involving foreign companies. New tenders for Sakhalin-3 have been held up by the slow progress of new legislation on the Sub-soil law and on foreign investment in Russia's strategic sectors. Thus, this second generation is essentially a 'lost generation' that should have followed on behind Sakhalin-1 and -2, providing the conveyor belt of projects needed to create an offshore oil and gas province.

The 'third generation' of projects represents the new generation.[27]

Table 6.4: The Current Status of the Sakhalin Projects (as of 1 May 2007)

Project	*Fields/Blocks*	*Participants*	*Recoverable Reserves*	*Project Status*
Sakhalin-1	Chayvo, Odoptu, Arktun-Dagi (oil & gas)	Operator: Exxon Neftegaz Ltd (US) 30% SODECO Ltd. (Japan) 30% ONGC Videsh Ltd. (India) 20% SMNG-Shelf (Russia) 11.5% Rosneft-Astra (Russia) 8.5%	307 mln t of oil 485 Bcm of gas	PSA Phase 1: early oil. First oil in October 2005. First exports October 2006
Sakhalin-2	Piltun-Astokhskoye (oil), Lunskoye (gas)	Operator: Sakhalin Energy Investment Co. Gazprom (Russia) 50% + 1 share Royal Dutch Shell (Netherlands/UK) 27.5% Mitsui (Japan) 25% 12.5% Mitsubishi (Japan) 10%	600 mln t of oil 700 Bcm of gas	PSA Phase 1: early oil in 1999 Phase 2: Gas/LNG in 2007-08
Sakhalin-3	Kirinsky	Operator: Pegastar ExxonMobil (US) 33.3% ChevronTexaco (US) 33.3% Rosneft (Russia) 33.3%	453 mln t of oil 700 Bcm of gas	Exploration project suspended due to loss of exploration rights
	Vostochno Odoptinski, Ayashski,	ExxonMobil (US) 66.6% Rosneft (Russia) 33.3%	167 mln t of oil 67 Bcm of gas	Project idle and now suspended due to loss of exploration rights
	Veninski	Operator: Venin Holding Ltd. Rosneft (Russia) 74.9% Sinopec (China) 25.1%	51 mln t of oil 578 Bcm of gas	First drilling season summer 2006

Table 6.4: *Continued*

Project	*Fields/Blocks*	*Participants*	*Recoverable Reserves*	*Project Status*
Sakhalin-4	Astrakhanovskii offshore structure	Rosneft (Russia) 51% BP (UK) 49%	89 Bcm of gas	Project stopped by Rosneft
Sakhalin-5	Vostochno-Shmidtovski and Kaigan/Vasyukan (and Zapadno-Shmidtovskii) blocks	Rosneft (Russia) 51% BP (UK) 49% Joint Venture Company Elvary Neftegas being created to develop the project	600 mln t of oil 600 Bcm of gas	Exploration First well drilled in summer 2004, second in 2005 and two planned for 2006.
Sakhalin-5	Lopukhovski	Gazpromneft (Russia) 100%	130 mln t of oil 5 Bcm of gas	Sibneft acquired block from TNK-BP. Gazpromneft was created in late 2005 when Gazprom purchased Sibneft
Sakhalin-6	Progranichnii	Urals Energy	200 mln t of oil	Acquired when Urals Energy purchased Alfa Echo.

Source: Table based on oil and gas industry press and oil company websites and press releases.

Map 6.1: Sakhalin Gas Fields, Pipelines and the Sakhalin 2 LNG Project

These are exploration projects being developed on the basis of normal tax and royalty, rather then a PSA. Initially, the Sakhalin-4 block, involving ARCO and Rosneft and SMNG was to be developed under a PSA; however, indifferent drilling results and no progress on the PSA resulted in ARCO losing interest in the block. When BP acquired ARCO in 2000, Sakhalin-4 became part of BP's long-harboured interest in Sakhalin. BP had long harbored an interest in Sakhalin; on two occasions BP had looked at joining the first generation projects, when McDermott left Sakhalin-2 in 1997 and when Rosneft and SMNG halved their share in Sakhalin-1 in 2001. In the case of Sakhalin-2 the other shareholders exercised their right to absorb McDermott's share and in the case of Sakhalin-1 the price for 20 percent of the project was considered too high by BP but the Indian State Oil Company ONGC acquired the 20 percent interest.

Instead, in 1989, BP signed an agreement with Rosneft to explore the Sakhalin-5 block. This alliance was later expanded to include Sakhalin-4 and most recently has become part of an alliance to develop projects in the Arctic and offshore Russia-wide. The Sakhalin-4-5 project was the first involving a foreign company to proceed without a PSA. Rosneft obtained the exploration licence for the Kaigansky–Vasukansky blocks of Sakhalin-5 in 2002 and in 2003 Elvary Neftgaz (a 51 percent Rosneft, 49 percent BP joint venture) was created. The drilling programme began in 2004 and is on-going.[28] The Sakhalin-5 project is further out, in deep water and in colder conditions than the other projects and its development will present many additional technical challenges. In discussion around the 11th Sakhalin Oil and Gas Conference, held on the island in early September 2007, it became clear that if the summer 2007 drilling season did not yield positive results the future of the project could be in doubt. Furthermore, both BP and Rosneft are clear that substantial offshore development would not be viable under the current tax regime and that a special regime is required to promote high-cost sub-arctic offshore exploration and development.

Two other projects can be considered part of this new generation involved in exploration activity on the continental shelf: the Sakhalin-6 project focused on the Progranichnie Block, which is now owned by the Russian oil company Urals Energy, which acquired 97.16 percent ownership from the Alfa Group in 2004; the Sakhalin Oblast Administration owns the remaining 2.84 percent.[29] Exploration of the block began in 2004 and in 2005 Urals Energy was granted a 5-year extension to its licence and plans to drill a minimum of three test wells. The final project that can be considered part of the new generation is Sakhalin-3 Veninskii block.[30] In 2003 Rosneft obtained an exploration licence for the Veninskii block and in 2005 entered into a Memorandum of Understanding with the Chinese State Oil Company Sinopec under which Sinopec would finance 75 percent of the exploration costs and Rosneft 25 percent. In autumn 2006 Venin Holding Ltd. was established as the sole shareholder of Venineft, the licence owner and operator of the Sakhalin-3 project in which Rosneft has a 74.9 percent ownership and Sinopec 25.1 percent.[31] The first exploratory well was drilled in the summer of 2006 and confirmed the presence of hydrocarbons. Further seismic surveys and drilling operations are planned. All three of the new generation of exploration projects are a long way from commerciality and all seem to be hoping to find oil, rather than gas. However, over the past decade, it is the development activities of the first generation projects that have been the focus of attention and that will establish Sakhalin as an exporter of natural gas to Asia.

Sakhalin-2. Sakhalin-2, which has been producing oil since 1999, is the most advanced project in terms of gas development and planning. The first phase of the project involved the installation of the Molikpaq production platform, but ice conditions restricted production to between 4 and 6 months. By late 2006, cumulative oil production from the Piltun–Astokhskoye field had reached 12 million tons and, while no gas had been produced, Phase 1 has established Sakhalin-2 as an exporter of hydrocarbons to the Asia–Pacific market.[32] In June 2001, Sakhalin-2's Supervisory Board approved the plan for development for Phase 2 of the project involving the construction of the infrastructure to enable year-round production and export of oil and gas as LNG. Described by Sakhalin Energy as the 'world's largest integrated oil and gas project', Phase 2 involves several elements:

- a second production platform on the Piltun–Astokhskoye field, in addition to tying the Molikpaq into the new pipeline system;
- a new platform on the Lunskoye field, capable of producing 17 Bcm/year of non-associated gas;
- an onshore processing facility to take oil and gas from both fields;
- two 800 km onshore oil and gas pipelines to the south of the island;
- the construction of Russia's fist LNG plant with a capacity of 9.6 Mtpa in two trains, and associate export facilities for LNG and oil;
- and substantial infrastructure upgrades to roads, bridges, rail, ports, airports and medical facilities as demanded by the project.[33]

At the time of commencement the cost of Phase 2 was put at $9.6 billion.

As part of Phase 1, Sakhalin Energy had obtained project financing from a group of international financial institutions (IFIs) led by the European Bank for Reconstruction and Development. Phase 1 had attracted some critical attention from the international environmental movement (hereafter ENGOs), but a modest loan package was agreed. Phase 2 has turned out to be a very different matter, the scale of the project has meant that it has become the target of a concerted global ENGO campaign to persuade the IFIs not to provide $4–5 billion in project finance.[34] The campaign focused on the impact of the construction of the offshore structures on the endangered western grey whale, the impact of the construction of the onshore oil and gas pipelines on salmon spawning rivers and the impact of the project on the island's indigenous peoples. Sakhalin Energy responded by re-routing the offshore pipelines and developed a Sakhalin Indigenous Minorities Development Plan. All of this added time and cost, but failed to placate the ENGOs. Soon the cost of the project was being described as 'more than $10 billion'.[35]

In addition to the external pressure applied by the ENGOs, Sakhalin Energy was also under increasing pressure to allow a Russian partner into the project. It was the only project without Russian equity participation. On 8 July 2005 Shell announced that it had agreed an asset swap with Gazprom to exchange a 25 percent share of the Sakhalin-2 project for a 50 percent share of the Neocomian formation of the Zapolyarnoye field in West Siberia. However, no sooner had this been announced than Shell was forced into having to disclose 'staggering' cost overruns and delays in the delivery of Phase 2.[36] In an official press release on 14 July 2005, Sakhalin Energy announced that Phase 2 costs were anticipated to be in the order of $20 billion and that the first LNG deliveries were to be postponed from the end of 2007 until the summer of 2008.[37] Malcolm Brinded, head of Shell's exploration and production division said that the cost increases were due 'to factors such as currency movements, rises in steel prices, problems with getting its onshore pipelines across rivers and the right environmental licences', he further stated that 'it is now clear that the Sakhalin project's budget and schedule were significantly underestimated'.[38] Not surprisingly, negotiations on the asset swap with Gazprom stalled.

In April 2003 the EBRD had declared the Environmental and Social Impact Assessment (ESIA) for Phase 2 'unfit for purpose'; subsequently the EBRD and its consultants worked closely with Sakhalin Energy to address areas of concern and in November 2005 declared the addendums to the ESIA as 'fit for the purpose of public consultation'. In early 2006 the EBRD held a 120-day public consultation with hearings in London, Moscow, Sakhalin and Hokkaido in Japan.[39] The EBRD promised a decision in the early summer of 2006, but events in Russia soon complicated matters. Growing Kremlin criticism of the financial terms of the Sakhalin PSAs and the increasingly high profile ENGO campaign against the project, which was now firmly focused on problems along the onshore pipeline corridor, gave regulators at the Ministry of Natural Resources the opportunity to put pressure on the project.[40] The campaign escalated over the summer and culminated in a statement by MNR Minister Yuri Trutnev threatening to revoke the State Environmental Expert Review, an action that would have brought substantial delays and cost billions. The Western media maintained that this 'administrative pressure' was being used to force Sakhalin Energy into accepting Gazprom into the project on favourable terms. The Kremlin and Gazprom denied this.[41] Nonetheless, the pressure continued to mount with prosecutions and fines being threatened.

On 21 December 2007 Shell, Mitsui and Mitsubishi announced that they had signed a protocol to bring Gazprom into Sakhalin Energy as the leading shareholder. Under the terms of the protocol, Gazprom would acquire a 50 percent plus one share stake in Sakhalin Energy for a total cash purchase price of $7.45 billion. To accommodate this, the other shareholders reduced their holdings by 50 percent: Shell retaining 27.5 percent, Mitsui 12.5 percent and Mitsubishi 10 percent. By the end of the third quarter of 2006, Phase 2 was 80 percent complete with investment to date of over $12 billion. It is reported that the foreign partners will absorb some $3.6 billion of the cost over-run and even pay the Russian government an annual royalty to stay in the project. On 18 April 2007 Gazprom formally took control of the Sakhalin-2 project. Gazprom's Deputy Chairman Alexander Medvedev said that 'joining the Sakhalin-2 project provides a powerful impetus for accomplishing a large-scale project in the energy supply sector to Asia Pacific Countries and North America. It will stimulate a stage-by-stage entry strategy on the world LNG market'.[42] The new shareholders have agreed a budget of $19.4 billion for Phase 2 and a plan to deal with the environmental problems has been agreed with the state regulators.[43] The shareholders are now considering building a third train at the LNG plant.

The Sakhalin-2 marketing strategy has always been to sell the LNG to Asia–Pacific markets (Japan, Korea, Taiwan and China) and initially Sakhalin Energy aimed to conclude sales contracts by July 2002 with the first LNG being produced in November 2006.[44] However, the delays in delivering Phase 2 inevitably impacted on the marketing process. Nevertheless, at the Sakhalin Oil and Gas Conference, held on the island in September 2006, Ate Visser, the Commercial Director for Sakhalin Energy, announced that the Sakhalin-2 project had sold all of its LNG.[45] At the conference he said that Sakhalin Energy had received more than 10 'expressions of interest', with opportunities in Japan, Korea, the USA, Taiwan, China and India. In the end, 63 percent (through 9 separate contracts) of the 9.6 Mtpa capacity will go to Japan, 16 percent to Korea, 19 percent to the US and Mexico and 2 percent will be retained for operational flexibility (see Table 6.5). The Sakhalin-2 project will then account for about 8 percent of the

Table 6.5: Sakhalin 2: Contractual Commitments for LNG Exports (June 2007)

Buyer	*Agreement*	*Term (years)*	*Quantity (Mtpa)*
Tokyo Gas	SPA/FOB	24 (2008~)	1.6
Tokyo Electric	SPA/FOB	22 (2008~)	2.0
Kyushu Electric	SPA/DES	22 (2009~)	0.5
Toho Gas	SPA/DES	24 (2009~)	0.6
Baja Mexico(SETL)	SPA/DES	20 (2008~)	1.8
Korea Gas Corp	SPA/FOB	21 (2008~)	1.5
Hiroshima/Saibu Gas	SPA/FOB	20 (2008~)	0.2
Tohoku Electric	SPA/FOB	20 (2010~)	0.4
Chubu Electric	HOA/DES	20 (2011~)	0.5
Osaka Gas	HOA/FOB	20 (2011~)	0.2
Boil-off gas for DES delivery			0.1
Total			9.4

* Contract Year of all the agreements is from 1 April to 31 March of the following year.

Source: Sakhalin Energy

Asia–Pacific LNG market. He also suggested that, by de-bottlenecking trains 1 and 2 and building a third train, capacity could be increased to around 16 Mtpa. However, at present, Sakhalin Energy does not have access to sufficient gas to enable such expansion to take place.

Sakhalin-1. During the second half of the 1990s licensing problems and the complex geology of the Chaivo field meant that the Sakhalin-1 project slipped well behind Sakhalin-2 in terms of time lines. This undermined the idea that the projects would share a common infrastructure. In October 2001, the Sakhalin-1 partners finally declared commerciality of Phase 1 of the project involving the start of oil production in 2005 with 'limited gas supplies available to help meet Russian domestic demand'.[46] The first oil deadline was met and in the autumn of 2006, despite problems with the Ministry of Natural Resources (MNR), the oil export terminal at De Kastri on the Russian mainland commenced year-round oil exports.[47] The project has made considerable use of Extended Reach Drilling (EDR) technologies from the world's largest land rig, the Yastreb, thereby reducing the need to build offshore structures. By February 2007, the project had supplied 35 billion cubic feet (1 billion cubic metres) of gas to customers in Khabarovsk Krai. As far as gas exports are concerned, the press release that announced commerciality phrased the position thus:

> Gas market development and sales remain a high priority for the project. An assessment of gas export markets in Japan and China has been completed and the Consortium has concluded that a pipeline, built to international design standards, would be the most cost-effective method to deliver gas to these export markets. Discussions with potential export customers are under way, along with a framework for developing commercial gas sales to Russian domestic purchasers. [48]

Thus Sakhalin-1 had a completely different marketing strategy compared with Sakhalin-2 focusing on pipeline gas. A feasibility study of pipeline exports to Japan was completed in 2004. According to a presentation by Jack King, then Vice President of Exxon Neftegas Project Services Inc., at the Sakhalin Oil and Gas Conference in London in November 2004, the results of the study demonstrated that building a pipeline to Japan is 'technically and economically viable'.[49] Since the completion of the feasibility study there has been no progress on the Japan pipeline option. At subsequent Sakhalin Oil and Gas conferences in 2005 and 2006 Stephen Terni, President of Exxon Neftgas Limited, refused to be drawn on Sakhalin-1's gas export strategy, he maintained that all options remained open. By this he meant pipelines to Japan or China, or even LNG. Now that Phase 1 is complete, Sakhalin-1 is under pressure to commit to a Phase-2 that would see substantial natural gas exports. In October 2006 ExxonMobil announced that it had entered into a preliminary agreement with the China National Petroleum Corporation (CNPC) to sell all of the natural gas production from Sakhalin-1 to China via pipeline. [50] At the same time, ExxonMobil said that it was also in talks with Gazprom 'regarding potential options for delivering pipeline gas to Northeast China'.[51]

As discussed below, the 2002 Presidential Decree gave Gazprom control over Russian gas exports to Asia and although the Russian Government has affirmed Gazprom's monopoly over gas export pipelines, from a legal perspective the terms of the Sakhalin-1 PSA gives it the right to build an export pipeline. The reality is that any pipeline to China would require Gazprom's approval, if not participation. It is also the case that both the Japanese and Indian shareholders in Sakhalin-1 would like the gas to be made available as LNG to supply their home markets. Speaking at the CERA Conference in Houston, in 2007, ExxonMobil CEO Rex Tillerson said that 'we do continue to look at all the other options on gas sales, because we have not concluded anything...with the Chinese. There's still great interest from the Japanese to purchase the gas. There's great interest from the Indians to purchase the gas. And we continue to have discussions with all those potential customers'.[52]

Following its acquisition of a controlling stake in Sakhalin-2, Gazprom has stated that it is interested in purchasing all of Sakhalin-1 natural gas production. Speaking to the Japanese media in late April 2007, Alexander Medvedev said that 'we are in consultations with Sakhalin-1 operators to purchase the total gas output. We are interested in enhancing the project's economic effectiveness'.[53] At the time of writing, the future of Sakhalin-1's gas remains unresolved. At the 11th Sakhalin Oil and Gas Conference the local Gazprom representative, Vladimir Kozlov, stated that Gazprom's view is that Sakhalin-1 is the only source of gas for the Russian Far East before 2016. Thus, it seems likely that one way or another Gazprom will gain control over the project's gas. In the past Gazprom has said that it would be interested in acquiring Rosneft's 20 percent share of the project. Rosneft has shown no signs of giving up participation in Sakhalin-1, but also shows no appetite to gain a larger share of the project. The future of the project seems to rest with ExxonMobil. In October 2006 the Russian Government approved the increase in the Sakhalin-1 budget by 33 percent to $17 billion, this includes all costs associated with Phase 1. The investment required for Phase 2 is over $10 billion, which represents a major commitment by the Sakhalin-1 shareholders. ExxonMobil has stated that it would like to see more clarity as to how Russia plans to treat foreign investors before it considers any new projects there. What is not clear is whether or not it considers Phase 2 of Sakhalin-1 a new project.

Clearly, the current uncertainty over gas marketing needs to be resolved before a commitment can be made to Phase 2. Two options appear to be on the table: first, a pipeline to China and, second, selling gas to Sakhalin-2 for export as LNG. There is insufficient gas to do both and it makes little sense to build an additional LNG plant at this stage. The most likely outcome is that Sakhalin-1 reaches an accommodation with Gazprom and Sakhalin-2, with or without Gazprom equity participation in Sakhalin-1, and that the additional gas permits the expansion of the Sakhalin-2 LNG plant, as well as expanded supplies to the mainland of the Russian Far East and to the southern regions of Sakhalin. This option makes LNG available to the Japanese and Indian shareholders and may also see exports of LNG to China. In order to work, the arrangement between Gazprom and ExxonMobil has to be commercially viable and ExxonMobil, as Sakhalin-1 project operator, has to be confident about the environment for foreign investment in Russia. Thus, a great deal of conditionality still hangs over Sakhalin-1's gas exports.

In the mid-1990s when the first generation Sakhalin projects were being developed, it was envisaged that they would share a joint infrastructure. As the time lines of the two projects slipped, Sakhalin-2 was forced into building Phase 2 alone. Then Sakhalin-1 went its own way with an oil export terminal on the mainland and a pipeline-based gas export strategy. There is a certain amount of irony that it is only with the arrival of Gazprom that shared gas export infrastructure now appears a real possibility. Gazprom has other means of delivering pipeline gas to China (see below), but only Sakhalin provides the immediate opportunity to sell LNG to the Pacific Basin. Gazprom clearly sees this as part of a global strategy of a stage-by-stage entry into the global LNG market. While there is still some uncertainty over Sakhalin-1's gas phase, the future of the first generation projects seems set. What remains unclear is what will happen next. The Sakhalin-3 projects should have been developing behind Sakhalin-1 and Sakhalin-2, but they were stopped in their tracks in 2005. The future now lies in the new generation of exploration projects, but they are at least a decade away from development. Thus, at present, Sakhalin's ability to export gas to Asia is limited to Sakhalin-1 and Sakhalin-2, most likely in the form of LNG.

The Kovyktinskoye (Kovykta) Project

The Kovyktinskoye (Kovykta) field, located 430km north of Irkutsk, was discovered around 1986–87 but little publicity was accorded to the find probably because of its remoteness from domestic or international markets.[54] In 1992, the licence to develop Kovykta (at that time with proven and probable reserves of 870 Bcm, and the much smaller Verkhnechonskoye field) was awarded to Baikalekogaz but within a year transferred to Rusia Petroleum which was then simply a consortium of local and regional Irkutsk government and state-owned company interests.[55] A lack of both tax exemption for the project and commitment to export markets led to the withdrawal of BP and Statoil which had developed some new geological data. Shortly thereafter, a government initiative saw the founding of Sidanko as a regional company controlling Rusia Petroleum. In 1994–95, Rusia became a subsidiary of Sidanco which thereby became the main equity holder in the field.

From the mid-1990s, the equity ownership of Rusia Petroleum changed considerably (see Table 6.6). The Korean East Asia Gas Company bought a 27.5 percent stake and, for a while, it seemed that the company would become the largest shareholder. But in late 1997, British Petroleum bought a 10 percent shareholding in Sidanco – specifically targeted at participation in Rusia Petroleum and oil and gas exports to China. A combination of Asian (1997) and Russian (1998) economic

Table 6.6: Rusia Petroleum Shareholdings 1997–2007. Percent

	Up to 11/97	*Post 11/97*	*2001*	*31/12/03-1/12/07*	*1/12/07*
Sidanco	37.82	57.82			
Angarsk Petrochemical Co					
BP (post-2001 TNK-BP)		26.02	30.84	62.89*	
East Asia Gas Co	27.5	7.5			
Irkutsk Oblast Administration	18.86	18.86		10.78	10.78
Irkutskenergo	12.75	12.75			
Others	3.07	3.07	4.44		
Interros			23.29	25.82	25.82
Irkutsk Property Fund			13.97		
Vitra Holdings (Kantupan)			13.0		
Naftrad (Rinko)			8.46		
Tyumen Oil Co.			6.0		
Minority Shareholders				0.51	0.51
Gazprom					62.89*

* Agreement was reached in June 2006 for the sale of TNK-BP shareholding to Gazprom; this was intended to be completed by October 2007 but at the time of writing had been delayed to December 2007

Sources: Wybrew-Bond and Stern, Table 6.8, p.248. http://www.rusiap.ru/invest/

crises wrecked the finances of the Russian and Korean shareholders and set back prospects for development. But in December 1997, 5-party talks – the Russian Ministry of Fuel and Energy/Sidanco, China National Petroleum Corporation (CNPC), Korean gas Company (Kogas), Japan National Oil Corporation (JNOC) and Mongolian Petroleum Administration – led to the launch of a feasibility study. However, perhaps the most significant development was that the meeting firmly established China as the principal market for Kovykta gas.[56]

The bankruptcy of Sidanco and the conflict between BP and the Tyumen Oil Company (TNK) involving major assets of that company, were finally settled leading to the position that by 2001, the remaining major equity holders in Rusia were BP, Interros and Kantupan, with others (including TNK) holding minority shares.[57] Up to this point, Gazprom had shown no interest in Kovykta and the company's post-Soviet leadership under Chairman Vyakhirev questioned whether reserves were sufficient for an export project and subsequently offered gas to China from the Yamal Peninsula in Western Siberia. In 2001, the corporate landscape changed with the arrival of a new Gazprom management under Alexei Miller who, as Deputy Minister of Energy, had already expressed the view that Gazprom should be involved in Kovykta. Following his appointment as CEO of the company, Gazprom became fully involved in the project. During the late 1990s, a three-stage development concept emerged, the first being regional gas development around Irkutsk, the second stage encompassing wider East Siberian development and only the third stage envisaging exports.

A trilateral agreement to conduct a feasibility study was signed in late 2000 by representatives from Rusia, Kogas and CNPC. Until 2000, three main pipeline routes were considered: a route through Mongolia, and two eastern routes[58] – one which followed the Russian–Mongolian border through Buryatia and Chita into north eastern China and then through North Korea to South Korea and another which traveled offshore through the Yellow Sea to bypass North Korea.

According to the feasibility study which was completed in November 2003, starting in 2008, the project would deliver 20 Bcm/year to China (building up to 30 Bcm in 2017) as well as 10 Bcm/year to Korea; 4 Bcm/year would also be delivered to local markets in Irkutsk, Buryatia and Chita.[59] The 2003 feasibility study estimated the project cost at $17bn. The selection of routes for the Kovykta project was strongly influenced by the extremely negative stance taken by the Chinese authorities to the Mongolian route. The Russian parties were well disposed towards the Mongolian route, which was around 1,000km shorter with a capital cost saving of around $1bn and provided opportunities for sales to Mongolia.[60] As recently as late 2004, the Governor of Irkutsk invited the Mongolians to participate in Kovykta, but the Chinese had decided to bypass that country and it proved impossible for other parties to oppose their wishes.[61]

Along with difficulties in the export phase, the domestic aspects of the field's development also began to receive greater scrutiny as the decade progressed and the future of the project therefore began to look increasingly precarious. There was a crescendo of announcements by the Federal Authorities about the potential violation of the licence condition requiring 9 Bcm of production by 2006 and a lack of any progress on the part of the Kovykta partners to bring Gazprom into the development as a partner

The original development plan for the project foresaw production of up to 5.5 Bcm by 2000.[62] In 1997, after the arrival of both TNK and BP (but before the merger of the Russian interests of those companies) Clause 6.11 of the licence was altered requiring the subsoil users to produce 9 Bcm of gas by 2006 in order to cover the needs of the Irkutsk region. The 9 Bcm figure resulted from a 1996 regional development forecast which was based on the idea that Irkutsk coal would be used to resolve the energy crisis in regions further east and gas would replace coal for regional energy needs. In the event, neither of these developments occurred. The production requirements for the local market would probably have been amended by the production sharing agreement (PSA) which had been envisaged for the project. But in the early 2000s, PSAs were rejected by the Russian government as unnecessary, and the field was developed under the original licence arrangements with a tax and royalty regime.

The first suggestion of problems with the licence came in early 2003 when the General Director of Rusia Petroleum (a former Deputy Minister of Natural Resources) publicly identified the issue of insufficient market but noted that there were no current problems. At that time it was still possible that the problem could have been remedied by a PSA.[63] After the Russian government's rejection of PSAs, the licencees were forced to continue under the original framework. The 2003 feasibility study envisaged sales of up to 4 Bcm in the Irkutsk, Chita and Buryatia regions, an estimate with which there was no public disagreement at that time.[64] In mid-2004, TNK-BP and the Irkutsk Regional Administration established the East Siberian Gas Company to deliver gas to local markets.[65] By the mid-2000s, there was general agreement that the regional market for gas would be around 2.5 Bcm in 2010 and 4 Bcm in 2015; thus the requirement to produce 9 Bcm for the local market by 2006 had become completely unrealistic. It is possible that this requirement could have been finessed had exports commenced, or at least been in sight by 2007, but this was not the case.[66]

Gazprom's participation had also been a major issue. In 2001, before their merger, BP announced that it was ready to offer Gazprom participation in the venture and the TNK Chief Executive stated that 'in order for the project to survive, Gazprom has to join'.[67] This may have reflected a judgment that the project needed a heavyweight Russian partner to proceed further, Gazprom having become the preferred partner of both the Russian government and the Chinese authorities, and Kovykta having fallen behind the West–East Pipeline (see Chapter 2) in the priorities of Chinese authorities. But, despite regular meetings between the two sides, there was no sign that Gazprom would be willing to take an equity position – even a controlling position – in Kovykta, and, as the 2000s unfolded, speculation increased that the company would simply wait for the withdrawal of the licence, to be followed by an auction which Gazprom would win.

In mid-2007 (as a decision on the Kovykta licence withdrawal seemed imminent) TNK-BP agreed to sell its stake in Rusia Petroleum and the

East Siberian Gas (distribution) Company to Gazprom for $700–900m, reflecting the investments that had been made by the company thus far.[68] This was announced in the context of a new strategic alliance between Gazprom, BP and TNK-BP with the latter allowed to exercise a call option to buy 25 percent plus one share of the Kovykta project following a significant joint investment by the companies or agreement on an asset swap.[69] While this was greeted in most western press and petroleum circles as a continuation (following the Sakhalin-2 events six months previously) of Kremlin expropriation of international energy companies, calmer voices noted that Gazprom had agreed to pay nearly double what TNK-BP had spent in developing the field, and that the strategic alliance provided BP with a more promising future in Russia than the outcome might have suggested.[70] Few commentators recognised the crucial distinction between BP and TNK-BP in the announcement, amid swirling rumours of strengthening pressure on the TNK shareholders in the joint venture to sell their shares to a Kremlin-approved company.

It was never a secret that, although the licence required the Kovykta partners to create local markets for gas, their principal aim had always been exports to China and Korea. The 2003 Energy Strategy reinforced the decree of the previous year (see below) establishing Gazprom's principle of the 'single export channel' (i.e., Gazprom's export monopoly) in the east of the country.[71] In 2004 it was formally agreed between TNK-BP, Interros and Gazprom that the latter would be in charge of all export negotiations in respect of Kovykta.[72] At this time, Gazprom began to question the validity of both the licence and the feasibility study conclusions, asserting its own authority over all gas resource development in the region and using alleged licence violations as a reason why it could not consider equity participation.[73] Gazprom and government sources also began to use the helium content of the field as a reason for delay.[74]

In March 2006 the announcement of the plan for two Russian gas pipelines to China gave renewed impetus to the project. Despite the fact that the Altai Pipeline (see above) was intended to come first, the second phase would be an 'eastern route'. A year earlier, a Deputy Head of the Ministry of Industry and Energy had suggested four options for export routes to foreign (Chinese, Korean and Japanese) markets: west, central, east and the Rusia Petroleum option based on Kovykta.[75] But, although the sources of gas and pipeline routes for this second phase remained undefined, the Kovykta field was the obvious choice, given its location, established reserves and corporate structure.[76]

After the Gazprom takeover of the field in mid-2007, sporadic announcements from both government ministries and Gazprom during the 2000s that the best use of Kovykta gas would be to pipe it westwards to join up with the UGSS (the main gas network) and deliver gas to the Russian (and European) market took on a different meaning.[77] On the face of it, such a plan (to pipe Kovykta gas west rather than export it to China) seemed so economically ridiculous that it could only be a negotiating tactic designed to force concessions from the Kovykta partners. To move Kovykta gas west would require a pipeline of at least 2200 km to connect with the UGSS. However, given what was said above about the cost, complexity and potential delays in developing new supergiant fields on the Yamal Peninsula, by 2007 it had become a serious option (although an expensive one) included in the Eastern Gas Programme (see below) that Kovykta could make a contribution to Gazprom's western supply portfolio until Yamal gas becomes available. The case for this was strengthened by the publication of Gazprom data showing 1.6 Tcm of proven and probable reserves around Irkutsk and a total gas resource of 6.2 Tcm.[78]

Sakha Republic (Yakutia)

As noted above, the development of Sakha Republic (also known as Yakutia) gas fields for export dates back to the Soviet period. Table 6.2 shows that the key fields are Chayandinskoye and Talakanskoye where the licences have been due to be auctioned for several years. The tender for the Chayandinskoye field, with more than 1.2 Tcm of reserves, has been pending since at least 2002 and the field was one of the first to be classed as 'strategic' and therefore not available for majority foreign ownership.[79] In late 2007, the timetable for the tender was still not fixed, but it was widely assumed that Gazprom would simply be awarded the licence.[80] Gazprom then suggested that it should be awarded the Chayandinskoye (and Sakhalin-3) licence without a tender and that it would start production from the field in 2014.[81] In response, the Minister of Natural Resources publicly rebuked the company: 'To simply give someone something without a tender, even a respected company like Gazprom, does not appeal to me, like any irregular decision.'[82]

Over the years, a number of possibilities for pipeline and LNG exports of Sakha gas have been discussed including several routes: to Japan either as part of a Sakhalin pipeline project or a free-standing project or as LNG; to DPRK and South Korea via Russia; east and south to China (possibly via Mongolia) with a possible extension to Korea; west and south to link with a pipeline from the Kovytinskoye field.

But by 2007, these options had been replaced by the Eastern Gas Programme (see below).

Regional Benefits: the gasification programme

Despite sitting on top of a huge oil, gas and coal resource base, during the early post-Soviet period the Russian Far East was badly affected by energy supply crises and environmental degradation due to coal-fired power generation.[83] The depth of this crisis during the winter 2000/01 in the Primorye region led to the replacement of both the Russian Energy Minister and the Governor.[84]

The Russian Energy Strategy to 2020, first published in 2001 with a 2nd edition in 2003, stated a number of objectives in terms of energy and natural gas development in Eastern Siberia and the Far East, in particular the development of all the major gas fields in the region. The 2001 version called for the creation of a major oil and gas export zone and the gasification of Sakhalin oblast, Primorsk and Khabarovsk Krai, and Kamchatka Oblast.[85] The 2003 version spoke of '…priority being given to deliveries to Russian customers and the creation of the best possible conditions for socio-economic development of Eastern Siberia and the Far East…increasing the reliability of gas supplies by means of a single system of regional gas supply to the East'.[86]

The Russian Government has become increasingly concerned about the economic situation in the Russian Far East. During the 1990s the region suffered from significant population decline and its economy has not rebounded as fast as the European regions of the country. Thus, gasification is part of a wider concern to maintain enough population and economic activity to insure the effective occupation of the region.[87] But as Table 6.7 shows, by 2004 there was little tangible evidence of gasification in Eastern Siberia and the Far East: gas was used in only 6 out of 15 districts in Siberia and four out of 10 districts in the Far East. The figures are for Federal Districts rather than individual regions but the Energy Strategy was relatively modest in its projections suggesting gas demand in 2020 of 9.2–9.8 Bcm for Eastern Siberia and 4.6–5.1 Bcm for the Far East.[88]

Since 2004, Gazprom has shown significantly greater interest in these regions undertaking two well-publicised week-long visits by a delegation led by an executive board member.[89] It is worth surveying Gazprom's gasification plans for regions in the eastern part of the country in order to get a feel for some specific regional expectations in the context of the 2007 Gazprom plan for an average of 62 percent gasification of Russia in 2007 (67 percent in urban areas and 44 percent in rural areas).[90] Gazprom's budget for regional gasification was substantially increased

Table 6.7: Gas Demand in Eastern Siberia and the Far East, 2004. Bcm

Siberian Federal District	14.84
Altai Krai	0.54
Krasnoyarsk Krai	3.54
Kemerovo Oblast	3.89
Omsk Oblast	1.60
Tomsk Oblast	2.42
Chitinsk Oblast	2.85
Far East Federal District	2.93
Sakha Republic	1.40
Khabarovsk Krai	0.8
Kamchatka Oblast	0.1
Sakhalin Oblast	0.63

Source: Rosstat Data

by 8bn Rubles to 60bn Rubles in 2007, Altai and Irkutsk being cited as two of five regions to benefit specifically from the increase.[91]

The Altai Republic is planned to benefit significantly from an export pipeline to China (see above) and a regional gasification agreement was signed in 2006.[92] Altai Krai is already minimally gasified (5.9 percent in urban and 1.2 percent in rural areas) and in 2006 11 new settlements were added to the network via the Barnaul–Biysk pipeline.[93] The first leg of the Barnaul–Biyisk–Gorno Altaisk pipeline was completed in late 2006 and branch lines will allow gasification of nearly 120,000 apartments in 300 towns and villages; more than 100 million rubles was allocated for this project in Gazprom's 2007 budget.[94]

In Irkutsk, gasification is a big issue as would be expected given the problems of the Kovykta licencees in creating a market (see above). Gazprom made a very public issue over regional gasification and had its own programme even before its takeover of the TNK-BP share in the East Siberian Gas Company, which started construction of a pipeline from Kovykta to Irkutsk in 2006 with the first gas to be delivered in 2008.[95] In addition to Kovykta, gas will be sourced from the much smaller Chikanskoye and Dulisminskoye fields (in cooperation with Urals Energy).[96] Gazprom projects that gasification of Irkutsk Oblast will exceed 82 percent overall, and more than 40 percent in rural areas.[97]

In Krasnoyarsk, the emphasis is on production with gasification getting only a passing reference in Gazprom's agreement with the region.[98] Gazprom intends to use associated gas for gasification particularly in the Evenkiya region where many of the fields are located.[99] In Tomsk

Oblast, 6.3 percent of urban and 3.8 percent of rural population have access to gas with Gazprom supplying 1.3 Bcm in 2006; the distribution network will be expanded in 2007.[100] In the Aginsk–Buryat Autonomous Region an agreement was signed for the first time in April 2007 to develop distribution infrastructure.[101]

In the Sakha Republic, Gazprom has significant aspirations for gas development in a region where gas is already supplied to 18.5 percent of the population but not by Gazprom.[102] As one of the four designated centres of gas activity in Eastern Siberia and the Far East, few definite commitments have been made for expansion of gas development despite the consortium with Rosneft and Surgutneftegaz and substantial numbers of meetings between the company's senior management and key regional politicians in 2006.[103] But in mid-2007, Gazprom signed a major cooperation agreement with the region for all phases of gas development, including exports, although this did not appear to include any definite commitments.[104]

Primorskii Krai currently has no gas supply but this is planned to change when the Sakhalin–Komsomolsk on Amur–Khabarovsk pipeline is extended to Vladivostok as part of Gazprom's five year cooperation agreement with the region.[105] Gazprom signed a regional gasification agreement with Khabarovsk in January 2006 where 12 percent of the population already has access to natural gas but (as in Sakha Republic) this is not supplied by Gazprom.[106] The Khabarovsk pipeline will clearly provide the means for a substantial expansion of supply to the region.

For the first time in 2007, Gazprom started to pay attention to gasification in Kamchatka and a press release suggested this was a 'top priority' for the socioeconomic development of the region.[107] The wording of the statement suggests some pressure from government to pay greater attention to the region, but greater urgency may also be connected with gas discoveries by a consortium of Rosneft and Korean companies in 2006.[108] Agreement will need to be reached on the possible merger of Kamchatskgazprom and Kamchatskenergo in order to finance the construction of a pipeline and convert the region power supplies from fuel oil to gas.[109] Finally, Gazprom's interest in Sakhalin has clearly been transformed by its controlling interest in Sakhalin-2 and most of Gazprom's attention is focused on resource and export development opportunities. But in 2006, the company signed a cooperation agreement with the Oblast which appeared to envisage (but did not explicitly include) a commitment to increase the level of gasification from the current 9 percent.[110]

From the perspective of the regions, there is substantial urgency to see the gas projects move ahead quickly as they provide promises of: local employment, increased tax revenues, and substantial foreign currency revenues, a proportion of which will find their way back to the local economy; as well as infrastructural development, secure and environmentally benign energy supplies. For regional actors therefore – both political and commercial – there is less concern about the detail of which export markets may be served and which pipeline or LNG routes may be followed, and more concern to ensure that these projects move ahead quickly and bring much-needed revenues to severely deprived regions. During the 2000s, the theme of gasification of East Siberian and Far East regions became increasingly visible in both government and Gazprom documents culminating in the 2007 Eastern Gas Programme (see below). This is a strategic priority for the Kremlin which wants to ensure the 'effective occupation' of these regions and increase Russia's economic interaction with the Asia–Pacific region in general.

While it is understandable that the regions should see energy developments as a potential source of economic development, Sakhalin's experience presents a cautionary tale. First, the under-developed nature of a region's infrastructure actually limits its ability to absorb the potential economic multiplier generated by energy-related investment. Consequently, much of the activity takes place outside the region. Second, the majority of the impact is associated with the construction phase. As the first generation Sakhalin projects are now nearing completion, local politicians on Sakhalin are worried that there will now be a lull in activity and the economy will experience the classic 'boom and bust' cycle often associated with oil and gas projects. Third, the Sakhalin-2 project, which is export-oriented, will make no contribution to regional gasification. This has been a major bone of contention for the local politicians and Gazprom has promised to address this problem. Fourth, the economic benefits of energy related investments can be tempered by the nature of Russia's federal fiscal structure. The substantial increases in Sakhalin's Gross Regional Product generated by oil and gas-related construction activity actually resulted in a reduction in the level of federal transfers to the region. Finally, the Federal Government has reneged on an agreement to give Sakhalin Oblast 50–60 percent of the royalty generated by oil and gas production; instead it will receive 5 percent.

The initial impression from a review of gasification plans for Eastern Siberia and the Far East was that they were riding on the back of larger, export-oriented projects. This would make sense given the problem that local customers will have to pay a price for gas (or gas-fired electricity) which will make viable the construction of major new infrastructure. Neither the government nor Gazprom has much to say about this problem. When asked about profitability, the general director of Mezhregiongaz (responsible for the gasification programme) repeated Gazprom's general principle of a 15 percent rate of return on investment but admitted that this would be very difficult to achieve and that there were problems with regional authorities who pressed for gasification without wishing to allow the necessary price increases to finance it. However, he also noted that: '...this is part of the social obligation of Gazprom'.[111] The 2007 Eastern Gas Programme in combination with Gazprom's commitments to individual regions would appear to create a political commitment to gasification from which it will be difficult to step back. However, if the pipeline export projects progress quickly it would be quicker (and less costly) to complete these regional programmes.

The Role of Government and Gazprom: the 2007 Eastern Gas Programme

In the early years of the post-Soviet era, neither the Russian government nor Gazprom demonstrated any substantial interest in Asian gas development and export initiatives. Part of the reason for this was the company's Soviet-era disillusion with Japanese gas buyers.[112] Another cause was a lack of top level management time given the domestic problems and challenges of the post-Soviet era and more immediate opportunities for increasing exports to Europe.

In the early 2000s, as the Russian economy and global oil and gas prices began to recover from their 1998 nadir, so Gazprom under new management reappraised its options and prospects in Eastern Siberia and the Far East. One of the first steps was to set up an alliance with Rosneft and Surgutneftegaz in 2004, foreshadowed at a meeting between the three CEOs in late 2003 and signed on 24 January 2004.[113] Under this agreement the three companies agreed to develop oil and gas fields and also a unified gas network in Eastern Siberia and Sakha Republic. While many at the time saw this as part of Gazprom's attempt to control all activity within the region, it appeared as if this might be a genuinely cooperative effort when, in May 2004, it seemed possible that TNK-BP would be invited to join.[114] While cynics saw their views vindicated by the subsequent lack of action, Gazprom continued to insist that the problem was mainly due to the delay in agreeing a comprehensive programme for the development of these regions.[115]

The institutional basis for gas development in Eastern Siberia and the Far East changed fundamentally in June 2002 when the Russian Government issued Decree No 975-r, which instructed the Ministry of Energy and Gazprom to draw up a programme for a unified system of production, transmission and distribution of gas in Eastern Siberia and the Far East, taking into account the possibility of exports to China and other Asia–Pacific countries. Critically, the Decree named Gazprom as the coordinator of the programme thereby placing the company in control of all future projects in the region.[116] Since then, the basis on which Gazprom proceeded was that, if a project does not fit into its programme for the region, it would not make significant progress. A major obstacle to planning future domestic and export development in the region was the delay in publication of the programme.

Occasionally, external observers were able to get a glimpse of how Gazprom's thinking on this Programme was evolving. Alexei Miller's speech to the World Gas Conference in Tokyo in June 2003 was one such opportunity.[117] Miller's speech contained an entirely different concept for Far East development and was accompanied by a map of the region in which the West Siberian pipeline network was extended east from Novosibirsk to Irkutsk, with the Krasnoyarsk fields (Yurubchenskoye and Sobinskoye) connected to that pipeline. At Irkutsk, a pipeline was projected from Kovyktinskoye north to Chayandinskoye and east to Khabarovsk, meeting a pipeline from the Sakhalin fields which appears to feed two LNG plants, one near Sovietskaya Gavan' and another at Vladivostok.[118] With hindsight, this representation – it was not a proposal and the text of the paper did not discuss any specific projects – signaled a fundamental change in policy towards regional development.[119]

This was also the first suggestion of parallel development of gas and oil pipelines. Following the development of competing oil pipeline projects in the early 2000s, the Russian government opted for the Taishet–Kozmino line to be built by the Russian state-owned oil pipeline company, Transneft, in three phases.[120] In Phase 1, a pipeline with a capacity of 30Mt will be built only as far as Skovorodino close to the Chinese border; this phase is under construction and is due to be completed in 2008.[121] With the acquisition of the Sibneft oil company in 2006, Gazprom became a million barrel per day oil producer.[122] It is not yet clear where Gazprom's oil ambitions – or perhaps more precisely, government ambitions for Gazprom in this sector – will lead, but a significant increase in oil production by means of Sakhalin developments (e.g., the Kirinskii block) is a possibility.

The Miller concept of a Far East gas pipeline network was further elaborated by another Gazprom board member:

> The mainstay of eastern Russia's future gas supply network is said to be the so-called Central Gas Main that will unite basic fields of Eastern Siberia and the Far East, to be linked with Russia's United Gas Supply System (UGSS). Based on the Central Gas Main, eastern Russia will see before 2020 already the construction of a United gas production and transmission system which may possibly be built in several stages, allowing the optimisation of capital investments for all the main gas fields for a diversity of regional markets while avoiding competition between the projects. [123]

While the 2002 Decree placed strategic control of regional gas development with Gazprom, lack of ownership of regional gas assets prevented any significant development. In early 2007, Gazprom Deputy Chairman Ananenkov pinpointed the major problem:

> Gazprom has been instructed to coordinate the implementation of the eastern program, which is impossible to do without its own gas sources. To establish efficient control and attain state interests, Gazprom needs to be involved in major extraction projects in the eastern part of the country. [124]

But as we have seen above, this situation changed from the end of 2006 with Gazprom taking a majority share of Sakhalin-2, a project which was substantially completed, followed by the purchase of the majority TNK-BP share of Kovykta. Having gained control of two major gas assets, this left others to be decided: Chayanda and Sakhalin-3 (Kirinskii) where (again as detailed above) it appeared that government seemed in no hurry to auction the licence for these resources; and Sakhalin-1 which has reached an impasse on gas sales; no quick resolution was in sight. Elsewhere in Sakhalin, Gazprom has demonstrated its intention to buy into existing projects – particularly when the Sakhalin-3 Kirinskii block is retendered. In this connection, if the Gazprom–Rosneft merger had succeeded, this might have accelerated gas development in Sakhalin projects 3 to 5.[125]

The acquisition of majority shares in Sakhalin-2 and Kovykta projects in 2006–7, followed by the adoption of the Gas Programme, has added asset ownership to Gazprom's extremely powerful political and institutional position in the region. It is clear that Gazprom can prevent any other company, with one possible exception, and certainly any foreign company, from pursuing any gas development plan which it does not approve. The major remaining tasks for the implementation of the Programme are the reauctioning of the Chayanda and Sakhalin-3 licences, although comments from Gazprom's top management already seem to assume that the company will acquire these assets.[126] The possible corporate exception to Gazprom's dominance is Rosneft which has assumed the position of Russia's national oil company and has significant gas assets particularly in Sakhalin. Gazprom and Rosneft's ambitions to

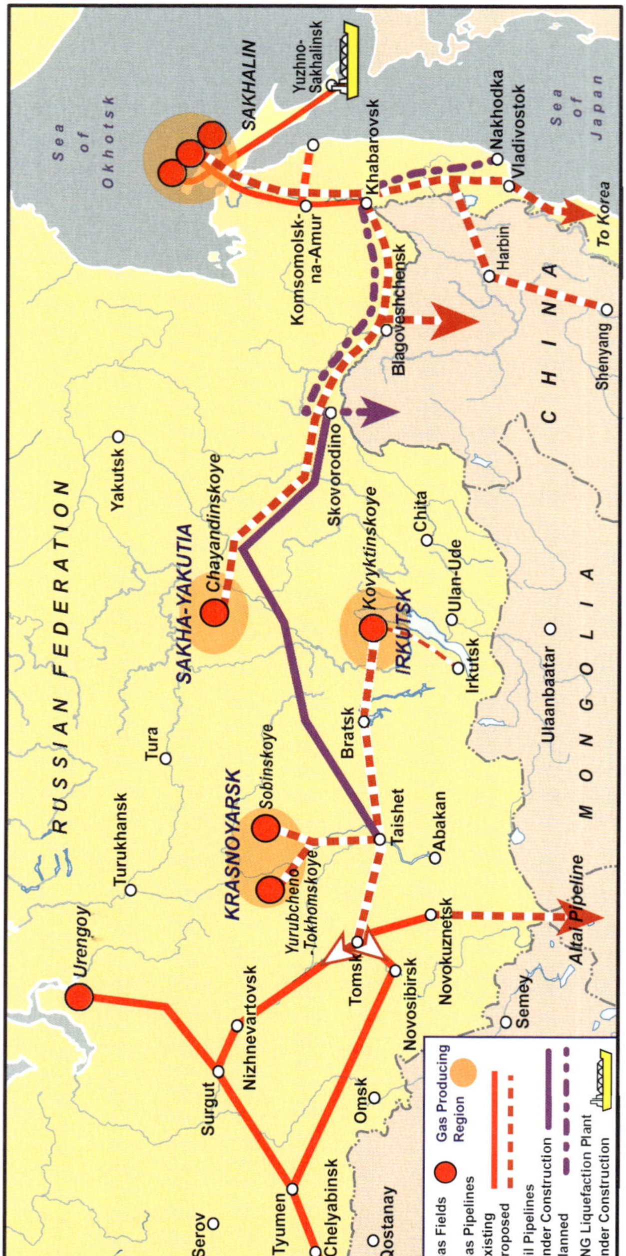

Map 6.2: The Eastern Gas Programme Plans for Pipeline Developments in Eastern Siberia and the Russian Far East

control gas (as opposed to oil) developments in the Far East will need to be resolved at presidential level and any decisions taken by president Putin before he leaves office cannot automatically be assumed to be followed by his successor.

The 2007 Eastern Gas Programme

In September 2007, after 5 years in the making, the adoption and main features of the Eastern Gas Programme were confirmed in a press release from the Ministry of Industry and Energy.[127] From 15 different options, 'Vostok 50' – hailed as Gazprom's favoured option – was selected.[128] The Programme outlines four main centres of gas resources and sets out how production from those fields will be allocated (see Map 6.2):

- the Sakhalin centre will provide supplies for the Sakhalin, Khabarovsk and Primorsk regions; the Jewish Automonous Republic and exports of pipeline gas and LNG to Asia–Pacific countries. Sakhalin-1 gas will be devoted to the gasification of Sakhalin and Khabarovsk, while deliveries to Primorsk and exports to Korea will be achieved via a new Sakhalin–Vladivostok pipeline system which could also take gas from Sakhalin-3 and Yakutia fields.
- the Yakutia centre, based on the Chayandinskoye field, will allow the gasification of the southern regions of the Sakha Republic, Amur Oblast and also exports of pipeline gas to Asia–Pacific countries;
- the Irkutsk centre will meet existing demand and future needs of industrial customers in the Irkutsk and Chitinsk Oblasts and also the Republic of Buryatia as well as, if needed, deliveries of gas to the UGSS (serving the west of Russia);
- the Krasnoyarsk centre will meet existing demand in Kransoyarsk Krai as well as, if needed, deliveries of gas to the UGSS.

The gas transportation network will be optimised by utilising the route of the East Siberia–Pacific Ocean (ESPO) Oil Pipeline which will also allow the utilisation of associated gas from oil production.

The gas market in Eastern Siberia and the Far East is expected to grow to 27 Bcm in 2020 and 32 Bcm in 2030. Taking into account gas demand from gas processing industries, e.g., petrochemicals, this figure could rise to 41 and 46 Bcm in 2020 and 2030 respectively. The share of gas in the energy balance of these regions should increase more than six-fold to 38.5 percent. Outside the region itself, deliveries to western Russia via the UGSS could be as high as 35 Bcm/year. Pipeline gas exports to China and Korea could reach 25–50 Bcm by 2020, and LNG exports to the Asia–Pacific region 21 Bcm by 2020 and 28 Bcm by 2030 which would mean a doubling of Sakhalin-2's 9.6Mt LNG export capacity. The programme confirms that the principle of a 'single export channel' – i.e., Gazprom's monopoly of exports – will be maintained. An earlier version of the Vostok 50 Programme, with lower domestic demand projections and no allowance for a connection to the UGSS, nevertheless makes clear that the vast majority of gas for domestic consumption and exports will be produced at Yakutia and Sakhalin, with Irkutsk and Kransnoyarsk having a relatively marginal role.[129]

The investment required for the whole programme up to 2030 is estimated to be in excess of 2.4 trillion rubles of which 290bn will be for exploration and 1.3 trillion for production and processing.[130] However, the Programme suggests that the these investments will be more than recouped by contributions to the Russian Federation budget by 2030 resulting from increased economic activity. The clear hope is that the Programme will contribute to an economic renaissance in the eastern part of Russia

Pipeline Export Markets: China and Korea

From 1997 onwards, Gazprom management began increasingly to mention Asian gas exports as part of their general export strategy.[131] Chairman Vyakhirev devoted an entire section of his speech to the 1997 World Gas Conference to the 'East-Asian Gas Market' mentioning all of the potential projects.[132] At the APEC Summit in November 1998, Gazprom's Chairman announced a plan to link the fields of the Yamal Peninsula with Shanghai, a distance of 7,100km, taking a route via Tomsk, Barnaul, Irkutsk and Urumchi. Gazprom officials also spoke of the Zapolyarnoye field as a possible source of gas for China (although this has not been mentioned again since the field started production in 2001) as well as fields in Krasnoyarsk region such as Bolshekhetskaya.[133] Two years later, Gazprom's Chairman was more specific about projects and timing:

> The Far East region, China, Korea and Japan is a new promising market for Russia's pipeline gas. Various options of Russia's involvement in this market development are under discussion. In the first phases, the Far Eastern gas transmission system project is supposed to be realised with gas pipeline supply up to 30 Bcm annually. Gas flow will go from the fields in the Irkutskaya Oblast and the Republic of Sakha. Gas produced from Sakhalin offshore (Sakhalin-2) will mainly go to liquefaction…as an alternative, construction of a gas transmission system to deliver gas to China and Korea via Khabarovsk is under consideration. Annual deliveries of Sakhalin gas could reach 12–15 Bcm by 2010 and 20–25 Bcm by 2020. [134]

This confirmed Gazprom's recognition that significant new export markets were opening in Asia and these could be combined with increased attention to gas development prospects in Eastern Siberia and the Far East. The creation of its Vostokgazprom subsidiary in 1999 which became involved throughout the gas chain with an annual production of 3.5 Bcm, was a tangible manifestation of this recognition.[135] In 2000, Vostokgazprom suggested a project which would bring 30 Bcm of West Siberian gas to join the planned Chinese West–East gas pipeline in Xinjiang and hence to Shanghai.[136] But at that time, such proposals were probably based on an expectation that Gazprom would take a role in the West–East Pipeline after the company had been selected as one of the preferred foreign partners to enter into negotiations with the Chinese authorities to participate in the project. When Gazprom (and all other foreign partners) dropped out of the West–East Pipeline the idea of West Siberian gas being exported to China was not spoken of for some years.

China.[137] The failure of Gazprom to participate in the construction and development of China's West–East Gas Pipeline in the early 2000s was a substantial setback to Russian export aspirations. Gazprom (and Shell with which it formed a partnership in the project) had imagined that equity in the West–East Pipeline would provide it with the opportunity to supply gas from Western Siberia into the line. With the collapse of all foreign investment in the West–East Pipeline, Gazprom was left to rebuild the relationship.[138] In October 2004, China National Petroleum Corporation (CNPC) entered into a strategic cooperation agreement of a very general nature with Gazprom. While there was general discussion of China needing gas imports of 20 Bcm by 2010, it was stressed by Chairman Miller that the discussions were about 'the resource base and its optimisation, but we did not talk about specific [pipeline] routes…the price formula for Russian gas to be imported to China has not yet been agreed upon either'. [139]

In March 2005, the Deputy Head of the Fuel and Energy Department at the Ministry of Industry and Trade, in a presentation which seemingly dealt mainly with the ESPO oil pipeline, proposed four new options for building a regional gas pipeline network with export options to East Asian countries.[140] However, these proposals were superseded by the March 2006 announcement by the Russian president of the Altai (West Siberian) and Eastern Pipelines (see above). Following the announcement and the unrealistically rapid timetable attached to completion of negotiations and construction of the pipeline, there has been limited information on progress. It is known that a major problem in bilateral negotiations has been the issue of price, with the Chinese favouring a price linked to coal, which is their major alternative fuel for power generation, and Gazprom insisting on a price linked to oil.

The outlook appeared to change dramatically with the announcement of a Memorandum of Understanding between Gazprom and CNPC at the meeting of the Chinese and Russian presidents in Beijing in March 2006 for two gas pipeline projects – one from Western Siberia and the other from gas fields further east with a projected 68 Bcm of Russian gas to be exported to China in 2020.[141] Somewhat surprisingly, the main priority project for gas exports became the Altai pipeline from Western Siberia to the Chinese border a distance variously estimated at 2,600–2,800km at an estimated cost of $13.6bn, linking up with the West–East Pipeline.[142] Even more surprising were the suggestions at that time, that commercial negotiations could be completed by the end of 2006 and the pipeline could be delivering its first gas by 2011 reaching 30 Bcm by 2020.[143] This always seemed extremely unlikely and so it proved. The only tangible progress was in discussions on the route of the Altai pipeline, over which environmental opposition was raised in relation to the crossing of the Udok Plateau which is a natural heritage site, but this has apparently been accepted in return for promises of employment and regional gasification (see above).[144]

During 2006, press releases charted regular contacts both in the context of the CNPC cooperation agreement and commercial negotiations.[145] One of the few commercial details given by Alexei Miller in his speech to the 2006 World Gas Conference was that '...China is ready to buy gas at the JCC-based price'.[146] It seemed unlikely that Gazprom would compromise on this position given that West Siberian gas can be delivered through existing pipelines to existing customers in Europe at prices linked to oil, and that the transportation system necessary to bring gas only as far as the Chinese border requires an investment in excess of $14bn.[147]

At the March 2007 meeting between the Russian and Chinese Presidents in Moscow, nothing was said about gas or gas pipelines, suggesting that no significant progress had been made.[148] In late 2007 therefore, the future of negotiations remained uncertain with no suggestion that the parties were close to an agreement. In late 2007, following new Chinese purchases of Australian LNG, Gazprom's chief negotiator noted that 'China paying market prices is good news for our discussions', suggesting that price remained a major stumbling block.[149] As far as Sakhalin-1 exports to China are concerned (see above), while Gazprom has not denied the right of the Sakhalin-1 partners to direct exports, it has stressed – backed by the Eastern Programme – that Sakhalin-1 gas is needed for domestic consumption in the region.[150] In the face of such implacable opposition from Gazprom and the government it must be doubtful whether a direct pipeline to China will be possible.

Korea.[151] In the section above on Kovykta it was noted that, during the 1990s, Korean investors were clearly aiming at delivering pipeline gas from the field to their home market. The withdrawal of those investors created a several year break in relations but in May 2003, Gazprom signed a five-year cooperation agreement with the Korean gas company Kogas to examine the prospect of Russian gas supplies to Korea. From the Gazprom side, the agreement:

> ... will lay the legal basis for intensive cooperation between the two companies in the pipeline gas deliveries to South Korea as well as for the realisation of joint projects in third countries...The agreement concluded today means that the principle of delivering gas via a single export route will be effected for the Korean gas market. [152]

It was anticipated that the Gazprom–Kogas Working Group would start negotiating a long-term contract for deliveries to Korea with Gazexport.[153] Since 2004, there have been a string of meetings between Gazprom and Korean representatives within the context of the Working Group, but no significant progress on gas deliveries has been reported.[154] The only exception was the delivery of a spot cargo of LNG to Kogas in October 2006 which had been purchased from Mitsubishi and Tokyo Electric by Gazprom Marketing and Trading.[155] Since 2005, nothing more has been heard of the plans to deliver pipeline gas to South Korea bypassing China, which will be necessary if negotiations with the Chinese fail.[156] The 2007 Programme (above) suggests that 12 Bcm of pipeline gas could be delivered from a combination of Sakhalin or Yakutia via the Sakhalin–Vladivostok network.[157] However, this would require the pipeline either to pass through North Korea, or to take an offshore pipeline or LNG route avoiding that country; but these options raise serious political and economic issues (see Korea chapter). Nevertheless, what is important from the Russian perspective is that, in sharp contrast to their Chinese and Japanese counterparts, both Kogas and the Korean government have appeared consistently and unambiguously to be in favour of Russian pipeline gas deliveries.

Russian Gas Exports: Europe versus Asia. From the late 1990s, government and commercial perspectives on Asian gas exports viewed from Moscow changed. The main export task of establishing a market share for Russian gas in Europe had been achieved. Current contracts (and likely renewals of existing contracts) will see sales of at least 180 Bcm/year of Russian gas in Europe. Although this volume may increase further during the 2010s, the task will be one of maintaining market share rather than capturing large new markets. European liberalisation initiatives, including the likelihood of gas to gas competition and the possible demise of traditional long-term take or pay contracts indexed to oil-related products, began to introduce new and unwelcome elements of risk into the European gas business for Gazprom.[158] In addition, worsening political and energy relations between Russia and Europe since the Ukraine crisis of 2006, have led to rising European concern regarding existing, let alone increasing, dependence on Russian gas.[159]

Thus there is no longer any significant growth story to tell for Russian gas exports to Europe and this means that Asian markets, with huge growth potential, have become correspondingly more interesting. Not only are gas markets in East Asia less well developed but they are less likely to introduce aggressive liberalisation and competition initiatives and less likely to attempt to introduce political and environmental conditionality in relation to gas export projects. However, East Asian markets present significant infrastructural constraints: currently Korea is the only country with a national pipeline network; the beginnings of such a network can be seen in China while Japan has rejected this option. The general lack of infrastructure to bring pipeline gas to major Asian markets appears to favour LNG exports. Yet LNG exports will not support regional and local gas networks planned in Eastern Siberia and the Far East which – given the level of economic disadvantage in these regions – would find it extremely useful if export infrastructure could subsidise their development.

However, the case of exports to Asia should not be seen as suggesting that Russia will turn its back on the European gas market. In mid-2006, bizarre stories circulated that Gazprom's exports to Europe could be at risk because of the company's new commitments to the Chinese market.[160] Not only were these firmly denied by Gazprom but in late 2007 the likelihood of any immediate pipeline exports to Asian markets seemed low.[161] While Gazprom is not averse to reminding Europeans that there will be other large scale export markets for Russian gas in future, this falls somewhat short of a 'China card' that can credibly be played to exert commercial pressure on Europe.

The Foreign Investment Climate

While the Russian legal and fiscal framework for development of energy projects and infrastructure became significantly more settled in the 2000s, key legal and regulatory instruments have yet to be finalised and implemented clearly and consistently. Equally worrying for foreign investors is the development of a political environment around energy assets in the 2000s which has seen Russian companies – and particularly the majority state-owned companies Gazprom and Rosneft – assume an overwhelmingly dominant position, with foreign companies correspondingly marginalised. The majority foreign opinion is that in the cases of Sakhalin-2 and Kovykta, this marginalisation has amounted to expropriation of assets.[162] The Russian view, strongly expressed by President Putin, is that Russian government actions were amply justified by the commercial agreements entered into during the 1990s:

> [Sakhalin-2] is a colonial treaty that has absolutely nothing to do with the interests of the Russian Federation. I can only regret that in the early 1990s the Russian officials allowed such incidents to take place, incidents for which they should have been put in prison. Implementing this treaty resulted in a situation in which, for a long period of time, Russia allowed its natural resources to be exploited and received nothing in return. Almost nothing at all.
>
> [Kovykta]... does not only affect BP but also ... Mr Wechselberg's company and Mr Potanin's company. They are all Russian economic residents. And for that reason the affair is not limited to BP, to a foreign partner, but to all shareholders that have committed to developing this deposit and, unfortunately, have failed to comply with the terms of their permit. They have not yet started to develop it. According to the permit's conditions they should have already begun extraction last year. And not simply begun but also extracted a certain amount of gas. Unfortunately, they have not done so.[163]

The Putin view of subsequent changes in ownership of these projects concluded in the 1990s was unrepentant: they were economically unfair to Russia and it was legitimate to correct this unfairness by any available means, making Russian, rather than foreign, companies majority owners of these assets. By contrast the international oil company view was that the original terms and conditions correctly reflected the risks inherent in reaching those agreements in the economic and political conditions of the mid-1990s, and the willingness to assume those risks should be respected.[164] Had those investors not been willing to take these risks, then these projects would not exist.[165] It will be interesting to see how the disagreement over Sakhalin-1 gas is resolved given that the Russian authorities have not discovered any violation of the project's terms and conditions.

Central Asia: Turkmenistan, Kazakhstan and Uzbekistan

In the immediate aftermath of the break-up of the Soviet Union, the newly independent countries of Turkmenistan and Kazakhstan lost no time in exploring the possibilities for new gas export markets in Asia. In late 1992, the Chinese pipeline company CNPC together with Japanese trading house Mitsubishi visited Turkmenistan to propose what became known as the Energy Silk Route pipeline bringing gas from Turkmenistan to China, Korea and Japan.[166] Turkmen and Kazakh gas would need to be piped across Central Asia to the east coast of China and from there to Japan and Korea. The distance from Turkmenistan to China's east coast is around 6,500km and a further 2,300km of offshore pipeline would be needed to reach Japan and Korea.[167] During the years 1993–96 a joint feasibility study was carried out on the project which apparently reached an unfavourable conclusion which was hardly surprising at then current energy prices given the distances involved.[168]

The exact dimensions of Central Asian gas resources are highly uncertain – an uncertainty which has increased in the 2000s – as increasingly determined attempts have been made by national leaderships to obscure specific data while issuing ever-larger estimates of resources. All three countries have a very substantial resource base, even using the restricted definition of proven reserves, let alone in more speculative categories. The figures for proven reserves in Table 6.8 are substantial but far smaller than will be found in press statements. The table does not sufficiently bring out that Turkmenistan is the resource giant of the region because in a country of four million people with little industrial base outside natural gas and cotton there is limited potential for domestic consumption.

Problems of accurate estimation of Turkmen reserves were highlighted in late 2006 when the newly discovered Yoloten field was said to have reserves of 7 Tcm – more than twice the size of country's total proven reserves (see Table 6.8) – followed in early 2007 by the discovery of the Osman field also with substantial reserves, which may be part of the same structure as Yoloten.[169] Press reports suggested that 1.3 Tcm

Table 6.8: Central Asian Gas Reserve Estimates. Trillion Cubic Metres

	Proven
Turkmenistan	2.9
Kazakhstan	3.0
Uzbekistan	1.87
Total	7.77

Source: Proven Reserves at the end of 2006 from: *BP Statistical Review of World Energy*, June 2007, p. 22.

of new reserves had been established in fields on the right bank of the Amu Darya River from which gas would be piped to China.[170] Turkmen reserves have been the subject of controversy since 2005 when the late President Niyazov, under pressure from Gazprom, commissioned an international reserves audit of the Dauletabad field and then refused to reveal the results.[171] Asked in 2007 whether Turkmenistan had sufficient reserves to meet all of its potential commitments to Russia, China, South Asia and Europe, President Berdymuhammedov said simply: 'Don't worry, there will be enough gas'.[172] Although this case needs to be made in greater detail, projections of domestic gas demand of 19 Bcm in 2020 (compared with 14 Bcm in 2006) will hardly make any significant dent in Turkmen resources.[173]

By contrast, Uzbekistan with its much larger population and more diversified economy already has a gas demand of around 40 Bcm/year which could increase to as much as 75 Bcm by 2020. Despite its significant resource endowment, this level of demand would probably not allow Uzbekistan to dedicate reserves to a baseload export project. By 2007, Uzbekistan appeared to have committed virtually all of its exportable gas surplus either to Kirghizstan, Tajikistan or Russia.[174]

Kazakhstan is in a more complicated position because of the geography of the country and the location of the major gas fields. The resource base is substantial and certainly in excess of the country's projected needs of 20–35 Bcm by 2020. But resources are concentrated in the Karachaganak field, in the north west of the country, and the Kashagan and Tenghiz fields in the Caspian Sea region.[175] The major fields are therefore extremely poorly located in respect of exports to Asia. By contrast, they are well located in respect of exports to Europe, being only a few hundred kilometres from existing pipelines to CIS countries.[176]

Following the enthusiasm of the mid 1990s, little was heard of grandiose plans to pipe Central Asian gas to China and other East Asian markets. Chinese companies became active in investments in Kazakh oil fields, refineries and possible export pipelines but gas exports seemed a longer-term prospect than exports from Russia. However in the mid-2000s, the situation changed principally with the completion of the Kazakhstan–China oil pipeline in 2005 which, despite only transporting relatively small volumes during its first two years, was accompanied by a plan to build a parallel gas export pipeline.[177] Kazakh estimates suggested that a 10 Bcm/year gas pipeline could be constructed by 2009 at a cost of $2–4 bn, which would subsequently be expanded to 30 Bcm/year by 2012.[178] But it was unclear from where the gas to fill such a large pipeline would originate.

A possible solution emerged in late 2005, when Turkmenistan signed an agreement to export 30 Bcm/year to China from new fields on the right bank of the Amudarya.[179] There was further confirmation in April 2006, when the Chinese and Turkmen presidents signed a general agreement for a pipeline to export 30 Bcm/year of Turkmen gas to China.[180] By late 2007, Chinese plans appeared to have crystallised on developing a pipeline from Turkmenistan to China via Uzbekistan and Kazakhstan;[181] and possibly a second line bringing gas from Kazakh gas fields.[182] While plans for such pipelines were still at an early stage, it seemed much more realistic to combine deliveries from all three countries (but principally Turkmenistan and Kazakhstan) into a single pipeline. These plans seemed to fit well with China's proposed second West–East Pipeline which would take gas from Central Asia as well as Chinese domestic fields on a route partly paralleling the existing pipeline (see Map 2.1).

Thus as this chapter was completed, all of the political building blocks seemed to be in place for a pipeline from Central Asia to China carrying a mix of (mainly) Kazakh and Turkmen gas. However, among the important commercial issues yet to be revealed were the sources of gas fields in Kazakhstan and Turkmenistan and how soon gas from those fields could be made available, the sources of investment for these developments, i.e., which parties would finance the development of gas reserves and pipeline construction, and the commercial terms on which the gas would be sold, in particular the price that Central Asian sellers would receive at their borders.

But in 2007 Turkmenistan was receiving \$100/Mcm from Russia at its northern border and Kazakhstan up to \$145/Mcm, with an expectation that by 2011 both countries will be receiving the equivalent of European prices netted back to their borders.[183] For the Chinese to match these prices through new pipelines over extremely long distances could be a problem. But despite these uncertainties, Central Asian pipeline gas supplies to China seemed set to commence by the mid 2010s.

Conclusions: Russian and Central Asian Gas Exports to Asia to 2020

The main rationale of Russian and Central Asian gas exports to Asia rests, and has always rested, on the obvious complementarity between exporting countries with surplus – and indeed 'stranded' – gas reserves, and importing countries with insufficient resources and huge market potential. One of the main conclusions of the first edition of this chapter (written in 2002) – that Russian LNG exports to Asia would commence by 2010 but volumes would be unlikely to exceed 10 Bcm/year – was slightly too pessimistic.[184] In the event, LNG exports should start in 2008 and should be more than 14 Bcm by 2010. But the other main conclusion – that large-scale pipeline exports of Russian gas to Asia on the scale of 20–50 Bcm/year seemed, as they have for the past 40 years, at least a decade away – remains robust.

As far as the pipeline gas export outlook is concerned, the major change since 2002 is not in relation to Russia, but Central Asia where a pipeline taking gas principally from Turkmenistan and Kazakhstan to China seems likely to be completed by the early 2010s. While commercial negotiations for Central Asian gas had not yet been finalised, there was great confidence that this will be achieved quickly in comparison with Russia which, in late 2007, appeared to have reached a commercial stalemate.

This is very surprising – and disappointing – given the duration of negotiations which have taken place between Russia and China. There is still time and an abundance of resources for Gazprom's projection of 68 Bcm of pipeline gas to be delivered to China by 2020, with additional pipeline gas to Korea. But in order to achieve such an outcome, both sides need to show much greater urgency and willingness to compromise than is evident in 2007. Since China has apparently selected the Central Asian option for the second West–East Pipeline, the prospects for the Altai Pipeline from Western Siberia have been seriously diminished. However, this is convenient for Russia since neither the economics of the project nor the idea of taking substantial gas from Western Siberia seemed sensible, given the vast reserves located closer to the Chinese market. But fundamental disagreement on prices – with Gazprom sticking to its European oil-related formula and the Chinese insisting on a formula reflecting local energy conditions – means that the two sides remain far apart. In addition to commercial problems, the Chinese clearly prefer to deal with producers who are willing to allow them direct access to resources and construction of gas infrastructure rather than to adopt the Russian model of selling the gas at the border of the importer.

With the publication of the 2007 Eastern Gas Programme, the sources of pipeline gas exports to both China and Korea – a Japanese pipeline export option was no longer even a subject for discussion – were firmly designated to be Sakhalin and Yakutia (Sakha). While the late 1990s and early 2000s mainly focused on the prospect of pipeline exports from Kovykta (Irkutsk) to both China and Korea, this option

– still seen by many as by far the most economically attractive – has been decisively rejected. Kovykta gas will be used for local demand and will possibly be piped west to connect with the main network carrying gas to western Russia and beyond. Pipeline exports to Korea, about which both the Russians and Koreans are enthusiastic will either have to be accomplished via North Korea or an offshore pipeline from Vladivostok, presenting difficult political and commercial problems.

It may therefore be that Russian exports to Asia will be easier to achieve in liquefied form. The Sakhalin-2 project will begin deliveries in 2008 but the Eastern Gas Programme only envisages a doubling of this capacity by 2030 which seems extremely conservative in relation to available resources. It would be possible for much larger-scale LNG exports based on already identified Sakhalin resources to be delivering to Asia by 2030. It might be that LNG could resolve the political and commercial barriers to pipeline gas exports. However, for this to happen, both Russian and Asian governments and companies have to show much more enthusiasm and urgency than has been the case thus far, with the single exception of Korea. On the Russian side, the government has played a key role in slowing down LNG development by annulling the Sakhalin-3 licences and failing to introduce the necessary legislation to hold new tenders. There is now a real concern on the island that the support industries developed for Sakhalin-1 and -2 will disappear as those projects are completed.

With the resources which have been identified, and the gas prices seen in international trade since 2003, there is no reason why Japan, China and Korea could not be importing at least 100 Bcm of gas a year from Russia and Central Asia by 2020; indeed much higher figures would be possible. But in 2007 only 14.5 Bcm of LNG from the Sakhalin-2 project was definite with an additional 30 Bcm from Central Asia seeming highly likely. There seemed no reason why political and commercial obstacles to substantially increased gas trade could not and should not be resolved, but the evidence of the period since 2000 gives no grounds for over-confidence.

Bibliography

Bradshaw 1990: Michael Bradshaw, 'Soviet Far Eastern Trade' in Alan Rodgers (ed.), *The Soviet Far East: Geographical perspectives on development*, London: Routledge, 1990, pp. 239–268.

Bradshaw 2003: Michael Bradshaw, 'Sakhalin Oblast', in Herd G.P. and Aldis A. (eds.), *Russian Regions and Regionalism: strength through weakness?* Richmond: Curzon, 2003 pp.141–163.

Bradshaw 2006a: Michael Bradshaw, 'Russian and transnational energy companies: conflict and cooperation in Pacific Russia', in Wenger A., Perovic J. and Orttung R.W. (eds), *Russian Business Power: The Role of Russian Business in Foreign and Security Relations*, London: Routledge, 2006, pp. 133–154.

Bradshaw 2006b: Michael Bradshaw, 'Sakhalin-II in the Firing Line: State Control, Environmental Impacts and the Future of Foreign Investment in Russia's Oil and Gas Industry', *Russian Analytical Digest*, Number 8, 2006, pp. 6–11.

Bradshaw and Kirkow 1998: Bradshaw, M. J. and Kirkow P, 1998, 'The Energy Crisis in the Russian Far East: Origins and Solutions', *Europe-Asia Studies*, Vol. 50, No. 6, pp. 1,043–1,063.

Loan Notes 2007: Open Joint Stock Company Gazprom, 'US$15,000,000,000 Programme for the Issuance of Loan Participation Notes', 18 May 2007.

Gazprom Press Releases can be found on the company's website where they are indexed by date: *http://www.gazprom.com/eng/archive.shtml.*

Miller 2003: Alexei Miller, 'Eurasian Direction of Russia's Gas Strategy', World Gas Conference, Tokyo, June 2003.

Miller 2006: Alexei Miller, 'Energy for the Planet', XXIII World Gas Conference, Amsterdam, June 2006, *http://www.gazprom.com/eng/articles/article19731.shtml.*

Miyamoto 1997: Akira Miyamoto, *Natural Gas in Central Asia: industries, markets and export options of Kazakhstan, Turkmenistan and Uzbekistan*, RIIA, 1997.

North East Asian Gas and Pipeline Forum 2007: 'A Long Term Vision of Natural Gas Infrastructure in North East Asia – 2007 Version', North East Asian Gas and Pipeline Forum, September 2007.

Paik 1995: Keun-Wook Paik, *Gas and Oil in North East Asia: policies, projects and prospects*, RIIA, 1995.

Paik 1996: Keun-Wook Paik, 'Energy cooperation in Sino-Soviet relations: the importance of oil and gas', *Pacific Review*, Vol 9, No 1, 1996, pp. 77–95.

Paik 1997: Keun-Wook Paik, *Tarim Basin Energy Development: Implications for Russian and Central Asian Oil and Gas Exports to China*, RIIA, CACP Briefing No 14. November 1997.

Paik 2005: Keun-Wook Paik, Pipeline Gas Introduction to the Korean Peninsula, Project on Energy and Environmental Cooperation on the Korean Peninsula, Chatham House,January 2005; *http://www.chathamhouse.org.uk/publications/papers/view/-/id/1/*

Paik and Choi 1998: Keun-Wook Paik and Jae-Yong Choi, *Pipeline gas in Northeast Asia: recent developments and regional perspective,* RIIA, Energy and Environmental Programme Briefing Paper No. 39, January 1998.

Pirani 2007: Simon Pirani, *The Ukrainian Gas Industry: domestic and international perspectives,* Oxford Institute for Energy Studies, 2007, *http://www.oxfordenergy.org/pdfs/NG21.pdf.*

Russian Energy Strategy 2001: *Osnovnie Polozheniya Energeticheskiye Strategii Rossii na period do 2020 goda*, IES Moscow, 2001.

Russian Energy Strategy 2003: *Energeticheskaya Strategiya Rossii na period do 2020 goda*, issued by the Government of the Russian Federation, August 28, 2003.

Rutledge 2004: Ian Rutledge, 'The Sakhalin II PSA – a Production 'Non-Sharing' Agreement: Analysis of Revenue Distribution', Sheffield Energy & Resources Information Service, 2004, available at *http://www.pacificenvironment.org/downloads/SakhalinPSA-www1.pdf.*

Sagers 1995: Matthew Sagers, 'Prospects for Oil and Gas Development in Russia's Sakhalin oblast', in *Post-Soviet Geography*, Vol.36, No.5, May 1995. pp. 274–290.

Simonia 2004: Nodari Simonia, *Energy Policy in Eastern Siberia and the Russian Far East*, The Energy Dimension in Russian Global Strategy, The James A. Baker III Institute for Public Policy, October 2004.

Stern 1999: Jonathan P. Stern, 'Soviet and Russian Gas: the origins and evolution of Gazprom's export strategy', in Robert Mabro and Ian Wybrew-Bond (eds), *Gas to Europe: the Strategies of Four Major Suppliers*, OUP/OIES, 1999, pp. 135–200.

Stern 2005: Jonathan P. Stern, *The Future of Russian Gas and Gazprom*, Oxford University Press, 2005.

Stern 2006: Jonathan Stern, *The New Security Environment for European Gas Trade*, Oxford Institute for Energy Studies, 2006, *http://www.oxfordenergy.org/pdfs/NG15.pdf.*

Vitsovskaia 1999: Galina S. Vitsovskaia, 'Lawlessness, Environmental Damage and Other New Threats in the Russian Far East', in Gilbert Rozman, Michael G. Nosov and Koji Watanabe (eds), *Russian and East Asia: The 21st Century Security Environment*, East West Institute/M.E.Sharpe, 1999, pp.179–199.

Vyakhirev 1997: Rem Vyakhirev, 'Russian Gas Industry in the 21st Century', World Gas Conference, Nice, June 2000.

Vyakhirev 2000: Rem Vyakhirev, 'The Perspectives of Russian Natural Gas. Role in the World Gas Market', World Gas Conference, Copenhagen, June 1997.

Wybrew-Bond and Stern 2002: Ian Wybrew-Bond and Jonathan Stern, *Natural Gas in Asia: the Challenges of Growth in China, India, Japan and Korea*, OUP/OIES, 2002.

Notes

1 Robert E. Ebel, *Communist Trade in Oil and Gas*, Praeger Publishers, 1970, p. 151 contains a map of three pipeline proposals for Sakhalin exports to Japan dating from the late 1960s, clearly showing options still being considered today. An early history of all the East Siberian and Far Eastern projects can be found in Peter Egyed, *Western Participation in the Development of Siberian Energy Resources*, Case Studies: East–West Commercial Relations Series Report 20, Institute of Soviet and East European Studies, Carleton, Ottawa, May 1983.

2 For details of the discovery process offshore Sakhalin, see Paik 1995, pp. 207–221.

3 Japan National Oil Corporation was the first company to become involved

and subsequently headed the SODECO consortium (see Table 6.4).

4 Western companies disputed this Soviet assertion about the volume of reserves which could, by western standards, be considered proven. For the background to the Yakutia project, see Jonathan P. Stern, *Soviet Natural Gas Development to 1990*, D.C.Heath/Lexington Books, 1980, pp.118–120.

5 In some respects Vostok, which envisaged the larger share of output being used for regional demand, with 20 (out of 48) Bcm allocated to exports, was a forerunner of the developments discussed later in this chapter. See *Oil and Gas of the Soviet Far East: how big a play?*, Petroleum Advisory No. 76, East West Center, Honolulu, September 3, 1991.

6 East Siberian gas reserves and discovery dates can be found in *Eastern Bloc Energy*, May 2001, pp. 22–24. We return to the development of the Kovyktinskoye field below.

7 The islands are Habomai, Shikotan, Etorofu and Kunashiri.

8 Gerald Segal, *Normalising Soviet-Japanese Relations*, RIIA Special Paper, 1991, pp. 36–38.

9 For a detailed discussion of these compensation agreements see: Bradshaw 1990.

10 Stern 1999.

11 The state of Japanese LNG trade in 1991 is summarised in: Sylvie Cornot-Gandolphe, *World LNG Trade, a new growth phase*, Cedigaz, November 1991.

12 The French drill ship carrying out the seismic surveys was using US-licensed technology.

13 Paik (1996) has a chronology of energy cooperation clearly showing that bilateral energy developments were well under way before the break-up of the Union.

14 Stern 2005, Chapter 1.

15 These figures are proven and probable reserves as certified by Degolyer and McNaughton taken from *Loan Notes 2007*, Appendix A. The division between producing areas and the Yamal Peninsula was assumed using the figures from Stern 2005, Table 1.1, p.3. These figures are somewhat lower than the Russian A+B+C1 reserve classification.

16 For example Gazprom purchase of producing assets Northgas and Itera during the period 2004–6 (Stern 2005, p.27 and p.190).

17 *Gazprom management committee resolves to initiate investment stage of Bovanenkovo field development*, Gazprom Press Release, October 6, 2007.

18 The problems with a 2011 start date are related to the lack of progress in creating the new pipeline network to evacuate the gas.

19 This seems increasingly likely to apply even to sales in the Russian domestic market where the government has announced that, by 2011, non-residential customers will be paying prices equivalent to European border prices. *http://www.government.ru/government/governmentactivity/government_meetings/* archive/2006/12/01/2833216.htm.

20 In 2005, the population of Sakhalin Oblast was 532,000, of Krasnoyarsk

Krai 2.9 million, of Irkutsk oblast 2.5 million and of Sakha Republic 951,000. The population of the entire Siberian Federal District (East Siberia)was 19.8 million and of the Far Eastern Federal District 6.6 million (*Rossiskyii Statisticheskii Yezhegodnik 2005*, Moscow: Rosstat, Table 4.3, p. 83)

21 In 2000 President Putin created 7 federal districts as part of a package of reforms to regain control over Russia's regions. These regions replaced the 11 economic regions previously used to report economic information. The definitions of West Siberia and East Siberia come from that scheme of economic regions. We use both regional schemes in this chapter. The Siberian Federal District was created by combining the old West Siberia and East Siberia regions, with the significant exception that Tyumen (including Khanty-Mansy and Yamal-Nenets) was moved into the Urals Federal District. This was a classic case of divide and rule as it split the major oil and gas regions off from the rest of Siberia. In the 1990s, under the banner of the 'Siberian Agreement', the Siberian regions had effectively lobbied the Yeltsin Government for greater autonomy. The Far Eastern Federal District is the only district that maps directly onto the old economic region. Maps of the economic regions and federal districts can be found at *http://www.geog.le.ac.uk/russianheartland/GeneralMaps/index.html* (Accessed 7 June 2007).

22 Russian reserve classifications A–D do not correspond exactly with general international reserve classification. Hence it is not strictly correct to refer to A+B+C1 reserves as 'proven'. Russian categories A+B equate to what are considered *proven* reserves while C1 roughly equates to *probable* reserves. Russian C2–D2 categories are generally considered as *resources* rather than reserves. For more details see Stern 2005, Appendix 1.1, pp. 60–63. International estimates of East Siberian and Far East Reserves (such as were quoted above for Western Siberia, see note 15) are not available other than for the Sakhalin-1 and 2 fields.

23 *Regiony Rossii Sotsial'no-ekonomicheskie pokazateli 2006*, Moscow: Rosstat, Table 1.5, p. 39.

24 Russian Energy Strategy 2003, p.55. Similar projections can be found in the 2008 Eastern Gas Programme discussed in Section 3.3 1 below.

25 Many of the project details in this section are taken from: Bradshaw 2003 and 2006a, and Sagers 1995.

26 For a detailed history of the Soviet period, see Matthew Sagers, 'Prospects for Oil and Gas Development in Russia's Sakhalin oblast', *Post-Soviet Geography*, Vol.36, No.5, May 1995. pp. 274–290. For a chronology of the Sakhalin tenders, see Paik 1995, pp. 255–60.

27 The Lophukovsky Block of Sakhalin-5 has also been subject to exploration. In 2003 TNK-BP did some initial seismic work. The block was then sold to Sibneft which was acquired by Gazprom in 2006 and in May 2007 Gazprom-neft announced that they were allowing the licence to expire and were no longer interested in the block. However, in July it was reported that

Gazprom Neft had asked Norway's Statoil to consider joining the project, *http://russogasoil.blogspot.com/2007/07/gazprom-neft-tenders-sakhalin-invite.html* (Accessed: 12 September 2007).

28 'Sakhalin-5: oil under the ice', 25 October 2006, *http://www.rosneft.com/news/today/9944.html* (Accessed 01 June 2007.

29 Further information on Urals Energy can be obtained at: *http://www.uralsenergy.com*.

30 Details from: *http://www.rosneft.com/printable/Upstream/Exploration/ExplorationAssets/Sakhalin-3/*. (Accessed 1 June 2007).

31 Initially, the Sakhalin Oblast Administration had a very small share, but it was removed from the project in 2006.

32 'Russia's first offshore oil production successfully completes another section', *http://www.sakhalinenergy.com/en/media.asp?p=media_page&itmID=197* (Accessed 1 June 2007).

33 'Sakhalin II- Meeting the Energy Challenge', *http://www.sakhalinenergy.com/en/ataglance.asp?p=aag_main&s=1* (Accessed 1 June 2007).

34 For a discussion of this campaign see *Squaring Off on Sakhalin: A 'Glocal' NGO Campaign Against Big Oil*, Centre for Applied Studies in International Negotiations, Geneva, 2005.

35 This was the phrase used during discussions with Sakhalin Energy on the island in early 2005.

36 It seems to be common knowledge that the timing of the announcement was forced on Shell as one of its Japanese shareholders informed the CEO of Shell of the preliminary results of an internal audit of Sakhalin-2. Jeroen van der Veer then had to disclose the information to the markets and to Gazprom.

37 *Sakhalin Energy confirms schedule and cost update'*, Press Release SEIC, Yuzhno-Sakhalinsk, 15 July 2005.

38 Ashley Seager, *Shell costs double at Sakhalin*, Guardian Unlimited, *http://www.guardian.co.uk/print/0.3858,5239718-103676,00.html* (Accessed 19 July 2005).

39 Full transcripts can be found at *http://www.ebrd.com/country/sector/natural/projects/sakhalin/*.

40 Much of the Russian position on the economic inequities of the Sakhalin-2 PSA was based on a report commissioned by western NGOs that was translated into Russian and widely distributed to policy makers; see Rutledge 2004.

41 For an analysis of the early stages of the campaign see Bradshaw 2006b.

42 *Gazprom Becomes Shareholder of Sakhalin-2 project*, Gazprom News, 18 April 2007. *http://www.gazprom.com/eng/news/2007/04/23260.shtml*. Accessed: 28 May 2007.

43 *Gazprom now controls SEIC*, The Sakhalin Times, 22 May 2007. *http://www.sakhalintimes.com/news/1/1/11-80.html* (Accessed 22 May 2007.

44 Steve McVeigh, 'Sakhalin II Project', a paper to the 5th Annual Sakhalin Oil and Gas Conference, London 13–14 November 2001.

45 A.S. Visser, 'Overview of Asia Pacific Energy Markets', a paper presented at the 10th Annual Sakhalin Oil and Gas Conference, Yuzhno–Sakhalinsk, 28 September, 2006.
46 Sakhalin-1 Consortium Press Release, Yuzhno–Sakhalinsk, October 30, 2001.
47 Details can be found at the Sakhalin -1 website: *http://www.sakhalin1.com/whatsnew/projectnews/pnw_fact_sheet.asp* (Accessed: 25 May 2007).
48 Sakhalin-1 Consortium Press Release, Yuzhno–Sakhalinsk, 30 October 2001.
49 J.B. King, 'Sakhalin-1 Gas Marketing Update', 8th Annual Sakhalin Oil and Gas Conference, London, 17 November 2004.
50 Greg Jones, 'Exxon Signs Prelim Gas Export Deal with NCPC–Exxon', Dow Jones News Service, 23 October 2006. *http://www.uofaweb.ualberta.ca/chinainstitute/nav03.cfm?nav03=51855&nav02=43661&nav01=43092* (Accessed 03 June 2007).
51 *Ibid.*
52 'Sakhalin I gas deal still up for grabs', Upstream Online, 13 February 2007, *http://www.upstreamonline.com/live/article127745.ece.* (Accessed 3 June 2007).
53 'Gazprom seeks exclusive purchase of Sakhalin-1 gas output', *Novosti*, 28 April 2007, *http://en/rian/business/20070428/64591972* (Accessed 30 April 2007).
54 The Rusia Petroleum website states that the field was located in the mid to late 1970s but officially announced only in 1987, *http://www.rusiap.ru/kovykta/*.
55 The original Rusia Petroleum consisted of the Irkutsk Regional Administration, Irkutskenergo, Angarsknefteorgsintez and Varieganneftegaz. The early paragraphs of this section are based on Simonia 2004.
56 Simonia 2004.
57 Ibid.
58 Paik 2005. Table 5 refers to these as Manzhouli Route 1 and 2.
59 *Ibid.*
60 These figures are taken from Luvsandorj Sukhbuyan, 'Why western route of natural gas pipeline from Kovykta?', Proceedings of the 7th International Conference on Northeast Asian Natural Gas Pipelines, Tokyo, December 3–5, 2001.
61 'China opts for eastern gas pipeline route'; 'Kovykta gas could be supplied to Mongolia', *Interfax Oil and Gas Weekly*, 26 July–1 August 2002, p.15 and 25 November–1 December 2004, pp. 25–6.
62 For the Original Development Plan see Wybrew-Bond and Stern 2002, Table 6.9. p.250.
63 'Rusia Petroleum Chief see no grounds for withdrawal of Kovykta licence', *Interfax Petroleum Report*, 28 February–6 March, 2003, pp. 14–15.
64 'Kovykta int'l feasibility study confirmed', *Interfax Petroleum Report*, 21–27 November, 2003, pp. 6–7.
65 TNK-BP, Irkutsk Region sign agreement establishing East Siberian Gas

Company, *Interfax Petroleum Report*, 12–18 March, 2004, pp. 19–20.

66 As late as 2002, press reports were suggesting that exports to China and Korea could start in 2008 – although this would still have been 2 years later than the licence required: 'Russia, China, Korea to discuss transportation from Kovykta in April', *Interfax Petroleum Report*, 5–11 April 2002, pp. 14–15.

67 'BP to offer Gazprom various forms of participation in Kovykta project'; 'TNK wants Gazprom to join Kovykta gas pipeline project', *Interfax Petroleum Report*, 2–8 November 2001, p.4; 12–18 July 2002, p.18; 'BP to offer Gazprom various forms of participation in Kovykta project', *Interfax Petroleum Report*, November 2–8, 2001, p.4.

68 The actual sale price is uncertain because of complex negotiations which were ongoing at the time of writing and not expected to be completed until December 2007.

69 Because Gazprom only owns 63 percent of Rusia, it is not clear whether the whole of the TNK-BP option would come from this share, or whether Gazprom intends to buy out the other partners in the project before TNK-BP exercises its option: 'Gazprom, BP and TNK-BP enter into major terms agreement', Gazprom Press Release, 22 June 2007; 'BP and TNK-BP plan strategic alliance with Gazprom as TNK-BP sells its stake in Kovykta gas field', TNK-BP Press Release, 22 June 2007.

70 Catherine Belton, 'Kremlin power grows in TNK-BP Siberian gas deal'; Ed Crooks, 'End of Dispute means BP can continue to do business in Russia', *Financial Times*, 23 June 2007.

71 In 2006, Gazprom was given a legal monopoly on exports.

72 'Outcomes of Alexey Miller and Viktor Vekselberg's working meeting'; 'Results of Alexey Miller and Vladimir Potanin's working meeting', Gazprom Press Releases, 4 June and 1 July 2004. Gazprom was subsequently given a legal monopoly of exports in 2006.

73 'Gazprom hedges with TNK-BP on Eastern Siberian fields', *Interfax Petroleum Report*, 1–9 June 2004, pp. 14–15.

74 Kovykta has a significant helium content which, it is claimed by government sources has immense value and requires a specific extraction and transportation programme to be developed at the same time as gas development. Development of a helium industry in Eastern Siberia and the Far East is specifically mentioned in the *Russian Energy Strategy 2003*, p.76.

75 'Energy Ministry proposes four options for sale of Far Eastern gas'; 'Ministry puts $16bn price tag on eastern pipeline', *Interfax Oil and Gas Weekly*, 17–22 February 2005, pp. 8–10 and 1–9 March 2005, pp. 10–11.

76 Press reports suggested that even Alexei Miller had agreed that Kovykta would be the most suitable field for the eastern route: 'Russia–China co-operation documents signed in Beijing', *Interfax Russia and CIS Oil and Gas Weekly*, 16–23 March 2006, pp.6–10.

77 'Gazprom to transport Kovykta gas to west'; 'Ministry says east option most effective for Siberian gas'; 'Gref says market to dictate Kovykta gas pipe

route'; *Interfax Russia and CIS Oil and Gas Weekly*, 9–15 September 2004, p.24, 14–20 April 2005, pp. 32–3, 15–21 September 2005, pp.28–9.

78 *http://www.gazprom.ru/documents/report_Medvedev_26.06.2007.pdf.*

79 'Yakutia to hold closed tender for Chayandinskoye gas field', *Interfax Petroleum Report*, 22–28 November 2002, p.18; 'Russia could restrict foreign access to four hydrocarbon projects', *Interfax Russia and CIS Oil and Gas Weekly*, 13–15 October 2005, p.11.

80 Chayandinskoye, like Kovykta, also contains helium and similar considerations apply, see note 74. The 2007 Eastern Gas Programme for Sakha includes underground storage of helium. For a discussion of potential helium export development from these fields see: 'On Meeting Devoted to Multicomponent Hydrocarbon Field Development in Eastern Siberia and Republic of Sakha (Yakutia)', Gazprom Press Release, 30 March 2006.

81 'Gazprom may launch Sakhalin-3 project in 2014, Chayandy in 2014', *Interfax Russia and CIS Oil and Gas Weekly*, 21–27 June 2007, p.37.

82 'Trutnev says Gazprom proposal on Chayanda, Sakhalin-3 licences irregular', *Interfax Russia and CIS Oil and Gas Weekly*, 14–20 June 2007, pp.10–11.

83 On the energy crisis see Bradshaw and Kirkow 1998; on environmental degradation in general see Vitsovskaia 1999.

84 Both Alexander Gavrin and Governor Yevgeny Nazdratenko officially 'resigned' their positions. In the case of the Governor there were also allegations of serious financial irregularities, but the President did a deal whereby in return for his resignation, Nazdratenko was appointed Chairman of the Federal Fisheries Committee. Astrid Wendtland and Andrew Jack, 'Putin orders shake-up over energy crisis in Far East', *Financial Times*, 6 February 2001; Martin Nicholson, 'Putin's Russia: slowing the pendulum without stopping the clock', *International Affairs*, Vol. 77, No.3, pp. 867–884.

85 Russian Energy Strategy 2001, pp. 69–70.

86 Russian Energy Strategy 2003, p.74.

87 For an analysis of the problems facing the regions see: Vladimir Kontorovich, 'Diagnosing the Russian Far East'. *Post-Soviet Geography and Economics*, Vol. 42, No. 6, 2001, pp 391–415.

88 Russian Energy Strategy 2003, Tables 5.13 and 5.14.

89 These were in 2004 and 2006. 'Gazprom Delegation's Visit to Far East Federal District Finalised'; 'Gazprom Delegation Ends Business Trip to Eastern Siberia and the Far East', Gazprom Press Releases, 19 November 19 2004 and 17 November 2006.

90 See the presentation of Kirill Seleznev at the press conference on the internal gas market and gasification, *http://www.gazprom.ru/documents/report_Seleznev_19.06.2007.pdf.*

91 'Gazprom says market relations with CIS by 2008', *Interfax Russian and CIS Oil and Gas Weekly*, 7–13 December 2006, pp. 25–7.

92 'Gazprom and Altai Republic Ink Cooperation Agreement and Gasification Accord', Gazprom Press Release, September 21, 2006.

93 'On Working Meeting Between Alexei Miller and Alexander Darlin',

Gazprom Press Release, 28 November 2006.

94 'On meeting of coordinating committee for Altai project', Gazprom Press Release, 10 August 2007.

95 'East Siberian Gas Co expects up to 12 bln rubles investment for gasification, *Interfax Russia and CIS Oil and Gas Weekly*, 15–21 March 2007, p.36; 'On Joint Session for Irkutsk Gasification Issues', Gazprom Press Release, 22 December 2006.

96 'Gazprom and Urals Energy sign Memorandum for Cooperation'; 'On joint meeting of Gazprom and Irkutsk Oblast Administration working groups', Gazprom Press Releases, 15 June and 29 August 29, 2007.

97 For a detailed map of Irkutsk Oblast showing all the fields and future distribution plans see, *http://www.gazprom.ru/documents/report_Medvedev_26.06.2007.pdf.*

98 'Gazprom Delegation is Taking part in the Krasnoyarsk Economic Forum', Gazprom Press Release, February 16, 2007.

99 'Gazprom Delegation Visits Evenk Municipal District in Krasnoyarsk Krai', Gazprom Press Release, 26 February 2007.

100 'On working meeting between Alexei Miller and Victor Kress', Gazprom Press Release, 19 February 19, 2007.

101 'Gazprom and Aginsk–Buryat Region Enter into Cooperation Agreement', Gazprom Press Release, 25 April 2007

102 'On working meeting between Alexei Miller and Vyacheslav Shtyrov', Gazprom Press Release, 13 December 2006.

103 'On working meeting between Alexander Ananenkov and Genady Alexeev', Gazprom Press Release, 25 October 2006; 'On working meeting between Alexander Ananenkov and Vyacheslav Shtyrov', Gazprom Press Release, 30 October 2006.

104 'Gazprom and Republic of Sakha (Yakutia) sign agreement of cooperation', Gazprom Press Release, 17 July 2007.

105 'Gazprom and Primorsky Area Enter into Cooperation Agreement'; 'Gazprom and Primorskii Krai sign gasification accord'; Gazprom Press Releases, 1 December 2005 and 15 November 2006.

106 'On Gazprom Delegation's Business Trip to Eastern Siberia and Far East', Gazprom Press Release, 16 November 2006.

107 'On Working Meeting for Gas Supply and Gasification in Kamchatka Oblast', Gazprom Press Release, 29 March 2007.

108 'Commission to discuss KNOC-Rosneft cooperation October 12', *Interfax Russia and CIS Oil and Gas Weekly*, 5–11 October 2006, pp. 22–23.

109 'Energy ministry backs UES initiative on Kamchatka gas producer', *Interfax Russia and CIS Oil and Gas Weekly*, 14–20 June 2007, pp. 12–13.

110 'Gazprom and Administration of the Sakhalin Oblast Signed a Cooperation Agreement', Gazprom Press Release, 14 September 2006.

111 'Postavka gaza na vnutrennii rinok. Rabota elektronnoy torgovoy ploshchadki. Realizatsiya programmi gazifikatsii Rossiskikh regionakh', Press Conference, 19 July 2007, *http://www.gazprom.ru/articles/article24067.shtml.*

112 See section 2 above.

113 'Ob itogakh robochey vstrechi Alekceya Millera c Sergeem Bogdanchikovam I Vladimirov Bogdanovym', Gazprom Press Release, 21 November 2003; 'Gazprom, Rosneft, Sugutneftegaz unite to develop Eastern Siberia', *Interfax Petroleum Report*, 26 December 2003–16 January 2004, p.5; 'Working meeting of the consortium for Eastern Siberia held', Gazprom Press Release, May 14, 2004.

114 TNK-BP may join East Siberia consortium, *Interfax Petroleum Report*, 20–27 May 2004, pp. 12–13.

115 'Gazprom expects East Siberian consortium revival', *Interfax Oil and Gas Weekly*, 26 May–1 June 2005, p.30.

116 'Komplekcnoe osvoenie mestorozhdenie Vostochnoy Sibiri I Dalnevo Vostoka', Gazprom Briefing, 3 June 2004.

117 Miller 2003.

118 The map is reproduced in Stern 2005, Map 3.6, p.158.

119 In sharp contrast to the map presented by the previous Gazprom Chairman at the previous World Gas Conference in which Sakha and Krasnoyarsk gas travelled through Irkutsk with the Kovykta field being the focal point of development (Vyakhirev 2000).

120 Kozmino is at least the third possible destination selected for this pipeline. For a history of the competing oil pipeline projects, see Simonia 2004.

121 Murad D. Muzhamedzhanov and Sergey S. Skaterschikov, *Transneft: new heights – new horizons*, March 2005. *http://www.indexatlas.com/Transneft_presentation_March_2005_Eng.pdf*; 'Khristenko says Russia will deliver oil to China', *Interfax Oil and Gas Report*, 20–26 January 2005, pp.5–7.

122 The Sibneft assets were consolidated with Gazprom's existing oil production into the newly created Gazpromneft, *http://www.gazprom-neft.ru/index-eng.php*.

123 'Alexander Ananenkov's report theses delivered at 4th All-Russian Oil and Gas Week', Gazprom Press Release, 26 October 2004. (Translated from the Russian by the author, the English press release talks about 'supplying targeted markets on an uncompetitive basis'.)

124 'Gazprom looks to control major gas fields in Eastern Russia', *Interfax Russia and CIS Oil and Gas Weekly*, 11–17 January 2007, pp. 28–9.

125 For the story of the failed Gazprom–Rosneft merger see Stern 2005, pp. 195-7.

126 See Deputy Chairman Alexander Medvedev's statement on the region: 'Gazprom na vostoke Rossii, vykhod na rinki stran ATP', Gazprom Press Conference, 26 June 2007, *http://www.gazprom.ru/articles/article24154.shtml*; see also Gazprom's statement on the region: 'Gazprom na vostoke Rossii, vykhod na rinki stran ATP', Spravochniye materiali, 26 June 2007, *http://www.gazprom.ru/articles/article24166.shtml*

127 *http://www.minprom.gov.ru/press/release/showNewsIssue?url=activity/energy/news/329/* The official name of the programme is: *Programme for the creation of a unified system of gas production, transportation and supply in Eastern Siberia*

and the Far East including possible exports of gas to China and other Asia-Pacific countries. This press release was the only official document available at the time of writing.

128 'Industry & Energy Ministry selects eastern gas strategy for Gazprom', *Interfax Russia and CIS Oil and Gas Weekly*, 6–12 September 2007, pp. 5–6; Yelena Mazneva, 'Vostochnuyu Sibir' razrabotayut po planu Gazproma', *Vedemosti*, 10 Sepember 2007.

129 'A Long Term Vision of Natural Gas Infrastructure in North East Asia – 2007 Version', North East Asian Gas and Pipeline Forum, September 2007, Figure 2.1.5, p.18.

130 2.4 trillion rubles is around $100bn at 2007 exchange rates.

131 Paik and Choi 1998 date Gazprom's interest from February 1997.

132 Vyakhirev 1997.

133 *The North East Asia Energy and Environmental Cooperation: Russian Approach,* Tokyo: NIRA Research Output, Vol 13, No. 1, 2000, p.47.

134 Vyakhirev 2000.

135 'Vostokgazprom celebrated 5-year anniversary of commercial gas production', Gazprom Press Release, 20 May 2004.

136 Vostokgazprom was established in 1999 to coordinate Gazprom's work in Eastern Russia and has developed to the point where the company has 30 production and transmission subsidiaries: S. Jvatchkine, 'The perspective energy projects of Vostokgazprom in China', proceedings of the Conference, *China Energy and Western Regional Economic Development*, 12–13 October 2000, Beijing, Royal Institute of International Affairs.

137 For a detailed account of the Chinese perspective on the Russian relationship see Chapter 2.

138 Shell and Gazprom intended to be partners in the project acting jointly. Reports on the withdrawal of the foreign investors did not make clear whether the companies had withdrawn from the project (given that the pipeline was virtually completed and no agreement had been signed), or whether Petrochina had told them that they were surplus to requirements. Carola Hoyos and Richard McGregor, 'PetroChina ends talks on pipeline venture with foreign companies', and 'Pipeline pullout embarrasses Petrochina', *Financial Times*, 4 August 2004.

139 'Gazprom, CNPC ink strategic gas cooperation deal', *Interfax Petroleum Report*, 14–20 October 2004, p.28.

140 'Ministry puts $16bn price tag on eastern pipeline', Interfax Oil and Gas Report, 3–9 March 2005, pp.10–11.

141 'Russia to supply 68 Bcm of gas per year to China by 2020', *Interfax Russia and CIS Oil and Gas Weekly*, 30 March–4 April 2006, p.31.

142 'Cost of Siberia–China Gas Pipe Estimated at $14bn', *Interfax Russia and CIS Oil and Gas Weekly*, 9–15 November 2006 p. 44.

143 'Gazprom and CNPC coordinating natural gas supply to China', *Interfax Russia and CIS Oil and Gas Weekly*, 11–17 May 2006 pp. 31–2.

144 'Russians approve Altai gas pipeline project'; 'Environmentalists criticise

Gazprom over Altai project', *Interfax Russia and CIS Oil and Gas Weekly*, 15–21 March 2007 pp. 31–32; 22–28 March 2007, pp. 36–37.

145 Gazprom Press Releases of 11 May, 7 and 17 June 2005 and 3 August, 6 and 18 October 2006.

146 JCC refers to Japan Crude Cocktail, a mixture of crude oils imported into Japan which determines the pricing of most of the LNG traded in the Pacific Basin. Miller 2006, *http://www.gazprom.com/eng/articles/article19731.shtml*.

147 The estimate of $13.6bn given above, made in 2006, may continue to rise due to the increasing cost of inputs, particularly steel.

148 In fact both Presidents made only passing references to 'energy' issues. See press statement following Russian–Chinese Talks, 26 March 2006; *www.kremlin.ru/eng/text/speeches/2007/03/26/1124*

149 Peter Smith, 'China's LNG deals "good" for Gazprom discussions', *Financial Times*, 10 September 2007.

150 'Gazprom na vostoke Rossii, vykhod na rinki stran ATP', Press conference, 26 June 2007, *http://www.gazprom.ru/articles/article24154.shtml*; see also the Chart in the presentation which accompanies that press conference which purports to show that Sakhalin-1 gas is insufficient to meet the regional gas demand of the Far Eastern Republics, *http://www.gazprom.ru/documents/report_Medvedev_26.06.2007.pdf*.

151 For more detail on the Korean perspective of the relationship see Chapter 5.

152 'Agreement on cooperation between Gazprom and the Kogas corporation (South Korea)', Gazprom Press Release, 12 May 2003.

153 'Outcomes of Alexey Miller and Jong Tae Ick's working meeting', Gazprom Press Release, 4 June 2004.

154 According to Gazprom press releases, meetings were held on 5 August and 3 September, 2004; 14 July 2005; 5 April, 29 July and 17 October 2006; 6 April and 7 September 2007. Meetings have also been held with other Korean representatives.

155 'Gazprom supplies first LNG cargo to Republic of Korea', Gazprom Press Release, 11 October 2006.

156 'Russia, South Korea prepare to sign gas agreement', *Interfax Oil and Gas Weekly*, 20–26 October 2005, p.8.

157 This is included in the preliminary programme for which more data is available, see North East Asian Gas and Pipeline Forum 2007, Figure 2.1.5, p.18.

158 The stress here is on the demise of traditional contracts with oil-related pricing and high levels of take or pay, rather than any general disappearance of long term contracts.

159 Stern 2006.

160 See the comments of presidential advisor Suvalov and Head of the Russian Union of Industrialists Shokhin in: 'Shuvalov says Russia's integration into global energy system will benefit all'; 'Shokhin warns Europe of gas supply

problems', *Interfax Russia & CIS Oil and Gas Weekly*, 18–24 May 2006 pp. 21–2, and 25–31 May 2006, pp. 9–10.

161 'Medvedev says Gazprom should be state-controlled', *Interfax Russia & CIS Oil and Gas Weekly*, 1–7 June 2006, pp. 14–16.

162 For example, John Gapper, 'We should learn to love foreign state-owned invaders', and Pierre Noel, 'How oil majors can thrive in an age of state monopolies', *Financial Times*, 28/29 July 2007 and 10 August 2007.

163 Inverview with Newspaper Journalists from G8 Member Countries, 4 Jun 2007, *http://www.kremlin.ru/eng/text/speeches/2007/06/04/2149_type82914type82916_132772.shtml.*

164 The principal risks being economic and political instability in Russia and a much lower oil and gas price environment.

165 This line of reasoning is more applicable to Sakhalin-2 than Kovykta where no substantial development had taken place prior to the Gazprom takeover.

166 Paik 2007.

167 Miyamoto 1997, pp. 71–74.

168 *Ibid*, Exxon was also involved in the feasibility study.

169 'Focus on Turkmenistan', *European Gas Markets*, 16 May 2007, pp. 4–5.

170 'Gas reserves at Amudarya right bank 1.3 Tcm – Turkmen president', *Interfax Russia and CIS Oil and Gas Weekly*, 30 August–5 September 2007, p.31.

171 'Turkmenistan audits reserves, won't tell Gazprom', *Interfax Oil and Gas Weekly*, June 2–8 2005, pp. 37–8

172 Press Conference Following a Trilateral meeting of the Presidents of Russia, Kazakhstan and Turkmenistan, 12 May 2007, *http://www.kremlin.ru/eng/text/speeches/2007/05/12/2059_type82914type82915_128357.shtml.*

173 Demand estimates in these paragraphs are 1998 estimates from the IEA, details in Wybrew-Bond and Stern 2002, Table 6.6, p. 241.

174 In 2006, Uzbekistan exported approximately 6 Bcm to Russia and 6.7 Bcm to Central Asian countries.

175 Principally associated gas from the Tenghiz field and the gas cap at the recently discovered Kashagan gas condensate field (not included in the figures in Table 6.5).

176 For a discussion of the relationship between Central Asian and Russian gas exports, see Stern 1999.

177 'Kazmunaigaz could start shipping gas to China by late 2008'; 'Rosneft not planning to use the Atasu–Alaskankou pipe in 2006', *Interfax Russia and CIS Oil and Gas Weekly*, 3–9 February 2005, p.30 and 13–19 July 2006, p. 22.

178 'Kazakh estimates gas pipe to China at $2bln–$4bln', *Interfax Russia and CIS Oil and Gas Weekly*, 28 September–4 October 2006, pp. 47–8.

179 'Turkmenistan to export 30 Bcm of gas per year to China', *Interfax Petroleum Report*, 24–30 November 2005, p.53.

180 The full text of the agreement can be found at *BBC Monitoring Service*, 4 April 2006.

181 The total length of the line will be around 6,500km of which 188km will be in Turkmenistan, 530km in Uzbekistan, 1,300 in Kazakhstan and more than 4,500km in China. 'Gas reserves at Amudarya right bank 1.3 Tcm – Turkmen president', *Interfax Russia and CIS Oil and Gas Weekly*, 30 August–5 September 2007, p.31.

182 Kazakh government sources have said that they would consider taking some of the gas from the Turkmen line to meet demand in the south of their country. 'Kazakhstan to allow Turkmenistan–China gas pipeline on its territory'; 'Uzbekistan, China to agree gas transit pipeline route by 2008', *Interfax Russia and CIS Oil and Gas Weekly*, 3–8 May 2007 pp.33 and 35.

183 For a discussion of Central Asian export prices see Pirani 2007, especially pp. 35–38.

184 Wybrew-Bond and Stern 2002, pp.273–4.

CHAPTER 7

SOUTH EAST ASIAN AND AUSTRALIAN GAS SUPPLY TO ASIA

David Ledesma

Introduction and Regional Overview

This chapter will discuss the gas supply countries of South East Asia and Australasia, which spans the region from Indonesia to Australia and Papua New Guinea. As Table 7.1 shows, this includes countries with wide variations in economic growth, population and politics all of which drive energy policy and the structure of the gas businesses which countries develop.

Table 7.1: South East Asian and Australian Gas Reserves and Economic Data

	Gas reserves Trillion cubic metres	*Gas reserves Change since 2000 (percent)*	*GDP US$ Billion*	*GDP per capita US$*
Indonesia	2.63	28	364	1640
Malaysia	2.49	7	157	5610
Australia	2.61	19	732	33120
Myanmar	0.54	145	12	208
Papua New Guinea	0.42	2	5	500
Brunei	0.34	–13	12	30928
Thailand	0.31	–9	206	3166
Vietnam	0.40	111	61	723

Source: ASEAN (2006 data for all countries except Australia and Papua New Guinea); World Bank (2005 data for Australia and PNG); BP, *Statistical Review of World Energy*, June 2007.

The region includes the Association of Southeast Asian Nations (ASEAN) founded in 1967 by Indonesia, Malaysia, Philippines, Singapore and Thailand.[1] ASEAN includes five gas exporters (Brunei, Indonesia, Malaysia, Myanmar and Vietnam) and two gas importers (Singapore and Thailand) and was established to encourage economic growth, social progress and cultural development amongst the member states through integration and working together. Energy cooperation between ASEAN members is of importance, but it could also become increasingly important to other countries should there be a change of policy across the region that encourages the use of hydrocarbons within the ASEAN region rather than their export to other countries. At the time of writing there is no indication that such a policy is being considered.

A full discussion of South East Asian gas supply must include, in addition to ASEAN countries, Australia, an existing exporter of liquefied natural gas (LNG), and Papua New Guinea, a country which is considering development of an LNG industry. Together this region exported 64 Mtpa (86.5 Bcm/yr) LNG in 2006 to seven countries, accounting for 40 percent of global LNG exports.

Singapore, the only ASEAN country without any indigenous energy resources, imports gas by pipeline from Indonesia and Malaysia, and, ever conscious of security of supply, is planning to develop an LNG import terminal to provide alternative gas supply sources. Thailand and Vietnam have considerable gas reserves, but they are targeted at the domestic market, specifically power generation. Myanmar has to date only exported gas by pipeline to Thailand, but recent offshore discoveries of just under 0.54 Tcm of gas may change this. The Philippines has developed its gas reserves from the Malampaya field which is used in power generation, and is exploring for further hydrocarbons. Cambodia is also exploring for gas and has made some small discoveries, but not sufficient for an export scheme.

Natural gas trade in the South East Asian region is therefore dominated by exports and particularly the LNG business which enables economic movement of gas over long distances, specifically to Japan, Korea and Taiwan under long-term contracts. LNG exports in the region started in the 1970s and many contracts have already been extended for additional twenty-year periods. Pipelines within the region are limited as key gas resource holders have favoured LNG exports to reach higher value markets. That said, there is, as discussed later in this chapter, a move in Indonesia to keep gas in the country for purposes of industrial development with resulting economic and political employment benefits.

ASEAN Cross-Border Linkages

The first trans-border pipeline in the region was completed in 1992 when Peninsular Malaysia's pipeline network was extended to Singapore. In 1998 the next trans-border pipeline was completed when the offshore gas field at Yaduna in Myanmar, followed by a second field

Yetagon, was connected to Thailand. In 2001, Indonesia's West Natuna field was linked to Singapore through a 640 kilometre (400 mile), 3.36 billion cubic metres per year (Bcm/yr), subsea pipeline giving a further cross-border regional linkage and in August 2003 a second 3.62 Bcm/yr natural gas connection to Singapore was opened when the South Sumatra–Singapore pipeline was completed. These small pipelines are seen to be the first stages of a wider Trans-ASEAN Gas Pipeline (TAGP), see Map 7.1, which envisions the establishment of a transnational pipeline network linking the major natural gas producers and consumers in South East Asia. Many in the region hope that TAGP will be completed and provide pan-ASEAN gas flows by 2020, the year set by the ASEAN energy ministers when they met in July 2002 as the target for the region to reach a stated level of economic growth. Achieving this aim will need key issues to be resolved between the different ASEAN members including cross border charges and operations, as well as the logistics of transporting power between countries, for instance, transmitting the abundant hydroelectric power from Laos to Thailand.

Whether the ASEAN pipeline network will be completed is questionable and there are two distinct views on the question. Some argue that efforts to develop the trans-ASEAN pipeline took place in an era when available gas supply was seeking markets, and so the gas sellers were interested in developing new routes to market. In 2007 however, the market situation was very different, with a tight supply situation allowing LNG exports to command significant premiums over domestic gas prices in most countries. Thus, the main driver to the development of the trans-ASEAN pipeline has been removed, the easy links have been constructed, and even though part of the pipeline has been constructed (from Bangkok to Jakarta through Malaysia, Singapore, then across Java to Surabaya) it is unlikely that the section from Java to Brunei and onto the Philippines will be developed due to limited stakeholder interest in supplying countries and high construction costs.

Others argue that the increasing linkage of energy policy within ASEAN, and the existence of physical power and gas grids within, and to some extent between, the countries mean that there are already sufficient connections. Gas can move from Malaysia to Singapore and to Indonesia and there is optimism that further interconnections of national networks will happen – it simply makes sense. It is worth noting that it is electric power which is of primary importance for ASEAN energy integration and gas is not necessarily the key to the growth of such interconnections, since gas is simply used to generate power. Map 7.2 shows the existing Asean Power Grid. The expansion

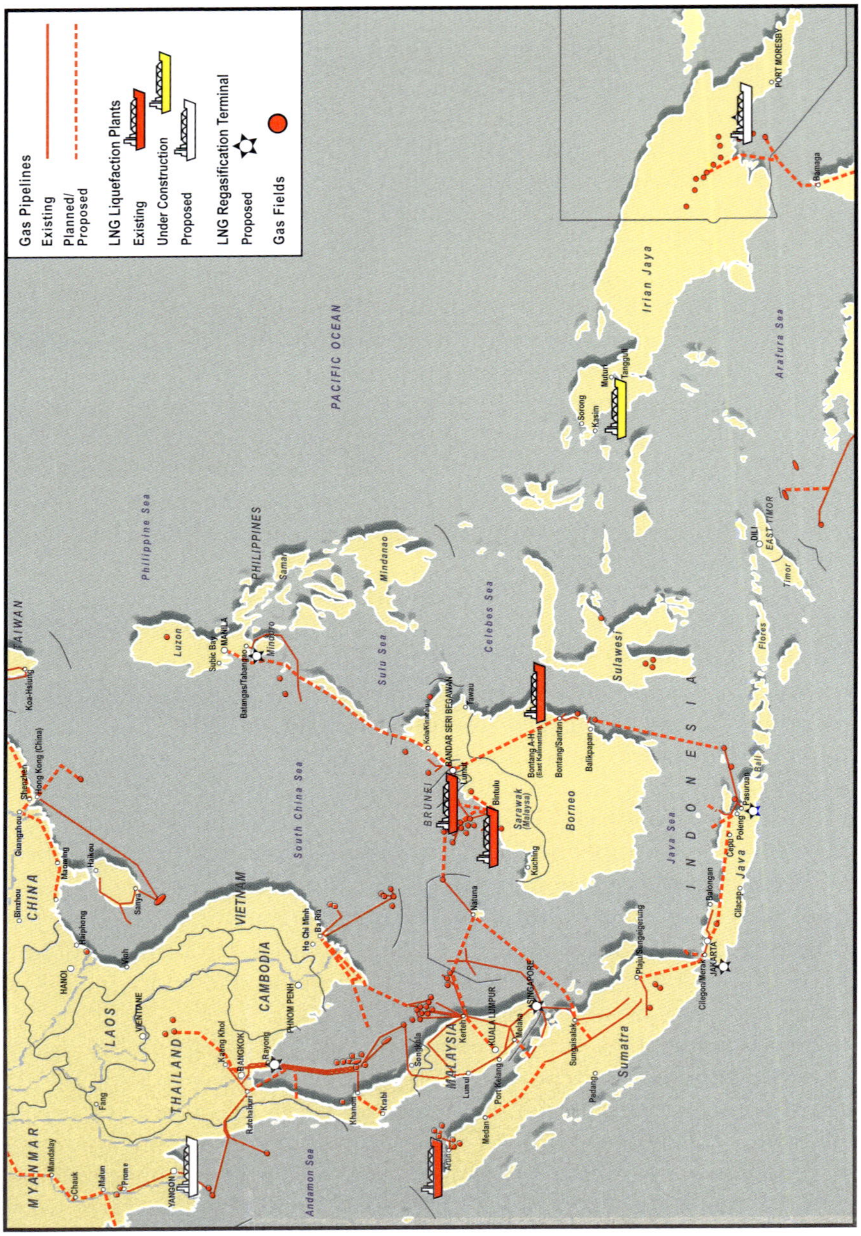

Map 7.1: Asian Gas Pipeline Network

Source: Compiled by author with input from www.aseaenergy.org and *Petroleum Economist.*

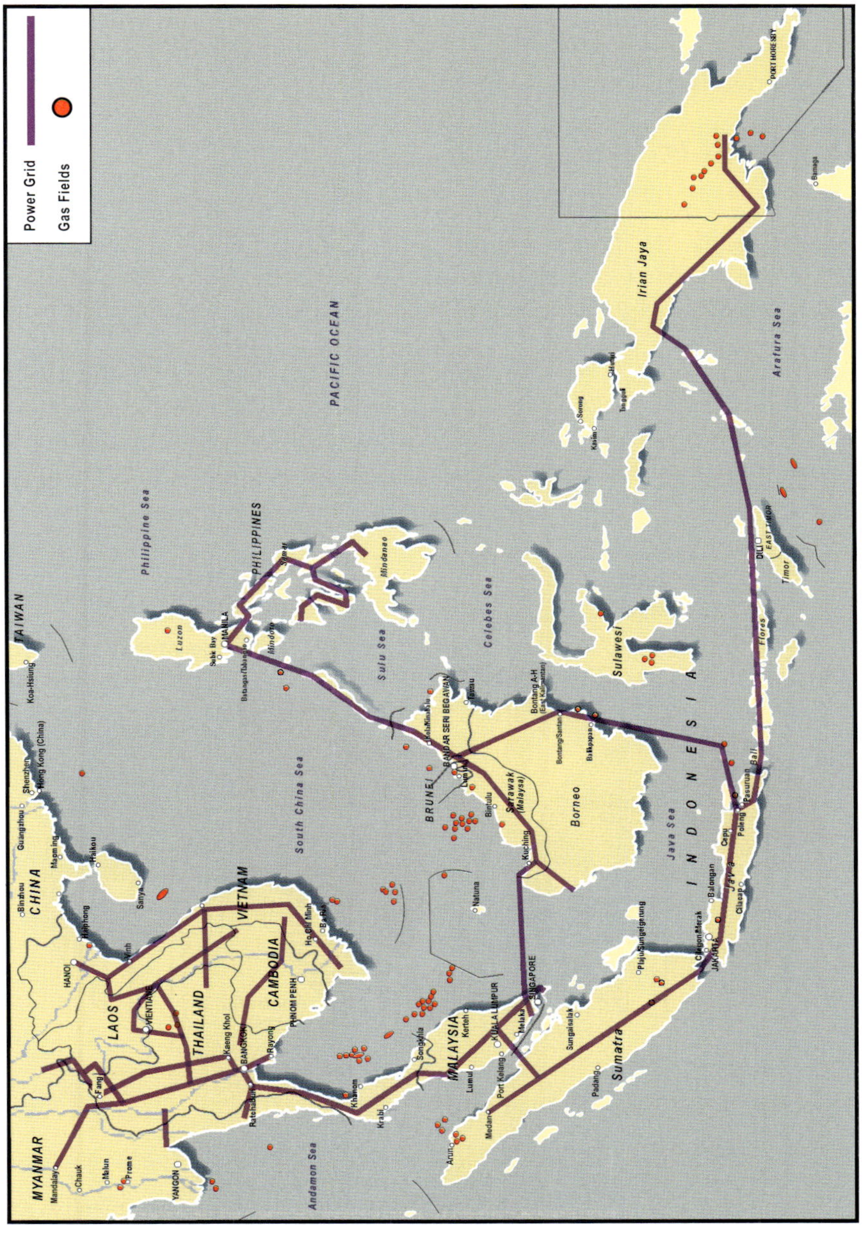

Map 7.2: ASEAN Powergrid

Source: www.aseanenergy.org

of power grids would be a function of the development of hydroelectric power rather than gas-fired generation, so is not reliant on additional gas reserves being found.

LNG Supply to Asia

South East Asian countries and Australia are, and will remain for some years to come, the largest exporters of LNG, primarily to markets in the region. As shown in Figure 7.1, in 2006 the Asian LNG market was 103.6 Mtpa (140 Bcm/yr), of which 62 percent was supplied from within the region compared to 2000 when the market was 72.5 Mtpa (98 Bcm/yr) and 76 percent from the region. This shows the rising importance of the Middle East as a supplier to Asia and the reduction in LNG volumes from Indonesia, due to reduced gas availability and a government policy of diverting gas to domestic markets (see Figure 7.2).

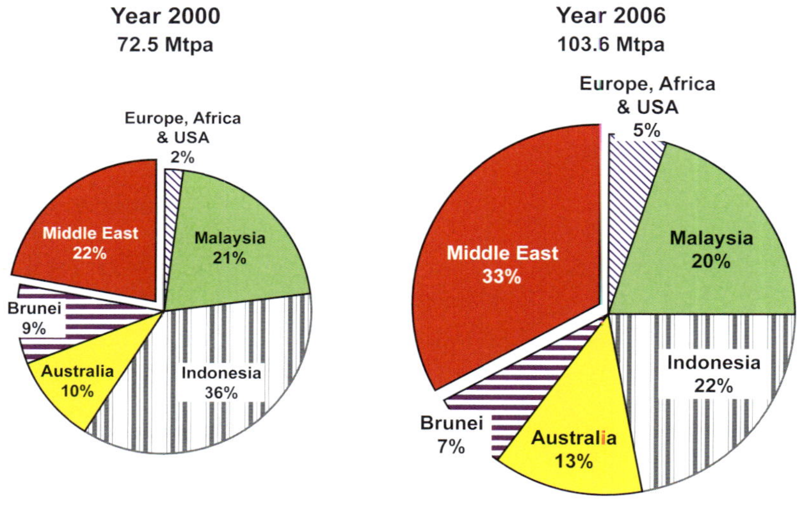

Figure 7.1: LNG Supply to Asia

Source: Cedigaz.

During this period, LNG exports from Australia have increased with the expansion of the North-West Shelf project and the start-up of Darwin LNG in February 2006. Australian LNG production is expected to increase further and the country should become the largest regional LNG supplier early in the next decade (see Australia section below).

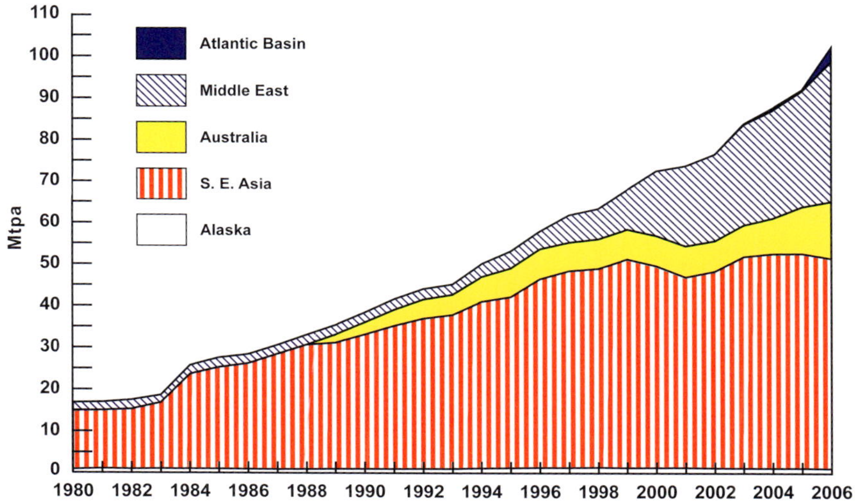

Figure 7.2: LNG Supply to Asian Markets 1980–2006

Source: Flower LNG.

Indonesian Gas Supply

Background

The Indonesian Archipelago has a population of approximately 234 million, the fourth largest in the world after China, India and the USA. The population is made up of several ethnic groups of which Javanese are the largest at 45 percent. The majority of the population (88 percent) is Muslim. Estimated 2006 GDP (at the official exchange rate) was $364 billion with GDP per capita at $1,640.

Indonesia has 2.63 Tcm of proven gas reserves and is the tenth largest holder of natural gas reserves in the world and the single largest in the Asia Pacific region. More than 70 percent of the country's natural gas is located offshore with the largest reserves found around Natuna Island, East Kalimantan, South Sumatra and West Papua (also known as Irian Jaya).

Under the Indonesian Constitution, following independence in 1945, it was stipulated that all natural resources would be controlled by the state and 'exploited for the greatest welfare of the people'. This was enshrined as a feature of the 1960 petroleum law which permitted foreign oil companies to operate in Indonesia, but only as contractors to the state enterprise. A subsequent law in 1971 identified that enterprise as Pertamina. Indonesia had an established oil industry even before independence and by the 1960s had a very significant oil trade with Japan.

It was a particular set of coincidental circumstances that made Pertamina take a new strategic look at the opportunities in LNG and then fully pursue it with maximum vigour to the extent of simultaneously developing two world-class liquefaction plants. Those circumstances included Japanese oil buyers, aware of the new technology (LNG), who advanced the idea of reducing the substantial pollution in Japanese cities and Japan's strategic decision, following the 1973 oil price increases, to change the energy portfolio of the country by importing natural gas. In parallel with this, gas reserves of about 0.65 Tcm were discovered by Mobil and Huffco in remote locations in Indonesia. It was the combination of these two factors that led to the development of Indonesian LNG exports.

While the Indonesian domestic gas supply position is still fragmented, lacking a national grid and any substantial domestic market penetration, Indonesia has developed a significant petrochemical industry requiring methane feedstock. Plants are located within the regions of the larger gas finds making bulk supply available and transportation easy for either export or indigenous use in the Indonesian islands. Examples are the urea and methanol plants in East Kalimantan and urea plants in Aceh. Gas feedstock prices to these plants are regulated, but are increasing gradually from $2/MMBtu in 2005 to $4–5/MMBtu in Q3 2007 for some of the newer gas buyers. The prices achieved by gas producers for gas sold within the country, compared with export prices, have been a cause of friction in the mid to late 2000s.

Pertamina describes how its LNG business evolved from a 'Golden Era' (1977–2000) and to a 'Challenging Period' (2001–2007).[2] During the Golden Era, Pertamina was one of the pioneers in the Pacific LNG market and steadily expanded, becoming the world's largest LNG exporter (losing the 'title' to Qatar only in 2006). Pertamina was also rightfully proud of its unblemished record as a reliable LNG supplier for more than twenty years. In those years Pertamina developed very close relations with its buyers and in many years was able to supply LNG volumes above the contractual requirements, while acting as the sole seller of Indonesian LNG. The LNG contracts agreed between the buyers and Pertamina, were matched by back to back contracts with producers who were assigned gas under 'packages'. A package comprised the consolidated gas volume from the different Production Sharing Contract (PSC) participants that made up the quantity of gas obligated under a contract.

LNG Export Plants

Arun Plant

Mobil (now ExxonMobil) acquired a PSC over two blocks in North Sumatra in 1968. After more than ten dry-holes, in 1971 it discovered the Arun gas field – some 0.48 Tcm of gas and condensate in a single reservoir. Based around these reserves an LNG project was developed with first deliveries of LNG from the initial three-train Arun plant, of 4.5 Mtpa (6.1 Bcm/yr) capacity, to Japan in 1978. The capacity of the plant was expanded in the 1980s and at its peak the Arun plant had six trains operational with a total of 12.4 Mtpa (16.8 Bcm/yr) capacity. The plant is wholly owned by Pertamina. The LNG is sold to Tokyo Electric (TEPCO), Tohoku Electric of Japan and Korea Gas Corporation (Kogas). Table 7.2 gives details of the current contractual position. The TEPCO and Tohoku contracts expire in 2009 and one of the two Kogas contracts for 2.3 Mtpa (3.1 Bcm/yr) at the end of 2007. The second Kogas contract for 1.0 Mtpa (1.35 Bcm/yr) expires in 2014. Reserves are, however, in decline and, despite significant exploration for gas, additional gas reserves in the area are not suitable for export because of high nitrogen content. Hence output from the existing trains is declining and is expected to stop altogether in 2010 when the 1.0 Mtpa (1.35 Bcm/yr) being supplied to Kogas from Arun is likely to be moved across to the Bontang plant. The only reason to keep the Arun plant in operation at such low production levels could be political, since state and federal governments are putting pressure on the operator to retain employment in the region.

Reduced production from Arun will no doubt have an impact on government budget requirements, countered by higher oil and gas prices and revenues, and by the start-up of the Tangguh plant in 2009.

Badak (Bontang Plant)

The second facility in Indonesia is the Bontang LNG project. In 1968

Table 7.2: Indonesia–Arun LNG Contractual Position (as at September 2007)

	Volume (BCMA)	*Primary Destination*	*Start Date*	*End Date*	*Delivery*
Tokyo Electric (TEPCO)	0.13	Japan	2005	2009	FOB
Tohoku Electric	0.83	Japan	2005	2009	FOB
Kogas	2.30	Korea	1986	2007	ExShip
Kogas	1.00	Korea	1994	2014	FOB

Source: Compiled by author with input from company sources.

the Sanga Sanga PSC was granted to Huffco over an area of the Mahakam Delta and after geological fieldwork and seismic surveys the first field, Badak, was discovered in 1972 with around 0.31 Tcm of gas and oil. As the area adjacent to Badak was a swamp and therefore unsuitable for either the footings of an LNG plant or the development of the necessary deep water harbour, a port location at Bontang Bay, some 35 miles north of Badak, was selected for siting the plant and harbour. First LNG deliveries from Bontang to Japan were made in 1977.

The Bontang plant is operated by PT Badak NGL Company and the current contractual position can be seen in Table 7.3. Since 1977 it has been expanded gradually and currently has eight trains with a capacity of 23 Mtpa (31.5 Bcm/yr). In 2006 it produced 18.5 Mtpa (25.3 Bcm/yr) LNG. The plant is wholly owned by Pertamina and the LNG is sold to Japan, Korea and Taiwan. Whilst the existing gas reserves of Badak are in decline, the area has proved prolific. Huffco, and its successor company Vico, had more finds onshore while other operators, Total and Unocal (now part of Chevron), had later success moving offshore into deeper and deeper water as technology developed

Table 7.3: Indonesia–Bontang LNG Contractual Position (as at September 2007)

	Volume (BCMA)	*Primary Destination*	*Start Date*	*End Date*	*Delivery*
ChubuElectric	2.15	Japan	1977	2010	ExShip
ChubuElectric	1.71	Japan	1983	2011	FOB
Hiroshima Gas	0.19	Japan	1996	2015	ExShip
Kansai Electric	2.56	Japan	1977	2010	ExShip
Kansai Electric	0.91	Japan	1983	2011	FOB
Kyushu Electric	1.55	Japan	1977	2010	ExShip
Nippon Gas	0.06	Japan	1977	2010	ExShip
Nippon Steel	0.62	Japan	1977	2010	ExShip
Osaka Gas	1.30	Japan	1977	2010	ExShip
Osaka Gas	1.10	Japan	1994	2013	FOB
Osaka Gas	0.46	Japan	1983	2011	FOB
Osaka Gas	0.10	Japan	1996	2015	ExShip
Toho Gas	0.57	Japan	1983	2011	FOB
Toho Gas	0.25	Japan	1977	2010	ExShip
Toho Gas	0.10	Japan	1994	2013	FOB
Tokyo Gas	0.80	Japan	1994	2013	FOB
Kogas	1.03	Korea	1998	2017	FOB
Kogas	0.97	Korea	1994	2014	FOB
CPC	1.84	Taiwan	1998	2017	ExShip
CPC	1.50	Taiwan	1983	2009	ExShip

Source: Compiled by author with input from company sources.

and supported the progressive expansion of the plant. However, in recent years the reservoir performance has been below expectation and there is currently insufficient gas to meet the contract commitments. As noted earlier in this chapter, Chevron and Total are also both exploring and developing gas reserves for supply to the Bontang plant. The existence of largely financially amortised plant onshore should permit the development of gas which, on a field stand-alone basis, may have been uneconomic.

Tangguh

The newest LNG development in Indonesia is the BP-led Tangguh project in Papua province. The Tangguh field contains just under 0.45 Tcm of proven gas reserves found both offshore and onshore in the Wiriagar and Berau blocks. The project received final approval from the Government of Indonesia in March 2005. The project shareholders are BP with 37.16 percent, China National Offshore Oil Corporation (CNOOC) 16.96 percent, Mitsubishi 16.3 percent, Nippon Oil 12.23 percent, Kanematu 10 percent and LNG Japan 7.35 percent. In early October 2007 it was announced that Kanematsu had sold its 10 percent interest to Nippon Oil, Mitsui and Mitsubishi & Inpex. It is interesting to note that Pertamina is not a shareholder in the Tangguh project as it was developed after the Indonesian government restructured the gas business in 2001 and BP Migas took on the role of the primary regulatory authority in the gas sector.

The project is being developed as a two-train, 7.6 Mtpa (10.3 Bcm/yr) plant, and is expected to start up in early 2009. LNG has been sold on a long-term basis to the Fujian LNG facility in China, K-Power & POSCO in Korea and Sempra Energy, for supply to Mexico, though it is understood that half of the volume can be diverted to Asian markets. Table 7.4 gives details of the current contractual position. The prices agreed with K-Power, POSCO and Fujian LNG are very low (with a price ceiling based on \$25/bbl oil) as they were negotiated in 2001 during the period of a 'buyers market' for LNG. With a shift to a 'sellers' market' after 2005, Indonesia has successfully sought higher prices and an increase of the oil price ceiling with Fujian to \$38/bbl so that the oil price cap has increased LNG prices from \$2.40/MMBtu to \$3.35/MMBtu, still well below 2007 regional market LNG price levels of around \$8/MMBtu.

There are plans for a third train but it will require further exploration and appraisal to prove up sufficient reserves. That said, it is expected that a third train will be operational by 2013–2014. If this expansion proceeds, there will be some considerable political discussion as to how much of the LNG will stay within Indonesia. In order for gas to be prioritised for the domestic market, domestic buyers will have to match international prices or, the Government of Indonesia may have to give subsidies to buyers to enable them to pay the required price levels.

Table 7.4: Indonesia–Tangguh LNG Contractual Position (as at September 2007)

	Volume (BCMA)	*Primary Destination*	*Start Date*	*End Date*	*Delivery*
CNOOC	2.60	China	2008	2032	FOB
POSCO	0.55	Korea	2005	2027	Ex-Ship
K Power	0.60	Korea	2006	2027	Ex-Ship
Sempra Energy	3.70	USA/Mexico and trading	2011	2035	Ex-Ship
Tohoku Electric	0.12	Japan	2010	2025	

Source: Compiled by author with input from company sources.

Senoro LNG Project (Sulawesi)

Located in Central Sulawesi, the currently named Senoro LNG project (though it has at times been referred to as the Donggi project), involves gas production and the construction of a 2 Mtpa (2.7 Bcm/yr) LNG plant. The project will be operated by a joint venture established by Mitsubishi (initially with a 60 percent share, Pertamina and Medco each holding 20 percent). The project will take gas from the Matindok and Senoro blocks and in May 2007 it was reported that the partners had agreed a gas price and plans and in August 2007, during a visit by the Japanese Prime Minister to Jakarta, Mitsubishi signed a contract with Pertamina and Medco to build the $1.13bn LNG plant, to start production in 2010. The finally agreed shareholding for the LNG plant is Mitsubishi 51 percent, Pertamina 29 percent and Medco Energi International 20 percent. All the production from the plant is understood to be sold to Japan.

The 'Energy Policy Dilemma' of the 2000s

Prior to 2001 Pertamina managed the Indonesian LNG business on behalf of the government. Since 2001, when Pertamina was forced to relinquish its monopoly status in upstream natural gas projects, BP Migas (no relation to the international oil company BP) has become the primary regulatory authority in the gas sector. Since 2001 Pertamina has suffered from gas supply shortages, as a result of an uncertain fiscal environment, and lack of clarity in the LNG business structure.

Pertamina also had a difficult management problem as its role was changed from being a regulator to a profit-driven limited liability independent company (albeit government-owned), combined with an obligation to supply domestic gas demand at considerably lower prices than could be achieved through exports. This, together with what production sharing (PSC) contractors see as low upstream profitability, has resulted in those contractors not wishing to invest, with resulting LNG supply shortages and contract renewal uncertainties.

Buyers became very concerned by this situation and in 2006 Pertamina reviewed its strategy and faced the issue of gas supply shortages using long-term relationships with LNG buyers to encourage cooperation and flexibility while the supply situation in the country was resolved. The key problem is that Pertamina and international companies are seeking international prices for their gas, but if the government insists that new gas is supplied domestically at lower prices, then new gas exploration may not take place and production will not increase. If, however, gas is exported to higher paying markets, then domestic industry may not be developed, with a resulting loss of employment. This, the 'Energy Policy Dilemma', is one faced by many countries but is particularly acute in Indonesia due to the political imperative to increase supplies to the domestic market resulting from the increase in economic growth and reduced availability of gas. To politicians the export of gas when local industry lacks energy supplies, even though it earns valuable revenue, is simply not acceptable. Politicians want to see job creation and growth as a means of ensuring country stability and this requires energy supplies. Stability is particularly important in Indonesia which has seen some local groups seeking independence, for instance, the Free Aceh Movement (Gerakan Aceh Merdeka, or Gam, founded in December 1976) which has been battling for the province's independence for the past 26 years.[3]

Pertamina on the other hand has contractual obligations to meet and is required to operate as a profitable organisation. It is therefore seeking to maximise its revenues and aspires to achieve this through increased LNG exports rather than meeting domestic gas supply obligations which provide much lower returns.

The price of gas sold to industry has seemingly been increasing steadily from an average of $3.50/MMBtu in 2005 to $5.50/MMBtu in Q3 2007, while gas export prices have been $3–4/MMBtu higher. Domestic price levels in the power sector have been even lower, resulting in a wider domestic/export price differential. Reports from BPMigas state that Indonesian companies have during 2007 concluded domestic gas purchase agreements in the range $3–5/MMBtu, with one at $3/MMBtu being fixed with only a 2 percent annual escalation up to 2022. Figure 7.3 shows this comparison which, although it seems to be narrowing, still creates pressure for gas to be exported and so earn higher revenues.

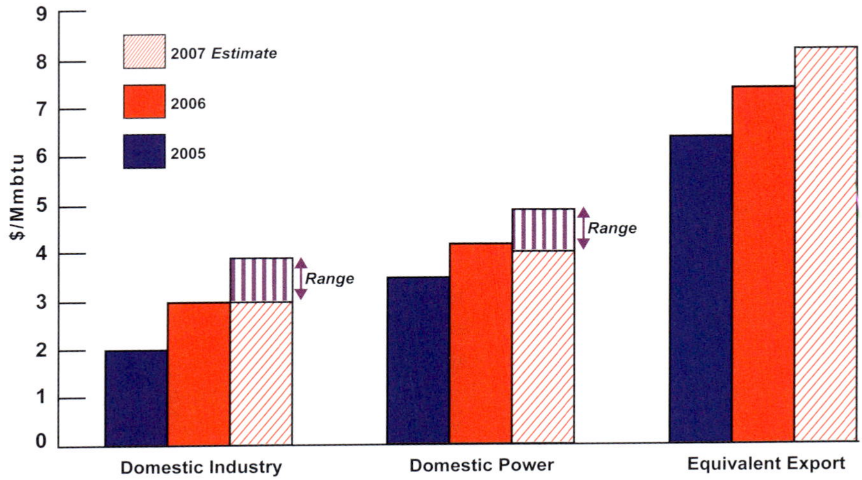

Figure 7.3: Estimated Gas Price Comparison, Domestic vs. Export

Source: Compiled by author with input from company sources.

From 2006 Pertamina asked LNG export buyers to take fewer cargoes and sought a common understanding on domestic/export gas policy from all parties involved in Indonesian gas and LNG management. Pertamina argues that if gas has to be supplied to domestic industry below international prices, then the government should refund the difference out of higher export revenues that Indonesia is able to enjoy. Also, there seems to be a perception in Indonesia that natural gas is a cheap source of energy compared to oil, which is not the case. This is Pertamina's view, but lack of clarity in government energy policy, together with extensive bureaucracy and complex decision making, seem likely to result in reduced upstream hydrocarbon exploration. If producers are unable to maximise the returns on their gas, then there may not be enough incentive to develop the resource, i.e., international oil companies will not invest.

New Domestic Gas Infrastructure

Within Indonesia there have been plans for a 7.24 Bcm/yr 1200 kilometre (752 mile) pipeline from the gas rich region of East Kalimantan to the populous demand centre on the island of Java (where there is a limited reserve base). This would link into the existing domestic gas pipeline network (see Map 7.1). If this project, reported to cost $1.25 billion in 2006, proceeds then gas may have to be diverted away from Bontang LNG exports. As of September 2007, the project was not proceeding due to high costs and insufficient gas reserves to support its development.

In order to meet the growing gas and power demand in Java, driven both by economic growth and rising oil prices, there are plans to supply the market from ExxonMobil's Cepu discovery located in Central and East Java provinces (0.05 Tcm gas supplying 4.14–8.27 Bcm/yr of associated gas as well as oil). As the development is based on oil, pressure to achieve international prices for the gas may be less than if it was solely a gas development. There are also plans to construct two LNG import terminals on the East and West side of Java. Supply would be expected to come from within Indonesia, but since there is no spare capacity in Bontang and Arun's closure is planned, LNG will have to be supplied from an expansion of Tangguh unless Bontang export volumes are reduced, but Japanese buyers would lobby hard against this. The view of Pertamina is that domestic users of gas from any future terminals in Java should pay international LNG prices to compete with export markets. But it is uncertain whether such a stance would be politically acceptable and therefore endorsed by the government.

Upstream Gas

More than 90 percent of the country's gas production is owned by Pertamina and six international companies – Total, ExxonMobil, Vico (a BP–Eni joint venture), ConocoPhillips, BP and Chevron (including the interests that company acquired through its purchase of Unocal).

The largest potential gas reserves are in the Natuna D Alpha gas field, located in the South China Sea, 683 miles north of Jakarta and 140 miles northeast of Natuna Island. This giant gas acreage remains undeveloped, with estimated reserves of 1.3 Tcm gas but with a 70 percent CO_2 content, is held 76 percent by Exxon and 24 percent by Pertamina under a production sharing contract. Exxon has stated that it has invested the resources necessary for development and is confident that it has commercial potential.[4] In 2006, Pertamina expressed public frustration that ExxonMobil had not developed the field and that BP Migas had said that the contract was cancelled. The reasoning was that ExxonMobil had forfeited its contractual rights to develop the field by not taking enough steps towards starting production, and further that ExxonMobil's deadline to establish economic feasibility of the block had expired and so it had become a public area.[5] In response, ExxonMobil argued that the current contract remains valid until January 2009. Whatever the position between the Indonesian government, Pertamina and ExxonMobil, there is no doubt that the development of Natuna D Alpha will be technically challenging and costly; but if a technical conclusion is found then there should not be a problem with financing its development.

More costly exploration will add support to the need for international prices for gas to ensure the continued development of projects. However, it is difficult to envisage domestic prices doubling to export equivalent levels, and hence the two-tier structure where projects are designated for export or domestic supply, or a mixture of both, looks set to continue. If companies are not going to be able to achieve the required higher prices then there are concerns, accepted by Pertamina, that they may not be willing to develop the reserves. The government should pursue consistent policies, streamline the decision making process and remove bureaucracy to promote investment and increase the country's reserves base.

If, as seems most likely, domestic price levels do not rise to export equivalent level then the government may require a percentage of all new gas discoveries to be sold to the local market at the lower domestic price levels. This could have an impact on LNG export volumes. In September 2007 the energy regulator BPMigas said that 'only if there is a deficiency in the domestic use of energy' could LNG export volumes be maintained to existing Japanese buyers, showing both that domestic demand will take priority and also that the future level of domestic demand is not clear and, if domestic demand is lower, the government will want to export the surplus gas. It is unclear how trade-offs between domestic and export supply will be resolved, but it is expected that political pressure will mean that, unless new gas is found, LNG exports will be reduced to meet domestic demand.

Indonesian LNG Supply Shortages

Reduced gas availability and the prioritisation of gas for domestic use has meant that since 2006, LNG exports under long-term contracts have been reduced. As a result, Pertamina has had to reduce the number of cargoes it can supply from Indonesian LNG production and source replacement cargoes from the spot market. Over 17 Mtpa (23.3 Bcm/yr) of Indonesian LNG contracts with Japan and other Asian buyers expire by 2012 (12 Mtpa – 16.2 Bcm/yr – with Japan) and even though the Indonesian President has said that he will honour these existing agreements for the remainder of their contract terms, there is still some uncertainty as to how much LNG will actually be available over the coming years. Even if only a small number of cargoes are not made available due to the tight LNG supply market, this can result in buyers being forced to acquire LNG cargoes on the global spot market at very high prices. The effect of this potential reduction of LNG supply will be to add to the uncertainty of an already tight supply market, to the discomfort of LNG buyers who contracted on a long-term basis to ensure a consistent supply. This situation has resulted in considerable political lobbying by buyer countries.

Reports in December 2006 suggested that Indonesia would not extend the 1.5 Mtpa (2.0 Bcm/yr) contract with Taiwan's Chinese Petroleum Corporation (CPC); Japan is being given preference because it helped to pioneer LNG trading and was Indonesia's earliest customer in 1973.[6] As of September 2007 the six Japanese buyers seeking to renew their contracts with Indonesia have held two negotiating meetings with Pertamina, the Indonesian Ministry of Energy and Mineral Resources and BP Migas, together with the three main production sharing contractors. The parties agreed that they planned to extend their accord by the end of 2007. Price is obviously one key negotiating point and Indonesia is seeking prices near to current market levels. But it is the volume of LNG that is a major sticking point since Pertamina is unable to agree to an extension of the full 12 Mtpa (16.2 Bcm/yr) volume, due to uncertainty over gas availability to the Bontang plant resulting from tight domestic gas supply and the pressure of demand.

Chevron and Total are planning to invest $14 bn in developing the Kutei Basin, which Chevron inherited through its purchase of Unocal and the Mahakam blocks.[7] These considerable upstream investments will result in additional gas availability to the Bontang LNG plant and Pertamina stated in September 2007 that up to 10 Mtpa (of the 12 Mtpa Japanese contracts) could be extended rather than the 3 Mtpa (4.1 Bcm/yr) extensions being discussed.[8] BP Migas are stating that the extension volume will apply to 3–6 Mtpa (4.0–8.1 Bcm/yr) though this figure could rise if domestic demand does not increase as planned.[9] The final Japanese extension volume is therefore not clear but the Senoro development (see above) will in part compensate Japanese buyers for lost volumes from the earlier projects.

Outlook

The outlook for LNG production in Indonesia will depend on the availability of gas, which will itself be decided by the success of upstream exploration and development, and the level of domestic demand. To make exact projections of LNG exports will be difficult, but this author believes that the Indonesian government and Pertamina will do their upmost to renew as much as possible of the 17 Mtpa (23.0 Bcm/yr) LNG contracts with Japan and other Asian buyers that expire by 2012. Figure 7.4 shows the decline in Arun and Bontang production and the effects of Tangguh and Senoro coming on stream. The 'Additional Bontang' bars on the chart identify the LNG volume that could be exported if feed gas is available and domestic demand permits.

Malaysian Gas Supply

Background

Malaysia, with a population of 24.8 million (2007 est.), is divided into two parts. West or Peninsular Malaysia, lying between Thailand to the north and Singapore to the south and East Malaysia, the states of Sarawak and Sabah on the island of Borneo, some three hours' flight from the capital Kuala Lumpur (KL).[10] These two states are themselves split by Brunei, a separate gas and oil rich country (see separate section). Malaysia, one of the tiger economies of the 1990s, had its economic growth trimmed through the 1997 Asian crisis and during the 2000s has experienced economic growth of 5 percent per annum with most GDP generated from the services and industrial sectors. 2006 GDP is estimated at $157 billion and GDP per capita at $5610. The population is made up of Malay 50.4 percent, Chinese 23.7 percent, indigenous 11 percent, Indian 7.1 percent and the majority of the population is Muslim.

During the period 2003–6, energy demand grew by 25 percent, met by gas, oil, coal, hydro and some renewables (with government policy to increase this source). Gas is sold to the power sector at $1.70–1.90/MMBtu and industry at $2.8–3.2/MMBtu (mid-2007 levels) considerably lower than export levels. Gas usage in the power sector has therefore increased considerably as the price of domestic gas is cheaper than oil products and half of power is generated from gas. In the financial year 2006/7 gas subsidies cost Petronas $4.5 billion (up 9.1 percent year-on-year), and it is understood that the company is expressing frustration at the subsidy level which is being reviewed by the government.

With increasing gas demand, Petronas has to import gas from Indonesia and the Joint Development Area (JDA – see below) and these

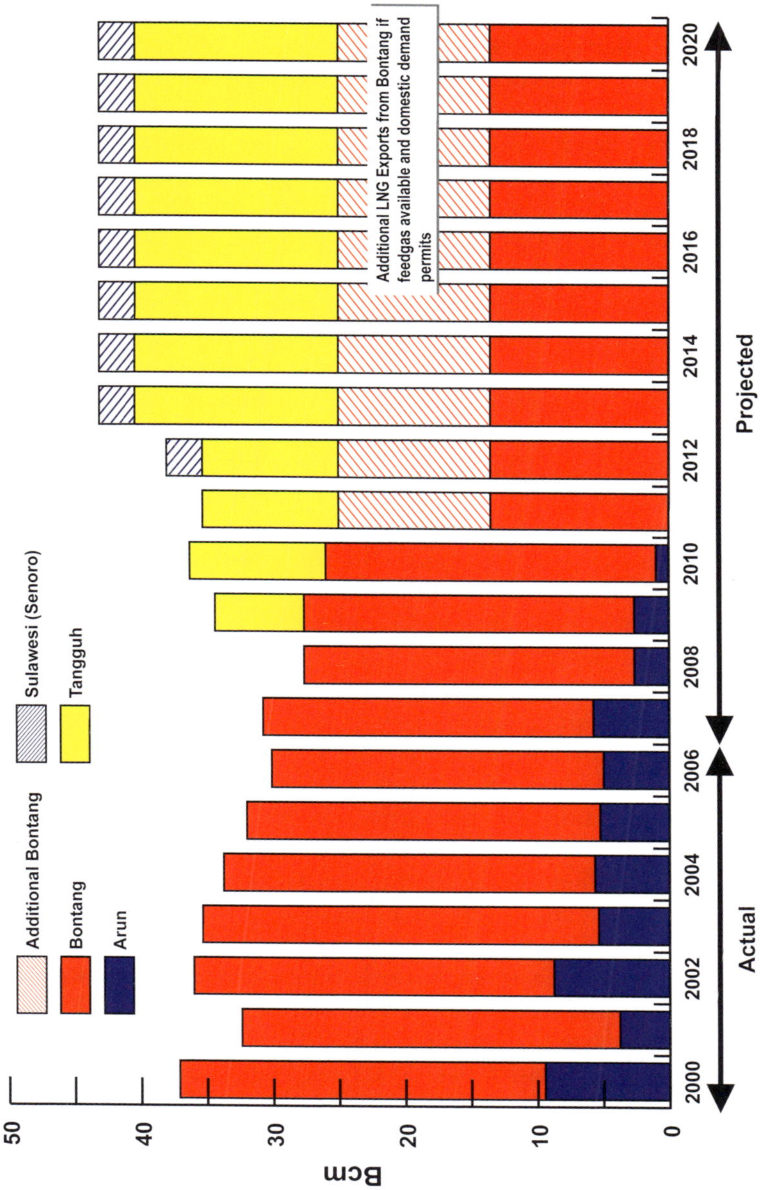

Figure 7.4: Indonesian LNG Production, Historical & Projected (as at September 2007)

Source: Actual figures, Cedigaz; projections, author with input from company sources.

imports increased to 5.1 Bcm in 2006. At the same time Petronas exported 1.4 Bcm in 2006 to Singapore, a key market that pays prices above $5.5/MMBtu (related to High Sulphur Fuel Oil). Import prices of gas from Indonesia are capped at $2.70/MMBtu making these imports cheaper than export prices. For comparison purposes, the average price of LNG sold from Malaysia to Japan in 2006 was $6.48/MMBtu. Malaysia is also considering importing coal to meet demand from new power plants since its price has been rising more slowly than that of gas.

Malaysia is a major player in the gas and LNG business. In 2006 it exported 28.4 Bcm of LNG (13.3 percent of global sales) and consumed 40 Bcm gas domestically. Gas is, therefore, an important energy source with 2.48 Tcm of proven reserves at the end of 2006. Malaysia's important gas position has been reflected in its being elected to the Presidency of the International Gas Union for 2009–2012. Malaysia also has 4.2 billion barrels of oil proved reserves and 747,000 barrels/day of oil production.

In line with the physical division of the country, the Malaysian gas business has developed on two different bases. Sarawak gas is produced offshore in relatively shallow water, although recent exploration and gas finds are in deeper water. As there is no ready domestic market (the Sarawak and Sabah population is only two million across a large landmass) gas commercialisation has been through development of LNG, gas to liquids and fertilisers for export.

Peninsular Malaysia, where around 80 percent of Malaysians live, is the centre of Malaysia's industrial, commercial and services sector growth. Gas is produced offshore the east coast town of Kuala Terengganu. Gas is supplied via pipeline to key consumers and power plants through the 2400 km (1580 mile) Trans-Peninsular Gas Utilisation Pipeline System (PGU) from the east coast Petronas gas processing plant complex down the east coast towards Singapore and then turning up the west coast to the Malaysian–Thailand border. This pipeline system is owned and operated by Petronas. In addition, there is gas production in the JDA offshore East Coast Malaysia and Thailand which is also supplied into the PGU.

The role of the international oil companies (IOC) has also tended to divide along the same regional lines. East Coast Peninsular Malaysia was historically the focus of ExxonMobil, while Shell interest concentrated on Sarawak and Sabah, on the island of Borneo. More recently, both companies have entered into the other's regional focus areas. Other companies such as ConocoPhillips, Amerada Hess, Murphy and Talisman have taken advantage of good geological prospects and commercial terms offered and there have been announcements of gas and oil finds (see below). Carigali, the 100 percent upstream company of Petronas has grown in importance and actively participates in exploration, development and production of oil and gas on a profit sharing and cost recovery basis.

Petronas

In 1974 the Government of Malaysia established Petroliam Nasional Berhad (Petronas) to be the owner and manager of petroleum resources in Malaysia. Petronas has since then grown into an integrated energy company, involved in upstream exploration and operations (through its subsidiary Carigali), gas processing and distribution as well as LNG development. Petronas also has a large complex oil refinery at Malacca, processing domestic and imported crudes as well as being an exporter of crude oil and products.

Since 1974, the Government of Malaysia and Petronas have set clear policies to put control and management of Malaysian energy resources under Petronas' control, and manage the depletion of national resources (National Depletion Policy, 1980) and fuel diversification (Four-Fuel Diversification Policy, 1981, then Five-Fuel Diversification Policy, 1999).[11] [12]

Petronas prides itself on well trained staff, operational efficiency, reliability and its ability to add value as well as being a global company. The average age of Petronas employees in 2007 was 35.7 years, young for an oil and gas company, and in April 2007, Tan Sri Mohd Hassan Marican, President and Chief Executive Officer of Petronas made a clear statement to the industry that 'scarcity of human resources will have a negative impact on the industry. What is required is a serious commitment to invest in human capital development. Poaching is not a solution'.[13] This message to industry shows that quality human resources are scarce and companies such as Petronas do not want to train staff just to see them move away to other countries or international companies, as occurred in the 1980s and 1990s when government companies were being established globally.

At the same conference Marican said that 'Resource nationalism is a reality and cannot be ignored, in that national energy companies are [developing] and have developed their own technical expertise to undertake gas and LNG projects, and so can therefore choose to carry out developments on their own or with international companies.' Petronas' strategy has been to bring new international companies into Malaysia to create competition between upstream PSCs to keep costs down, particularly as new potential reserves are located in the deep-water where exploration costs are high. Also, showing its commercial acumen, Petronas brought forward exploration rigs when rates were lower and this has enabled them to carry out exploration programmes as planned and enjoy good margins. As part of PSC terms, Petronas requires international companies to use local industry, which must have the necessary skills to avoid the international companies going overseas. This supports local industry and the Malaysian economy, but also allows international companies to obtain the necessary services and equipment at Malaysian rates, which in general are lower than international rates, thus again supporting good margins for Petronas.

Since the early 1990s Petronas has had a clear mandate from the Malaysian government to expand overseas and today has energy interests in over thirty countries, both within ASEAN and globally. In July 2006, Petronas purchased US$1.1 billion worth of shares in Rosneft. At that time Marican said 'Petronas aspires to be a global player and Russia is in a central position of providing access to markets in the East and West for oil and gas players....The next step is to widen the presence and, at the same time, reduce, dilute and spread the risk through this investment.' This move allows Petronas to spread its risk and seek hydrocarbons for the time when Malaysian resources will be no longer able to support domestic demand and shows a move by an Asian country to go global to manage depletion of its local energy sources.

Petronas acts de facto as the Malaysian energy ministry and regulator, setting a clear, attractive and stable investment environment. In its role of managing the country's energy resources, it promotes Malaysia as an upstream opportunity. The company has to be careful to manage the interplay between attracting international companies, their technology and resources, with developing Carigali, the local company as an international upstream player. To date Petronas has managed this potential conflict well, though some argue that the domestic regulatory functions should be taken away from Petronas and an independent regulator appointed. Petronas is a highly respected regional and global gas and LNG player, balancing the will to manage Malaysia's national gas and oil resources and developing profitable onshore and offshore operations in the company and the Malaysian national interests. This energy policy mandate is clearly given to it by the Malaysian government.

Gas Supply

Peninsular Malaysia

In 1968 Exxon and Conoco were awarded the first concession in

Peninsular Malaysia and since then have developed gas offshore East Coast Malaysia for supply to Peninsular Malaysia markets. To the north-east side of Peninsular Malaysia, bordering Thailand the Malaysian-Thailand JDA (Joint Development Area) was created as a means to resolve the joint claims by Malaysia and Thailand over hydrocarbon resources. The area is structured as three offshore blocks and is managed by the Malaysian-Thailand Authority (owned 50/50 by Malaysia and Thailand).

Production commenced from block A-18, and is expected to rise to 4.03 Bcm/yr which will be sold to the two countries on a 50/50 basis though initial volumes will be taken primarily by Malaysia. Linked to this gas production is the development of the 374 km Trans-Thailand-Malaysia Gas Pipeline System to link JDA gas with the Malaysian PGU at Kangar on the west coast of Peninsular Malaysia. These linkages provide security of supply to Malaysia and connect Malaysia, Singapore, Thailand and Vietnam, forming a potential segment of the planned ASEAN Gas Grid. Malaysia's physical location at the centre of such a grid, gives it the natural role of becoming a hub for pipeline trades in the region.

East Malaysia (Sarawak and Sabah)

Primarily the domain of Shell, Petronas' policy has been to diversify development of natural gas reserves in Sarawak and Sabah initially though its own subsidiary Carigali and then through the introduction of international oil companies. Recently new exploration has focused on the deepwater with eight major deep-water discoveries being made offshore Sabah and Sarawak. These include Murphy's Kikeh field (due to come on-stream in 2007/8) with gas sales agreed to the MLNG Bintulu LNG Complex. Also, ConocoPhillips has found oil and gas in the Kebabangan field. These finds are reported to have reserves of 0.21 Tcm of gas and it is expected that there is more gas to be found as other blocks have been awarded to international companies. Petronas also plans to move the gas to market by investing in new infrastructure – oil and gas facilities in Sabah, which will lead to gas being sent 480 km (320 miles) by pipe to the MLNG Bintulu complex. There is a methanol project at Labuan using local gas as feedstock, and there was a wish by some local politicians that the new gas found offshore Sabah would be used in developing a gas export project in Sabah as the local state would enjoy considerable revenues. Energy policy, however, is a Federal matter and the decision made was to focus on LNG at Bintulu and secure the long-term gas supply for LNG exports, ammonia and power. Petronas is keen to find more gas and, in September 2007, a decision was taken not to proceed with the de-bottlenecking of MLNG Tiga (see below) until more reserves had been found.

Between Sarawak and Sabah is Brunei. In 2003 Malaysia and Brunei ceased gas and oil exploration in certain blocks as a result of a dispute over offshore and deepwater exploration areas. Negotiations seemed to have reached stalemate and this may lead to international arbitration.

Ownership of the Spratly Islands, a group of approximately 100 reefs and islands in the South China Sea with possible potential for hydrocarbons, is also disputed. The extent of oil and gas reserves is not known and ownership is disputed between China, Taiwan, Vietnam, Brunei, Philippines as well as Malaysia, although military confrontation has thus far been avoided.

Bintulu MLNG Complex

The Bintulu LNG Complex consists of three LNG projects Satu, Dua and Tiga (meaning one, two and three in Bahasa Malaysia), with the following capacities and ownership structure:

Satu

8.1 Mtpa (10.9 Bcm/yr)

Petronas 90 percent, Mitsubishi: 5 percent, Sarawak State Government: 5 percent.

Dua

7.8 Mtpa (10.5 Bcm/yr), rising to 9.3 Mtpa (12.6 Bcm) following de-botttlenecking

Petronas: 60 percent, Shell: 15 percent, Mitsubishi: 15 percent, Sarawak State Government: 10 percent.

Tiga

7.4 Mtpa (10.0 Bcm/yr)

Petronas: 60 percent, Nippon Oil Corporation: 10 percent, Shell: 15 percent, Mitsubishi: 5 percent, Sarawak State Government: 10 percent.

The ownership diversification strategy of Petronas is evident in the managed ownership of the LNG projects. When in 2002/3 the initial twenty-year joint-venture expired, Shell (which previously had 17.5 then 15 percent) decided not to purchase any of the new shares, while Mitsubishi decided to reduce its shareholding from 15 to 5 percent. The Petronas shareholding consequently increased its holding from 65 percent to 90 percent. In all cases the strategy is that the local Sarawak state government enjoys revenues from the LNG projects (and also royalty from upstream gas and oil production). These payments give the state revenues to assist local development. Petronas is also not afraid of making tough decisions. In 2000, when a final investment decision was taken on the development of the Tiga project, Petronas had to persuade the shareholders to proceed without first selling the LNG.[14] As a result, not only did the project enjoy considerable capital investment savings, but it was subsequently able to sell the LNG without any difficulty. This shows how the national oil company has matured over the past twenty years increasingly taking over strategic decision making and project development from the IOCs.

LNG exports in 2006 were 21.0 Mtpa (28.4 Bcm/yr) and there is currently a de-bottlenecking/upgrading project underway to increase LNG production by 1.5 Mtpa (2.0 Bcm/yr). The three projects are located next to each other on one site. LNG exports are through Bintulu port, sold on a long-term basis currently to Japan (57 percent), Korea (26 percent), Taiwan (15 percent) with the balance (~2 percent) sold on a spot basis to other LNG import countries. In November 2006 MLNG finalised a 3.03 Mtpa (4.1 Bcm/yr) LNG sale to China State CNOOC from 2009. This transaction, together with recent sales to Osaka Gas and Chubu Electric, has been made by MLNG for supply from the 'Bintulu Complex', not from any specific project. This will give MLNG the flexibility to operate the plant and shipping as one unit to optimise output. MLNG is selling into the Atlantic Basin as well as to Asia.

MLNG prefers to commit its volume on a long-term basis and has developed excellent relations with its customers and shown itself to be a very reliable supplier since it started production in January 1983. Given Indonesian LNG supply problems (see above), the Japanese buyers are particularly concerned about their contracts with MLNG Dua that are due to expire in 2015–2017. Malaysia respects the long-term relationships that it enjoys with Japan and when the MLNG Satu's LNG contracts came up for renewal in 2003 they were all extended on substantially the same basis. It would be expected that the 2015–17 expiring contracts will be extended, albeit at prices reflecting the changed LNG market conditions. Details of the current contractual position are shown in Table 7.5.

Petronas subsidiary Malaysian International Shipping Company (MISC) is the world's largest LNG fleet operator, with 21 ships in operation and others on order. MISC also owns 89 other energy, FPSO and liner vessels. In line with Malaysia's wish to move overseas, MISC is developing its LNG ship management business offering to manage the shipping of other countries.

Table 7.5: Malaysia LNG Contractual Position (as at September 2007)

	Volume (BCMA)	*Primary Destination*	*Start Date*	*End Date*	*Delivery*
TEPCO	Japan	4.1	2003	2018	FOB/Ex-Ship
Tokyo Gas	Japan	2.1	2003	2018	FOB/Ex-Ship
EPCO/Tokyo Gas Me	Japan	1.2	2003	2018	FOB/Ex-Ship
Hiroshima Gas	Japan	0.08	2006	2013	Ex-Ship Via SendaiTe
Osaka Gas *	Japan	up to 0.92	2009	2024	Ex-Ship
Saibu Gas	Japan	036	1992	2013	Ex-Ship
Kogas	Korea	0.4	2005	2008	Ex-Ship
Tokyo Gas	Japan	0.80	1995	2015	Ex-Ship
Toho Gas	Japan	028	1995	2015	Ex-Ship
Osaka Gas	Japan	0.60	1995	2017	Ex-Ship
Kansai Electric	Japan	0.42	1995	2018	Ex-Ship
Saibu Gas	Japan	036	1993	2013	Ex-Ship
Sendai Gas	Japan	0.14	1997	2016	Ex-Ship
Shizuoka Gas	Japan	0.45	1996	2016	Ex-Ship
Tohoku Electric	Japan	050	1996	2016	Ex-Ship
CPC	Taiwan	225	1995	2015	Ex-Ship
Kogas	Korea	2.00	1995	2015	FOB
Tohoku Electric	Japan	0.90	2005	2024	Ex-ship
Tokyo Gas	Japan	034	2004	2023	Ex-ship
Toho Gas	Japan	022	2004	2023	Ex-ship
Toho Gas	Japan	052	2007	2027	Ex-ship
Osaka Gas	Japan	0.12	2004	2023	Ex-ship
Japex	Japan	050	2003	2023	Ex-ship
Tokyo Gas,Toho Gas,	Japan	0.92	2004	2023	FOB & Ex-ship
Kogas (Short-term)	Korea	0.71	2004	2008	Ex-ship
Kogas (Medium-term)	Korea	1.5-2.0	2003	2010	Ex-ship
Kogas (Long-term)	Korea	2.00	2008	2027	Ex-ship
Chubu Electric *	Japan	054	2004	2023	Ex-ship
Shanghai LNG *	China	31)3	2009	2029	Ex-ship

* Contract signed by MLNG Complex (assumes that supplies come from any project)

Source: Compiled by author with input from company sources

Outlook

The combined capacity of the Bintulu complex is 23.3 Mtpa (31.4 Bcm/yr) as at September 2007. Following the de-bottlenecking of MLNG Dua, production of LNG from the Bintulu Complex will rise to 24.8 Mtpa (33.5 Bcm/yr). However, the Bintulu Complex is operating below capacity due to upstream gas supply shortages and in June 2007 Petronas said that a fourth train would not be built, and that the de-bottlenecking of Tiga will have to wait until more gas reserves are found. Assuming that such reserves are found, LNG output from Tiga could increase from 7.4 Mtpa (19.1 Bcm/yr) to 8.4 Mtpa (11.3 Bcm/yr). This could mean that by 2013–14 total LNG exports could reach 25.8 Mtpa (34.8 Bcm/yr). It is not expected that exports will rise far above this figure unless large additional quantities of gas are found. The recent trend by MLNG to sell LNG from the Bintulu Complex, rather than from a specific project, is expected to continue, giving MLNG maximum planning and operational flexibility.

Australian Gas Supply

Background

Australia is richer than any of the other countries discussed in this chapter. With 2006 GDP of $732 billion and GDP per capita of $33,120 the economy is far larger than Indonesia or Malaysia. Its population is 20 million and in 2005 over 95 percent of its labour was employed in the commercial and industrial sectors; corresponding figures for Indonesia and Malaysia are 87 percent and 57 percent respectively.

With over 2.61 Tcm proven gas reserves (according to *BP Statistical Review of World Energy 2007*, though the Department of Industry and Resources of Western Australia estimated 3.17 Tcm at the end of 2005), and substantial potential for discovering additional reserves, Australia has grown into the largest gas resource holder in the Asia–Pacific region and one of the potential major gas resource growth areas globally. Recently additional reserves have been proved up, though further offshore in deeper water and thus more expensive to develop. The location of these gas resources is offshore the north-west of the country, well away from the energy demand centres of the East and South East coastal areas. Aside from export markets (LNG and fertiliser), the nearest demand centre is at Pilbara 219 km (136 miles) away, via the Pilbara Energy Pipeline, and Perth which is 1400 km (920 miles) and is linked by the Dampier to Bunbury pipeline; but this demand only represents a small offtake of the available gas supply. There are gas reserves in south-east Australia, in the Cooper Basin (onshore) and Bass Straight (offshore), which are used domestically in the high energy demand areas around Melbourne, Sydney and Adelaide. Small gas reserves in Central Australia (Amadeus Basin) are piped to Darwin. There are also substantial coal reserves in these areas from which coal bed methane is being developed (see below).

It is the export of gas as LNG which has driven developments of the remote gas reserves in Australia's North West; 13.4 Mtpa (18.4 Bcm/yr) LNG was exported in 2006 (21 percent of Asean and 8.5 percent of global supply), and plans for this to rise to over 20 Mtpa (27.4 Bcm/yr) by 2010. As a result of higher gas prices and increased gas finds, there are plans for more project development and it is possible that exports could reach over 70 Mtpa (95.9 Bcm/yr) by 2020. These LNG export estimates have increased over the past five years, since the last edition of this book, in response to the increased demand for LNG and available reserves in Australia. The final investment decision taken by Woodside on the development of Pluto LNG in July 2007 has also shown that LNG projects in Australia can be developed despite environmental difficulties and increasing capital costs.

Unlike the other big LNG exporters of Brunei, Indonesia, Malaysia, development of gas resources in Australia is less centrally controlled: neither State nor Federal Governments have any shareholding in LNG export projects, nor is there a national energy company as in Indonesia or Malaysia. There are therefore no production sharing contracts with companies operating within established legal and fiscal frameworks. This results in competition between companies to develop projects in a cost effective manner.

No single individual or government department is responsible for permits and development permissions and licensing of gas and LNG investments. Petroleum is regulated by both State (and Territory, which for all practical purposes, means a State) and Federal (or 'Commonwealth') governments. State governments have jurisdiction over onshore assets and facilities whilst the Federal government has jurisdiction over facilities more than three nautical miles offshore. As most gas activity is offshore, this means that the Federal government issues most of the necessary permits. It should also be noted that where the responsibility lies with the Federal government, it often delegates its responsibilities to the State authorities.

Environmental Issues

Both State and Commonwealth governments have to meet clear environmental requirements, in the form of submitting an environmental impact assessment or an environmental management plan. Although the State and Federal government requirements are different, the two authorities generally cooperate and there is usually one submission which fulfills both purposes. Under both relevant acts, there is due process to be followed, allowing for public comment and objection, responses, departmental review and recommendation and decision. This process takes time and can delay project development. It is the onshore environmental issues that tend to raise concern, as seen on Barrow Island and the rock art on the Burrup Peninsular, with companies seeking to develop projects having to carry out extensive environmental protection studies in order to gain the necessary approvals. In some cases companies enter into agreements with the local indigenous population, often with compensation, in order to ensure that a project proceeds. In July 2007, after several years of negotiation, the Japanese energy company Inpex and France's Total, the developers of the Browse LNG plant on the Maret Islands, signed a Heritage Protection Agreement with the island's traditional owners, the Wanambal Gamberra people. The developers hope that this agreement will enable the necessary approvals and speed up the development of the Browse LNG project.

There is a growing awareness of environmental and climate change issues and groups are becoming more organised and sophisticated in imposing their requirements on projects, thereby making infrastructure development more difficult. As a result, the due process in developing a gas or LNG export project can be lengthy and costly (additional surveys, reports, responding to public comment and trying to avoid conditions which will entail greatly increased capital or operating costs). Major gas developments are unlikely to be prevented, but they may well be delayed, by environmental concerns.

Following considerable public discussion between the pro- and anti-LNG lobbies, it was announced in July 2007, ahead of the project final investment decision, that the aboriginal rock art on the site of Woodside's Pluto LNG project would be protected by the Australian government. The Environment Minister Malcolm Turnbull said that the decision showed that environment and heritage protection could be balanced with economic and industry development.[15] In addition, the Pluto project faced environmentalist challenges related to offshore coral loss near the plant. These examples highlight the need to balance the environment protection and preservation of culture and historical artifacts with the need to develop industry – a dilemma for many governments.

The Domestic Market

The Australian gas industry is based on a free trade system with third-party access to pipelines. To manage this structure there are regulators in a number of jurisdictions, but these deal only with domestic markets; there is no export control. There is, however, a debate over the keeping of reserves for domestic use, particularly in Western Australia, where the State government has said that its resources should not all be exported without further benefit for the locals and has announced that 15 percent of all developed reserves for LNG export projects be allocated for domestic use 'if commercially viable'.[16] In Western Australia, domestic gas prices have historically been considerably lower than export prices, in the range \$1.50–2.00/MMBtu (~A\$1.70–2.30/MMBtu) but recent gas sales (summer 2007) have seen this rise to \$5.50–6.50/MMBtu (~A\$6.30–7.50/MMBtu) while East Coast prices have been in the range ~\$2.50–3.00/MMBtu (~A\$3.00–3.50/MMBtu); compared with an average Japanese (FOB equivalent) purchase price in 2006 of \$5.70/MMBtu.

As domestic prices are generally lower than export prices, any requirement to retain reserves, or to produce them for the domestic market, would be seen by developers as having a negative effect on profitability and increasing the commercial complexity of projects. At the Federal level there is a belief that imposition of these requirements would discourage investment and damage the sovereign standing of Australia. The revenue sharing between Commonwealth and State is further exacerbated by the situation whereby royalties from petroleum production go to the Federal government if the production is from more than 3 nautical miles off-shore. For the North West Shelf project Western Australia receives about 50 percent of the royalties.

Australia is the world's fourth largest producer of coal, producing 340 million tonnes per annum. In mid-2007 the Australian companies Arrow Energy and Santos announced two different plans to build LNG plants in Queensland, Eastern Australia fed by coal bed methane. These would be the first such plants in the world. Santos plan to make an investment decision in 2009 and start LNG production in 2014. Such projects would face environmental challenges, but as they are to be developed in a region that already has, and accepts, coal projects, they may find it easier to get local public support and the necessary approvals. The economics of such projects will depend on the cost of the coal-derived gas, but relatively low domestic east coast gas prices should support such developments, together with the reduced carbon footprint from using coal gas for LNG compared with traditional coal production. New power stations are also being planned in Queensland fed by coal bed methane.

LNG Export Projects

Map 7.3 gives details of pipelines in Western Australia. The following LNG projects are operational and under development in the country.

Map 7.3: LNG and Domestic Gas Pipelines in Western Australia

Source: Compiled by author with input from company sources.

North-West Shelf (Operational)
The North-West Shelf (NWS) project, operated by Woodside, commenced commercial operations in 1989 and in 2007 had a capacity of 12.2 Mtpa (16.5 Bcm/yr) which will rise to 16.6 Mtpa (22.4 Bcm/yr) in 2008 when the fifth train starts production supplying China's Guangdong LNG terminal. The NWS partners are Woodside, Shell, BP, BHPBilliton, Chevron and Mitsui/Mitsubishi each with a 16.67 percent shareholding. The LNG plant is located on Withnell Bay, Burrup Peninsular and it is estimated that the NWS area has sufficient reserves for gas supply to meet the existing LNG and domestic contracts plus an extension of the original LNG contracts for nine years. Table 7.6 gives details of the current contractual position. Further contract extensions will require more gas to be found. Sales of LNG are to Japan, Korea and China and since 1989 the project has delivered over 2300 cargoes of LNG. The NWS has been a success for Australian LNG and has responded to customer needs in the different regional markets, accepting that they offer very different risk and reward profiles.

Darwin (Operational).
The Darwin project, operated by ConocoPhillips commenced commercial operations in 2006 and in 2007 had a capacity of 3.5 Mtpa (4.7 Bcm/yr). There are plans for a second 5–6.5 Mtpa (6.8–8.8 Bcm/yr) train by 2012/13, but more reserves will have to be firmed up before the expansion can take place. Table 7.7 gives details of the current contractual position. The partners are ConocoPhillips 57.15 percent, ENI 10.99 percent, Santos 11.39 percent, INPEX 11.27 percent and Tokyo Timor Sea Resources 9.20 percent.

The Japanese companies have taken a stake in the project – a strategic move to invest in upstream activities to secure gas supply for the country and this could become the model for future LNG project developments.

Pluto (Investment decision taken)
On 27 July 2007 the Woodside board gave approval for the development of the Pluto LNG project being developed by Woodside (the 100 percent shareholder). The 4.8 Mtpa (6.5 Bcm/yr) capacity plant will be located on the Burrup Peninsular close to the existing NWS project. Gas will be supplied from the Pluto and Xena fields with combined reserves estimated at 0.14 Tcm. The project plans to start production at the end of 2010 or early 2011. The LNG will be sold to Tokyo Gas and Kansai Electric for 15 years (with an option for Woodside to extend this for a further five years). Table 7.8 gives details of the current contractual

Table 7.6: North West Shelf LNG Contractual Position (as at September 2007)

	Volume (BCMA)	*Primary Destination*	*Start Date*	*End Date*	*Delivery*
Chubu Electric*	1.42	Japan	1989	2009	Ex-Ship
Chugoku Electric*	1.50	Japan	1989	2009	Ex-Ship
Kansai Electric*	1.53	Japan	1989	2009	Ex-Ship
Kyushu Electric*	1.42	Japan	1989	2009	Ex-Ship
Osaka Gas*	1.07	Japan	1989	2009	Ex-Ship
Toho Gas*	0.31	Japan	1989	2009	Ex-Ship
Tokyo Electric (TEPCO)*	1.59	Japan	1989	2009	Ex-Ship
Tokyo Gas*	1.07	Japan	1989	2009	Ex-Ship
Korea Gas	0.68	Korea	2003	2010	Ex-Ship
Osaka Gas	1.35	Japan	2004	2034	FOB
Tokyo & Toho Gas	1.85	Japan	2004	2029	FOB
Shizuoka Gas	0.19	Japan	2005	2029	FOB
Tohoku Electric	0.62	Japan	2005	2020	FOB
CNOOC	4.46	China	2006	2031	FOB
Kyushu Electric	0.68	Japan	2006	2025	Ex-Ship
Chubu Electric	0.81	Japan	2009	2023	Ex-Ship
Kansai Electric	0.68	Japan	2009	2014	Ex-Ship
Osaka Gas	0.68	Japan	2009	2014	Ex-Ship
Tohoku Electric	0.68	Japan	2010	2018	FOB
Tokyo Electric (TEPCO)	0.41	Japan	2009	2017	Ex-Ship

* Original contracts

Source: Compiled by author with input from company sources.

Table 7.7: Darwin LNG Contractual Position (as at September 2007)

	Volume (BCMA)	*Primary Destination*	*Start Date*	*End Date*	*Delivery*
Tokyo Electric (TEPCO)	2.00	Japan	2006	2023	Ex-Ship
Tokyo Gas	1.00	Japan	2006	2023	Ex-Ship

Source: Compiled by author with input from company sources.

position. As part of the sales agreement, both companies will take 5 percent equity each in the project. Woodside and the Japanese buyers have also agreed joint shipping arrangements.

Woodside was keen to develop an LNG project on its own, without the involvement of any international companies, and not to use Pluto and Xena gas for a further expansion of the North-West Shelf project.

This decision no doubt helped with the fast development of the project (approximately two years from discovery of the gas reserves to investment decision). There should be some additional volumes available over the contracted amount and it is expected that LNG will also be sold into the US market, where Woodside is looking to develop the Oceanway LNG import terminal offshore California.

Table7.8: Pluto LNG Contractual Position (as at September 2007)

	Volume (BCMA)	*Primary Destination*	*Start Date*	*End Date*	*Delivery*
Tokyo Gas*	1.5–1.75	Japan	2011	2026	Ex-Ship
Kansai Electric*	1.75–2.0	Japan	2011	2026	Ex-Ship

* Both agreements give the seller the option to extend the contract period for five years

Source: Compiled by author with input from company sources.

Woodside has proposed the development of a gas hub with its Pluto project acting as the LNG aggregation hub – which may facilitate some project development challenges, but would also be of benefit to Woodside. This could also be a political move by Woodside to ensure its involvement (and control) in the development of LNG in North-Western Australia.

Greater Gorgon (Planned)

The Gorgon project is being developed by Chevron which has a 50 percent share, and ExxonMobil and Shell with 25 percent each. The original plans were for a two-train plant based on the Gorgon field. The shareholders have been trying to reduce the project's high unit costs and (as of September 2007) they may achieve this through the economies of scale of a three-train project, each one 5 Mtpa (6.8 Bcm/yr). The LNG plant is planned to be located on Barrow Island with the start of operations expected no earlier than 2013. Unlike the North-West Shelf project, each of the partners will market its own share of LNG: Chevron to Japan, Shell to China and its own trading organisation; ExxonMobil to India, Hong Kong and Korea (although this had not been finalised by September 2007). Details of the current contractual position are given in Table 7.9.

Browse (Planned)

Another LNG project is driven by Woodside (with 50 percent) based on BP, Chevron, BHPBilliton, Shell and their own gas reserves,

Table 7.9: Gorgon LNG Contractual Position (as at September 2007)

	Volume (BCMA)	*Primary Destination*	*Start Date*	*End Date*	*Delivery*
Shell (entitlement 5 BCMA)					
China National Petroleum Corp (CNPC)* (Shell volume)	1.35	China	n/k	n/k	n/k
Shell System	1.15	USA/Mexico	2011	2040	Ex-Ship
Chevron (entitlement 10 BCMA)					
Tokyo Gas*	1.2	Japan	2011	2035	FOB
ChubuElectric*	1.5	Japan	2011	2035	FOB
GS Caltex Corp*	0.25	Korea	2011	2035	FOB
ExxonMobil (entitlement 5 BCMA)					
Discussions with CLP (Hong Kong) and Petronet (India)					

* Heads of Agreements signed
Note: Each shareholder lifts and sells its own equity volumes

Source: Compiled by author with input from company sources.

estimated at 0.57 Tcm. Plans are for a 7–15 Mtpa (9.5–20.2 Bcm/yr) plant. Browse is the largest and most complex of Woodside's planned LNG projects with two development options: to build on Scott (coral) Reef or onshore with gas being piped over 400 km (260 miles). The onshore site at Kimberley also has serious native land rights issues and carbon sequestration problems (due to the CO_2 content of the gas), which could add to both timing and cost. These complexities have meant that construction costs are believed to be very high and the project may be locally unpopular. The WWF is also exerting huge pressure against construction on the offshore coral reef. Woodside is therefore examining other options, including piping the gas 920 km (570 miles) to the Burrup Peninsula, location of the proposed Pluto LNG plant.

Despite these difficulties, Woodside was pushing hard for the development of the project and in September 2007 signed a preliminary agreement to sell 2–3 Mtpa (2.7–4.1 Bcm/yr) LNG to China National Petroleum Corporation (CNPC) – details in Table 7.10. Reports suggest that Woodside made this commitment without the agreement of the other project partners, thus showing the confidence that it has in the development of LNG in North-Western Australia. With this commitment made, it can be expected that the project will proceed within the 2013–2015 timeframe set by Woodside unless the other partners do not support such a fast development.

Table 7.10: Browse LNG Contractual Position (as at September 2007)

	Volume (BCMA)	*Primary Destination*	*Start Date*	*End Date*
China National Petroleum Corp (CNPC)*	2.7–4.0	China	2013–5	2028–2035

* Key terms agreement signed 6/9/07

Source: Compiled by author with input from company sources.

Icthys (Planned)
With an estimated 0.27 Tcm of gas, Inpex (76 percent) and Total (24 percent) are seeking to develop a 12 Mtpa (16.2 Bcm/yr) LNG Project on Maret Island on North-West Australia for start-up in 2013/4, though it is expected that this date may slip. Total's involvement in the project brings hands-on LNG development experience to complement INPEX's Japanese market position. The project has been awarded 'major project facilitation' status by the Australian government which means the fast-tracking of bureaucratic processes as far as possible. There are expansion possibilities as Shell is exploring in adjacent acreage. If Shell were to become involved, it would certainly give the project a greater opportunity of success (although having two large IOCs involved could also create complications). This project has come up against major environmental hurdles concerning native wildlife which will no doubt delay its development, though political pressure from the Japanese will be aimed at keeping the project moving forward.

Pilbara (Planned)
BHPBilliton is planning to develop a 6 Mtpa (12.2 Bcm/yr) LNG project based on its (and ExxonMobil's) 0.23 Tcm Scarborough fields (further gas discoveries in mid-2007 may have increased this figure). The plant would be developed onshore at Onslow in Western Australia with possible start-up in 2012, according to BHPBilliton, but more likely in 2015. This would be the first LNG project developed by BHPBilliton and again would suffer from the engineering and human resource shortages and environmental challenges common to other projects. Success of this project may well be linked to whether BHPBilliton is able to develop a US regasification terminal. Its plans for the Cabrillo Port terminal offshore California have been rejected, so it will most probably be seeking to secure alternative terminal capacity as part of any development of Pilbara. As with Browse, development of this

project will depend on the project prioritisation of the other partner and how quickly ExxonMobil wishes to develop the project.

Sunrise (Planned)

The Sunrise project is based on the Sunrise, Troubador and associated gas fields (0.23 Tcm gas) located approximately 450km (300 miles) offshore Darwin and 80 km (50 miles) south of Timor-Leste.[17] The plan would be either to develop a floating liquefaction plant or to pipe the gas to an onshore plant in Northern Australia. Development of this project has been further complicated by the need to agree how revenues would be split in the Joint Development Area, between Australia and Timor-Leste. The treaty setting out the revenue sharing was signed in April 2003 and in February 2007 a maritime treaty covering the Timor Sea and a separate international unitisation accord setting out how offshore royalties will be split, were ratified by Australia and Timor Leste[18].

The gas is rich in liquids, and has a relatively low CO_2 content, making it commercially attractive to develop, which could allow development prior to 2015. The shareholders are Woodside (22.44 percent), ConocoPhillips (30 percent), Shell (26.56 percent) and Osaka Gas (10 percent). Though all companies would like to move the project forward as fast as possible, each company has its own agenda as to how the project will be developed. ConocoPhillips would like the gas to feed an expansion of Darwin LNG (see above), while Woodside, as the developer of Pluto and Browse, may not have sufficient financial and human resources to develop the three projects concurrently. Osaka Gas would like development as fast as possible to secure LNG for its own market.

There are also reports that Timor-Leste would like to site the liquefaction plant onshore, but the project sponsors are reluctant due to the uncertain legal and fiscal regime, and technical challenges of bringing the gas onshore through deep water. The companies may find it difficult to develop a project in Timor-Leste, and if this option is pursued, then sovereign risk issues may delay development further.

Outlook

As noted earlier there are several factors that will determine whether projects move forward to final investment decision and then construction. This author's view is that global LNG demand will drive project development in Australia and, as a result, exports will increase in volume. Figure 7.5 gives a view which would result in LNG production being over 90 Bcm/yr by 2020 and Australia rising to become the largest regional LNG producer. For this to become a reality, investment

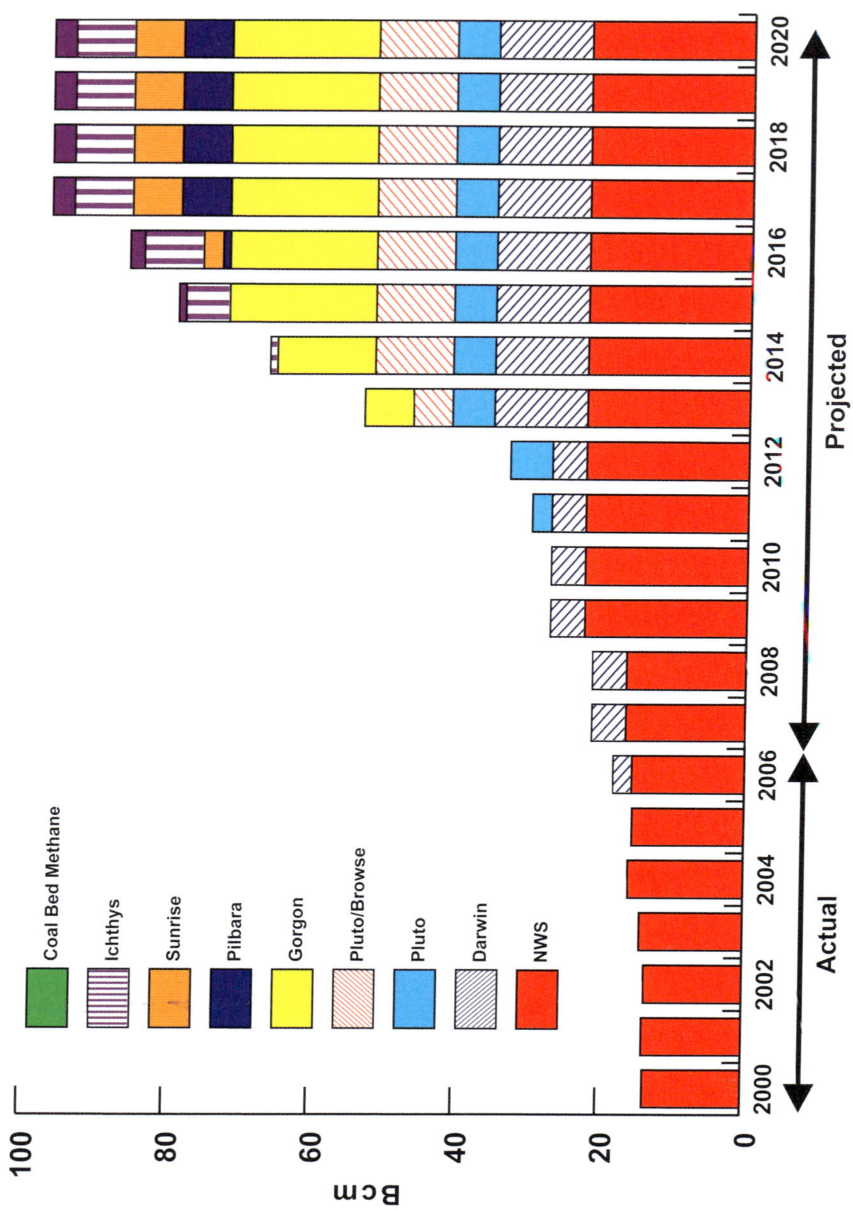

Figure 7.5: Australian LNG Production, Historical & Projected (as at September 2007)

Source: Actual values, Cedigaz; projection, author with input from company sources.

decisions would need to be made on four of the projects currently in the planning stage prior to 2012 – a challenging, but possible, timetable.

Project Development

There are three key reasons why development of gas export projects in Australia could be slower and more expensive than similar projects in other countries:

- The first is the shortage of human resources to manage the development of so many projects. Other countries, such as Qatar, bring lower cost labour from India or Philippines to carry out the construction, but Australia's restrictive labour laws mean that only Australian labour can be used. But there is not enough Australian labour to develop all the projects that are being planned. Australia has a limited pool of skilled labour which is not sufficient to construct more than one new greenfield LNG project at a time. Therefore, the first project that gets sanctioned will have a big advantage over its competitors. This may be the reason why Woodside decided to move forward so fast on the Pluto project.
- The second reason is technical considerations. Some of the new projects are technically complex, with potentially high CO_2 and deep offshore developments. For example, the cost of CO_2 removal for the Gorgon project is expected to be high either as a result of transporting the gas sub-sea in an acid-gas capable pipeline or due to the physical cost of removing the CO_2.
- The third reason is environmental factors. The development of infrastructure projects on greenfield sites has resulted in high costs of works to counter the environmental considerations. For example, the cost to Woodside of moving Aboriginal artworks from rocks at the Pluto site will be millions of dollars, and the World Wildlife Fund (WWF) is putting considerable pressure on project sponsors about where to site the Browse LNG and Ichthys projects. These considerations may drive project development offshore, but this is a more costly option with its own set of environmental problems.

In addition, the development of Australian projects will be affected by rising engineering costs. The overall effect is that Australian LNG projects will be expensive and take time to develop. To balance those problems, high commodity prices and the huge global thirst for LNG mean there has been no better time to develop LNG projects. Australia's geographic location will allow it to capitalise on growth forecasts for LNG in Asian (as well as US West Coast) markets, suggesting that Australian LNG will be a significant contributor to future global LNG supply.

Brunei

Largely as a result of its extensive petroleum and natural gas fields, Brunei has one of the highest per capita GDP levels in Asia at $30,928. A small population with an average age of 27 receives substantial government subsidies: free medical care and education to university level, subsidised rice and housing benefits. Crude oil and natural gas production account for just over half of GDP and more than 90 percent of exports. The country is very reliant on oil and gas-related income and is looking to strengthen the banking and tourist sectors to promote diversification.

At the end of 2006, Brunei had 0.33 Tcm of gas reserves and, given its very small domestic market (the population of the country is 375,000), has therefore developed its gas reserves and industry based on LNG. LNG exports from Brunei started in 1972 and by 2006 had risen to 7.3 Mtpa (9.8 Bcm/yr) from five LNG trains, 89 percent being sold to Japan and the balance to Korea under long-term contracts that expire in 2013 (details in Table 7.11). The shareholders of the project are the Brunei government (50 percent), Shell (25 percent) and the Mitsubishi Corporation (25 percent). There has been some discussion about the possible development of a 4 Mtpa (5.4 Bcm/yr) sixth train, but no further development has been possible due to the lack of gas reserves. The 35-year-old plant has recently been refurbished to support planned contract extensions with the existing Japanese and Korean buyers.

Table 7.11: Brunei LNG Contractual Position (as at September 2007)

	Volume (BCMA)	*Primary Destination*	*Start Date*	*End Date*	*Delivery*
Tokyo Electric (TEPCO)	5.45	Japan	1999	2013	Ex-Ship
Tokyo Gas	1.68	Japan	1999	2013	Ex-Ship
Osaka Gas	1.00	Japan	1999	2013	Ex-Ship
Kogas	0.95	Korea	1997	2013	Ex-Ship

Source: Compiled by author with input from company sources.

Other Regional Gas Suppliers

Myanmar

In 2006 Myanmar's estimated GDP was $11.95 billion and GDP per capita $208, one of the lowest in the Asian region, with an estimated 70 percent of the population working in agriculture. The reclusive authoritarian regime in Myanmar is subject to US and European economic sanctions over human rights issues and the ongoing house arrest of the National League for Democracy leader, Aung San Suu Kyi. The impact of these sanctions has to a certain extent been reduced by the eagerness of neighbouring China, India and Thailand to access Myanmar's natural resources to help fuel their growing economies. It is not clear whether their policies will change following the further demonstrations and civil unrest in September 2007.

Myanmar began export of natural gas by pipeline to Thailand in 1998, with gas production from two fields – the Yetagun (operated by Petronas, Nippon Oil and Thailand's PTTEP) and Yadana (operated by Total, Chevron and PTTEP). Myanmar's gas reserves were reported to be just under 0.54 Tcm as at the end of 2006. Exports of gas to Thailand (9 Bcm in 2006, accounting for 20 percent of Thailand's natural gas demand) has helped Myanmar's economy considerably and economic growth for the year ending April 2007 was 7 percent, up from 4.5 to 5.5 percent in recent years. The price for this exported gas was reported to be $6.60–7.00/MMBtu in August 2007. However, the country's longer-term prospects may be constrained by high inflation, economic isolation and unpredictable economic policies.[19]

In 2006 new gas finds were independently assessed, which again highlighted Myanmar as a potential gas exporter. A consortium of South Korean and Indian Companies (Daewoo, ONGC, GAIL and Korea Gas) found an estimated 0.28 Tcm (0.13–0.22 Tcm internationally certified) gas in the Shwe, Swe-Phyu and Mya offshore fields (blocks A-1 and A-3). The ownership percentages of each block vary between the companies and this could affect decisions as to how the gas is developed. The project sponsors want different commercialisation options and in early 2007 the Myanmar government selected Korea Gas and Japan's Marubeni Corporation to develop a proposal for an LNG export project. This is the option preferred by the Korean companies which wanted the LNG for their own market. The Indians however wanted to construct a 1400 km (920 mile) pipeline from Sittwe in Myanmar to the Jagdishpur–Haldia pipeline in India. Some of the initial gas, about 0.06 Tcm, is dedicated to the domestic market, only leaving sufficient resources for a small LNG export project. This means that there is considerable competition to secure additional exploration blocks around the Swe-Phyu and Mya fields.

While discussions continue on these options, the Myanmar government in April 2007 indicated that it preferred the option of selling the gas to CNPC of China via a 2380 km (1570 mile) pipeline from Kyaukphyu to Ruila in China's Yunan Province. A memorandum of understanding was signed in March 2007. CNPC has also proposed to build spurs along the line's 990km (615 mile) length inside Myanmar to help the country to develop a domestic gas market. The announcement about switching the gas to the Chinese market provoked the fury of the Korean and Indian governments and companies. Some see this decision as being linked to Beijing's veto, in January 2007, of a United Nations resolution critical of the Myanmar generals,[20] thus showing how the development of energy projects can be driven at least as much by political as economic considerations. These events also show the importance of developing the domestic market as well as earning export dollars, a message that has been clearly indicated elsewhere in the region. In August 2007 a representative of the Ministry's Energy Planning Department said that China was offering a price equivalent to the levels Thailand is paying for gas (reported in August 2007 to be $6.60–7.00/MMBtu as at August 2007) which is the highest price offered for the gas.[21]

Whichever export route is chosen, it is expected that gas exports from Myanmar will increase over the coming five years, but the speed of developing export options will no doubt depend on the political situation within the country and the ability of gas/LNG buying countries to resist international pressure to impose sanctions on Myanmar in response to government actions during the 2007 civil disputes. On this basis, exports to China, which are less likely to be swayed by such threats, would seem to be the most likely outcome.

Papua New Guinea

The eastern half of the island of New Guinea, the second largest island in the world, makes up Papua New Guinea (PNG) which achieved independence from Australia in 1975. Over the last ten years the country has been developing economically. PNG is richly endowed with natural resources, but the rugged terrain has hampered development of infrastructure and resources. Mineral deposits, including oil, copper, and gold, account for nearly two-thirds of export earnings. With 2006 GDP at $5 billion and GDP per capita of $500 the country is not as rich as its neighbour Indonesia. PNG is still an agrarian economy with 85 percent of the 5.7 million population working in agriculture which accounts for just under 36 percent of GDP. Australia annually supplies over $200 million in aid, which represents nearly 20 percent of PNG's national budget.

At the end of 2006, PNG's natural gas reserves were 0.44 Tcm, but with more exploration there is some expectation that this figure could rise to over 0.71 Tcm. PNG therefore is a country that is attracting considerable interest from gas buyers and investors. Over the past twenty years various projects have been proposed to commercialise the reserves. Of late, the main export project has been a gas export pipeline to Australia's eastern states, promoted by ExxonMobil and Oil Search; but due to its cost of over $4 billion the project has not moved forward. In September 2006, Australian Gas Light, which was leading the venture for the Australian portion of the pipeline, withdrew from the project. In 2007, high global energy prices and the tightening of gas supplies led to the pipeline coming under consideration again, together with the development of LNG export projects. In September 2007, three projects were under development:

- Liquid Niugini Gas with the participation of Merrill Lynch (33.3 percent), Canada's InterOil (33.3 percent) and Pacific LNG (controlled by Clarion Finanz AG) (33.3 percent). This project is based on gas from Interoil's Elk field and the intention is to develop a 9–10 Mtpa (12.1–13.5 Bcm) LNG project at InterOil's existing oil refinery (NapaNapa) near the capital, Port Moresby. A 36 inch diameter pipeline would transport 16.5 Bcm/yr of gas from the Elk field to the LNG plant. Gas to liquids development is also under consideration. The Government of Papua New Guinea will be included as a partner in the LNG project at a later stage.
- a project with the participation of ExxonMobil (49 percent), Oil Search (30 percent), Santos (17 percent) and Nippon Oil (2 percent) is based on the Hides, Angore and Juha fields and (according to some reports) the Moran and Kutubu fields (with additional participation by Australian Gas Light and the Mineral Resources Development Company). The six companies control 0.25 Tcm of gas, sufficient to support a 6.5–7.5 Mtpa (8.8–10.1 Bcm/yr) LNG project, to be located at Port Moresby. A pre-engineering design study was carried out in the summer of 2007.[22]
- an Australian company, LNG Limited, has proposed a 1 Mtpa (1.35 Bcm/yr) plant supplied with gas from the Forelands area (Uramu and Korobosea fields).

Malaysian companies as well as the China National Petroleum Corporation (CNPC) have also expressed interest in developing an LNG project. In September 2007 there were market rumours that CNPC was considering the purchase of Oil Search. This reported approach shows China's interest in securing gas reserves to supply to its growing domestic market.

In all cases the development of an LNG project, like the planned gas pipeline project to Australia, will face considerable technical, political and commercial challenges. In many cases the gas is located in a mountainous area, with substantial drilling depths, and it will be necessary to transport the gas across this mountainous terrain to the coast. Also, there are already reports that different regional interests are lobbying for the LNG plant to be located in their area with resultant economic benefits. In this author's view these problems should be overcome by the overwhelming thirst for LNG from Asian buyers and for that reason, projects will be developed. The first gas could be delivered around 2014–15 with a second project following later in the decade, but limited resources and country capacity may mean that only one project can proceed initially. Which of the projects will proceed is not clear. In September 2007 Liquid Niugini Gas seemed the most advanced, but if ExxonMobil were to put its full weight behind its project then this might be developed first.

Vietnam

In 2006 Vietnam produced 7 Bcm gas, all for domestic consumption. At the end of the year, it had 0.40 Tcm gas reserves. Vietnam uses gas for power generation as well as hydro and coal.

Conclusions

Energy Policy Frameworks

A variety of factors determine energy policy decision making for countries within the region as can be seen in Figure 7.6:

Gas Reserves
The volume of gas reserves is obviously crucial. Some countries are fortunate enough to be endowed with substantial oil and gas reserves; others are not so need to make tough decisions as to how to develop their gas businesses.

Location of Gas Reserves
For those countries with reserves, some are located at remote locations, away from markets and this determines how they are monetised. Malaysia for example exports gas produced in Sarawak, on the island of Borneo, as LNG as there is a relatively small population on the island, while gas produced on Peninsular Malaysia, where most population and industry is located, is used domestically. Australian gas reserves are located in the relatively remote North West region. Apart from a modest volume of gas supplied to Perth, gas is exported as LNG.

Environmental Considerations
Many countries operate in particularly sensitive environments; LNG projects in Australia are especially subject to environmental scrutiny. While Woodside is considering the construction of Browse LNG on Scott Reef, since that would be cheaper than building on the Australian mainland, the World Wildlife Fund (WWF) has responded that Scott Reef must be 'off limits' for LNG plants. Such environmental pressures are becoming stronger for all LNG developers, irrespective of whether their projects are new or expansions of existing capacity.

Figure 7.6: Regional Energy Policy Considerations

Source: Author

Prioritisation of Domestic Markets over Exports

Decisions by Indonesia and Myanmar to keep gas in the country rather than export it may be the start of a trend throughout the region. Malaysia has always had this policy, encouraging the development of industry and power in Peninsular Malaysia based on gas resources offshore. Even Australia is setting limits whereby 15 percent of new gas reserves should be diverted to the domestic market ('where economic to do so') because each State's politicians think that 'its' resources should not be exported without further benefit for the local population. This prioritisation of gas for domestic use constitutes a policy change compared with the early 2000s. By 2007, governments were focusing more on domestic gas development, driven by higher oil and gas prices (and therefore reduced need for increasing export volumes), and by a political imperative to support domestic and regional economic and energy demand growth.

The choice of market for gas depends on domestic industrial development, political decision making related to employment creation and the price local industry will and can pay for gas. In 2006, when Indonesia started reducing its LNG export volumes, gas availability was only part of the reason. Other factors included the government's need to maintain political stability within the country including the need to meet growing domestic energy demand in the diverse islands of the archipelago. By contrast, new countries entering the LNG business, such as PNG, are seeking export earnings as a means to develop the domestic economy.

A key question is whether other Asian gas-exporting countries will introduce similar policies and reduce export volumes. Will governments demand that a percentage of all new gas is reserved for the domestic market at discounted prices to the detriment of exports? The author believes that, due to the remote location of reserves used for LNG export, new industrial development would most probably be based on new reserves and fed by pipeline gas or domestic LNG supply.

In all countries in the region, domestic prices are lower than export prices for historical reasons, and because domestic prices have not risen as fast as Pacific LNG prices since 2000. Keeping domestic prices lower than export levels is not a sustainable policy; whether it continues depends on national policies and politics and on the ability of the local economy to sustain price rises.

Resource Nationalism

The region is seeing a shift, in line with global trends,[23] towards project development by state-owned and/or controlled national energy companies, rather than international energy companies. If national energy companies fail to invest profits in new oil and gas development (due to governments' financial need for their revenues), there will be a negative effect on the development of new projects. This was the case in Indonesia during the first years of this century and the country is currently experiencing the consequences in the form of reduced gas supply. Higher gas prices should support the development of new LNG projects, but given much higher technical and development costs (see Appendix 1 of this volume), governments may increasingly question whether to develop gas for export or keep it for their own use.

The structure of the national company and its relationship with central government is also relevant. If the national company delivers the agreed dividends to its government, then it will have a greater level of freedom and flexibility to manage the country's resources and develop new domestic and overseas projects. If the state continually presses the national company for funds, then the freedom to invest and operate independently is reduced. The success of Petronas is in part due to its clear mandate from the Malaysian government as well as to its ability to deliver the necessary dividends to the country.

Regional LNG Availability

Figure 7.7 shows full details of regional LNG production. The level of LNG exports in 2006 was 64 Mtpa (86.5 Bcm/yr) and, based on the author's estimates, could rise to 115 Mtpa (155 Bcm/yr) by 2015 and double to 137 Mtpa (185 Bcm/yr) by 2020. The timing of new LNG volumes coming on stream will depend on many of the issues discussed in this chapter, including technical development costs, environmental considerations, government policy in relation to the development of new gas reserves, the country's capacity to develop such projects and whether reserves will be used to meet domestic demand or LNG exports. This is particularly uncertain in the case of Indonesian exports and for that reason 8.5 Mtpa (11.5 Bcm/yr) of the Bontang volumes have been included in the graph as 'Additional Bontang?' since the gas volumes may, or may not, materialise as LNG exports.

LNG sellers will not have a problem placing any additional LNG volumes as and when projects are developed, but buyers must expect to pay global market prices or LNG sellers may sell outside the region. Sellers will continue to seek long-term contracts for 80–90 percent of project volumes to enable financing and the necessary support for project development. Spot and short-term LNG supply will become increasingly important as a means of managing variations in seasonal gas demand

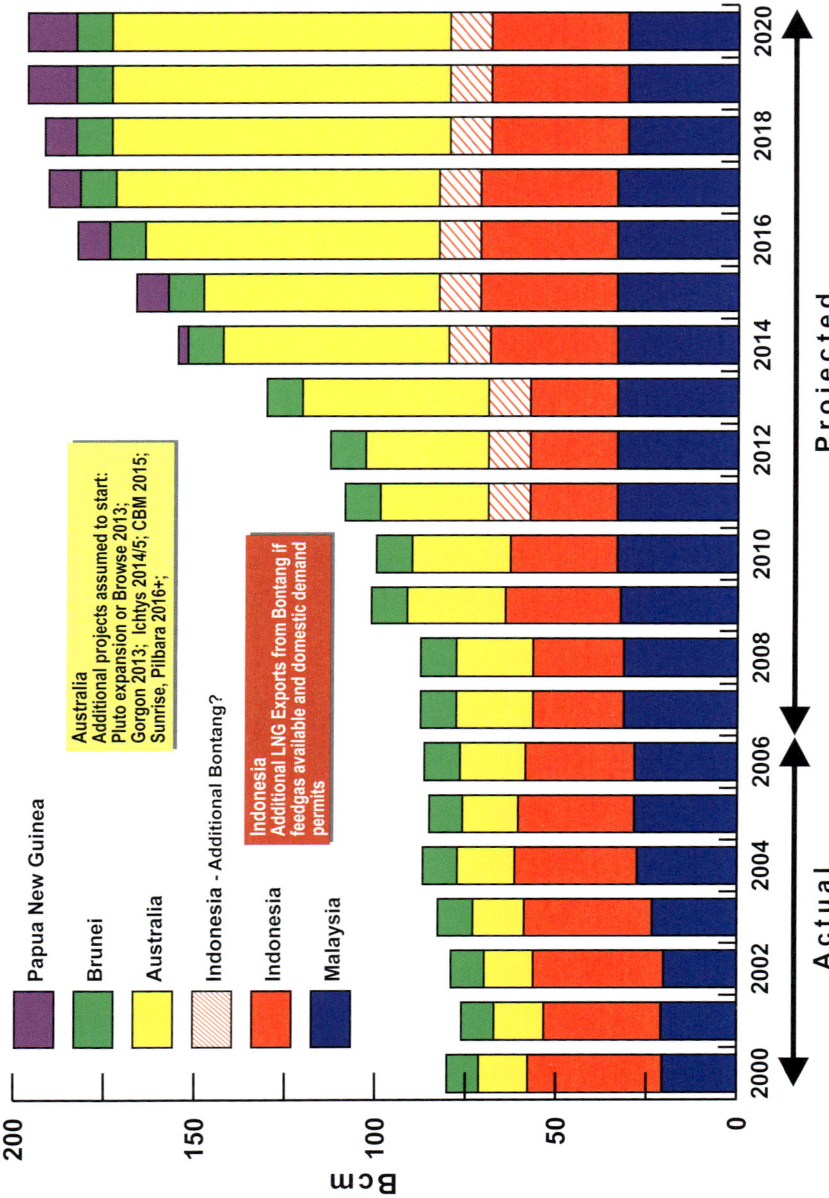

Figure 7.7: Regional LNG Production, Historical & Projected (as of September 2007)

Source: Actual figures, Cedigaz; projections, author with input from company figures.

and short-term shortages. The Asian way of doing business, based on 'trust', and long-term relationships will hopefully continue, but the introduction of new buyers, together with a shortage of LNG supply, may unfortunately put pressure on this traditional style of business.

Other Factors Facing the Region

Gas and energy demand, particularly in South East Asia, is growing rapidly while at the same time there are gas resource limitations. Such a situation would be expected to drive the development of additional gas supply capacity to meet this growing demand. But there are limitations to such growth: on one hand ASEAN is saying that it is looking for joint cooperation and a regional infrastructure so that its members can share resources; on the other hand countries are looking to retain a larger share of their gas for the domestic market and employment creation, and only export surplus gas. Therefore there will be no expansion of the existing ASEAN gas interconnections in the foreseeable future as there is insufficient surplus gas available for export to neighbouring countries. If additional gas should become available then it would be more likely to be used domestically.

Higher energy prices since 2005 have undoubtedly enabled the development of new gas export projects. When the first edition of this book was written in 2001, export of LNG from Papua New Guinea was not seriously considered. Seven years later, in an environment where Japanese LNG buyers are willing to pay over $8–10/MMBtu for LNG, such projects have become commercially realistic.

Upstream gas and LNG project developments all require stable economic, fiscal and government regimes. It is interesting to compare the approach taken in Malaysia, where Petronas has managed the country's energy resources for over thirty years, with that of Indonesia, where the government moved this responsibility from Pertamina to BP Migas, causing significant problems. Some part of the current shortages of gas in Indonesia may be due to these changes.

From a buyer's perspective, the recent LNG shortages and reduction in cargoes from Indonesia will have renewed the concerns, raised in the first edition of this book, that the degree of risk they can accept in taking LNG supplies from specific countries needs to be managed. Involvement of buyers in the upstream part of the chain will continue as they seek to secure gas reserves for their markets. The ability of Mitsubishi to secure additional gas for Japan, through the development of Senoro LNG in Indonesia, needs to be contrasted with Daewoo's apparent difficulties in developing an LNG project to move the gas it found in Myanmar to its home market of Korea. Gas buying countries will also continue to secure upstream gas reserves and LNG through acquisition, as seen by the reported approaches by CNPC in mid-2007 to acquire Oil Search as a means to gain access to gas reserves in Papua New Guinea. This emphasises both the importance of governments in the development of gas export projects and the political manoevres that buyer countries will carry out to secure energy supplies.

Projecting future regional gas and LNG supply is difficult due to the multitude of variables set out in this chapter. It is the view of the author that the region will remain short of gas and the political conflict between using gas domestically and exports will continue. Where economic, LNG will stay in the region, and sellers may prioritise exports to ASEAN countries (but not at 'discounted' price levels), LNG only being exported out of the region when high gas prices in the Atlantic Basin or US West Coast 'pull' cargoes away and when regional supply balances permit. Given growing regional and global energy demand and a tight gas supply, it is expected that competition to secure regional gas and LNG supplies will continue, and that events both inside and outside the region will affect the development of new gas and LNG supply sources.

Notes

1 The members of ASEAN are Brunei, Cambodia, Indonesia, Laos, Malaysia, Myanmar, Philippines, Singapore, Thailand and Vietnam.
2 Pertamina presentation to the Oil & Money Conference, London, September 2006: 'Indonesian LNG Update: The Changing Role of Pertamina'.
3 *http://news.bbc.co.uk/1/hi/world/asia-pacific/2558165.stm.*
4 *World Gas Intelligence*, 18 October 2006.
5 Heren LNG Markets, 13 July 2007.
6 *World Gas Intelligence*, 3 January 2007.
7 *World Gas Intelligence*, 19 September 2007.
8 *World Gas Intelligence*, 19 September 2007.
9 Heren LNG Markets 21 September 2007.
10 'Malaysian gas prospers as Petronas goes international', *Gas Matters*, August 2006 (other references in chapter).
11 Dr. Samsudin bin Hitam (Director General of Economic Planning Unit), 'Sustainable Energy Policy and Strategies: A Pre-Requisite for the Concerted Development and Promotion of the Renewable Energy in Malaysia'.
12 Abdul Rahman Mohamed and Keat Teong Lee, 'Energy for sustainable development in Malaysia: Energy policy and alternative energy'.
13 Tan Sri Mohd Hassan Marican (President and CEO Petronas), Keynote Address to the LNG 15 Conference, Barcelona 24 April 2007.

14 It is normal practice for large LNG projects to sell a large percentage or all of their LNG on long-term contracts prior to taking an investment decision.
15 'Art Science to co-exist as Australian state approves Pluto and protects paintings', Heren LNG Markets, Energy Focus.
16 'Australia's LNG future cloudy as cost issues persist', Heren LNG Markets 20 April 2007, Energy Focus.
17 In 2002, East Timor became the sovereign state of the Democratic Republic of Timor-Leste.
18 'Greater Sunrise Gets New Lease on Life', Poten & Partners LNG in World Markets, March 2007.
19 A report by the International Monetary Fund has found that increased gas exports are boosting economic growth in Myanmar. See *Financial Times* (London): *http://www.shwe.org/docs/gas-exports-boosting-economic-growth-in-myanmar-2013-report.html.*
20 *World Gas Intelligence*, 11 April 2007.
21 www.platts.com, September 2007.
22 In October 2007 Oil Search announced that it had agreed with BG not to proceed with the planned LNG project based on the MoU signed in 2006. Oil Search preferred to develop the LNG project with ExxonMobil.
23 'Nationalism and State Ownership seen as main threats to oil supply', *Financial Times* (London), 9 May 2007.

Other references and useful sources include:

BP Statistical Review of World Energy (www.BP.com)
Cedigaz (www.cedigaz.org)
Heren LNG Markets (www.heren.com)
World Gas Intelligence (www.energyintel.com)
Poten LNG in World Markets (www.poten.com)
LNG Focus (www.gas-matters.com)
Economic data from the Energy Information Administration Country Analysis and https://www.cia.gov (World Factbook).

Thank you to all those individuals, friends and colleagues who agreed to be interviewed and assist in the development and drafting of this chapter.

CHAPTER 8

NATURAL GAS FROM THE MIDDLE EAST

Andy Flower

Introduction

The countries of the Middle East collectively hold around 40 percent of the world's proven natural gas reserves. However, limited local demand and the region's distance from the main centres of gas consumption in Europe, North America and Asia have resulted in production in 2006 accounting for only 11.7 percent of the world total. Many countries around the world are facing the need for increased natural gas imports to meet growing demand because their own reserves are limited or indigenous production is declining. The Middle East is increasingly seen as a source of long-term supply, mainly as LNG but long distance pipelines to Europe or Pakistan and India are also possibilities. The growing interest in natural gas imports from the region has come at a time when local consumption is on the increase in support of growing economic activity.

The Middle East is here defined as the countries of the Arabian Peninsula (Saudi Arabia, United Arab Emirates, Oman, Yemen, Bahrain, Qatar and Kuwait), Iran, Iraq, Syria, Lebanon and Jordan. However, the chapter will focus on the current and potential gas-exporting countries around the Gulf – Qatar, UAE, Oman, Yemen and Iran – plus Saudi Arabia, which, after Iran, is the largest gas-consuming country in the region.

The countries of the Middle East have adopted very different policies for the development of their gas reserves. These range from Saudi Arabia, which has shown no interest in the export of natural gas, to Qatar, where LNG is now the country's main source of revenue. In between these extremes is Iran, which has ambitious plans for the export of LNG and pipeline gas but currently only supplies a small amount of gas by pipeline to Turkey and, in 2006, was a net importer of gas.

The countries of the Middle East are even more dominant in their share of world oil reserves, accounting for 62 percent of the total. The production and export of oil has had a major influence on the development of the gas industry in the region. In Saudi Arabia, for example, oil exports have largely met the country's revenue needs and there has been no interest in natural gas exports. Indeed, in Saudi Arabia, natural gas is used as an alternative to oil in the industrial sector and in power generation, releasing oil for export. At the other end of the scale, Qatar's gas reserves are much larger than its oil reserves and the Emirate has turned to LNG exports to provide a major source of income for the country.

Proven Natural Gas Reserves

The proven reserves of natural gas in the Middle East are unevenly distributed around the region as is shown in Table 8.1. Nearly three-quarters of the reserves are in just two countries, Iran and Qatar, which are, after Russia with 47.6 Tcm, the world's second and third largest holders, respectively, of natural gas reserves. A single field, known as the North Field in Qatar and the South Pars field in Iran, which straddles the maritime border between the two countries, is estimated to contain nearly 40 Tcm of gas, close to 55 percent of the regional total. The field is the largest single accumulation of natural gas, accounting for over one-fifth of the world's proven reserves. It is the basis for Qatar's LNG activities and for the most advanced of Iran's planned LNG developments.

Table 8.1: Proven Gas Reserves in the Middle East

Country	*Proved Reserves in Tcm*	*2006 Production in Bcm/a*
Iran	28.13	105.0
Qatar	25.36	49.5
Saudi Arabia	7.07	73.7
UAE	6.06	47.4
Oman	0.98	25.1
Yemen	0.49	-
Other Middle East	5.38	35.2
Total	73.47	335.9

Source: *BP Statistical Review of World Energy* 2007

Saudi Arabia and the UAE's reserves are less than one-third of those of Qatar and Iran, but are still large enough to place them fourth and fifth on the list of the world's natural gas reserve holders. The reserves in Oman and Yemen are more limited in comparison but are sufficient to support LNG exports. Iraq is the main contributor in the countries grouped together as 'Other Middle East', with proven reserves amounting to 3.1 Tcm.

The production of gas in the countries shown in Table 8.1 is small in comparison with the reserve base. Iran leads the way, in terms of current production, followed by Saudi Arabia. The UAE and Qatar are also significant natural gas producers. Oman's output is more modest while production in the Yemen will commence with the start-up of its LNG project in early 2009.

The reserves-to-production (R/P) ratio for most of the countries listed in Table 8.1 exceeds 100 years and for the region as a whole it is over 200 years, Qatar's reserves will last for over 500 years at the production rate in 2006 but the R/P ratio will fall to around 100 years as LNG production expands and other uses for its gas are developed. Although there is no doubt that the Middle East has enormous reserves of natural gas and the reserves-to-production ratios are, as a result, very high, some caution has to be exercised over the figures. The estimates have generally not been subject to the rigorous independent evaluations that are required by oil and gas companies, buyers or regulatory agencies. There are suspicions that in some countries reserves are over-stated for political or other reasons. On the other hand, lack of interest in natural gas may have led to the reserves, in some cases, being under-estimated. A significant part of the natural gas reserve base is only available for production in the long term because the gas has been injected into oil reservoirs to enhance production or is contained in a gas cap which maintains pressure in the reservoir. These reserves can only be produced when the oil is fully depleted.

Table 8.2 shows that the share of natural gas in total oil and gas reserves ranges from 14 percent for Saudi Arabia, the region and the world's largest oil producer and exporter, to 91 percent for Qatar, the region and the world's largest LNG producer and exporter. The oil reserves of the UAE are significantly larger than its gas reserves. While it has been an LNG exporter since 1977, its policy has been largely focused on oil exports and there has been no expansion of Abu Dhabi's LNG production since 1994. In the Yemen and Oman, the reserve base is split almost equally between oil and gas and both countries have adopted a policy of developing oil and LNG exports. In Iran, natural gas accounts for just over 50 percent of reserves and, while oil is the main source of export revenues, LNG and pipeline gas exports are being pursued as long-term development options.

Table 8.2: Middle East Proven Gas and Oil Reserves

Country	*Total Proven Oil and Gas Reserves in billion Barrels oil equivalent (boe)*		*Percentage Share of Gas*
	Gas	*Oil*	
Iran	171.2	137.5	55.5
Qatar	154.3	15.2	91.0
Saudi Arabia	43.0	264.3	14.0
UAE	36.9	97.8	27.4
Oman	6.0	5.6	51.6
Yemen	2.9	2.9	50.4
Other Middle East	32.7	219.4	13.0
Total	447.2	742.7	37.6

Source: *BP Statistical Review of World Energy* 2007

Gas Production

Data on gas production in Middle Eastern countries are often inconsistent leading to questions over their reliability. Variations occur because measurements are made using different units and the conversion factors depend on the quality of the gas. The way in which natural gas liquids and the gas used in treatment and compression are accounted for also varies between sources. Furthermore, in some cases gas re-injected into oil reservoirs to enhance production is included in the gas production data while in others it is excluded. The volumes of gas involved can be significant.

Figure 8.1 shows gas production in the Middle East over the period 1990 to 2006 from the *BP Statistical Review of World Energy*. During that period the growth rate in gas production has averaged 7.8 percent per annum, i.e. a doubling every nine years. This compares with a growth rate of 2.3 percent for gas production globally over the same period.

The Middle East's share of global gas production has increased from 5.1 percent in 1990 to 11.7 percent in 2006. The rate of increase has varied considerably amongst the countries, with Saudi Arabia registering an average growth of 5 percent per annum and Qatar 13.8 percent per annum.

Figure 8.2 shows the uses of the gas produced in the six countries that are the focus of this chapter (Iran, Saudi Arabia, Qatar, UAE, Oman and Yemen) over the period 1975 to 2006. Local use is divided into power generation, industrial uses, the residential and commercial sector and others. The latter category covers energy sector uses, includ-

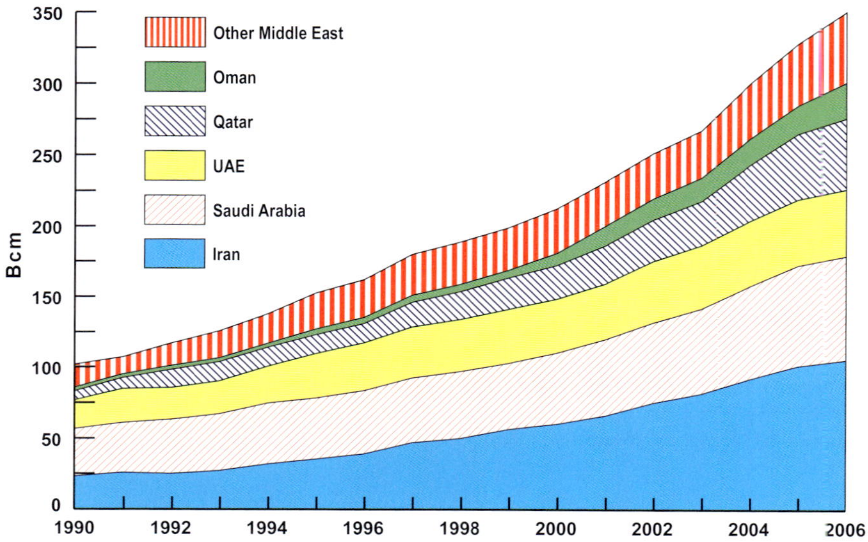

Figure 8.1: Natural Gas Production in the Middle East 1990–2006

Source: *BP Staistical Review of World Energy* 2007

ing the gas consumed in liquefaction plants, and consumption in the transport sector. Exports are separated into LNG and pipeline gas. The final category shown is flaring, shrinkage and losses, which includes natural gas liquids separated from the gas and the non-hydrocarbons, such as hydrogen sulphide, removed during the treatment of the gas. However, the main contributor in this category is flared gas.

In 1975, natural gas was widely treated as a by-product of oil production and assigned little value. As a result, flaring accounted for over 85 percent of the total production. Over the next two decades, flaring was reduced and, by 1995, the flaring, shrinkage and losses category had declined to under 20 percent of production. In 2006 its share was down to 15 percent of total production.

Between 1975 and 1995, the reduction in flaring was generally achieved by redirecting gas to local uses, principally power generation and the industrial sector. By 1995, each of these sectors accounted for around 30 percent of gas production. The share of power has remained at around 30 percent since then but the use of gas in the industrial sector has grown more slowly, and its share of production fell to 18 percent in 2006. Iran, which has by far the largest population of the countries included in Figure 8.2, is the only country where the residential and commercial sector accounts for a significant share of gas production.

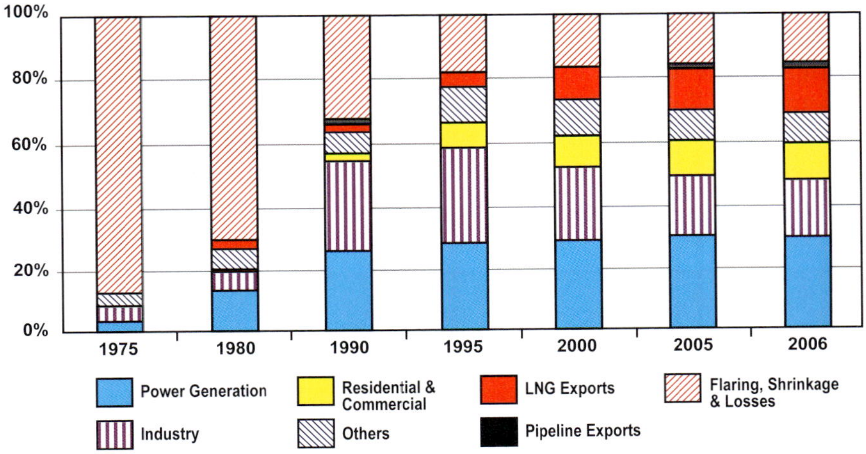

Figure 8.2: Uses of Gas Produced in Iran, Saudi Arabia, Qatar,UAE, Oman and Yemen

Source: FGEnergy

A small volume of gas was exported as LNG from the UAE (Abu Dhabi) from 1977, but exports did not take a significant share of gas production until the late 1990s when Qatar became an exporter. By 2006, LNG exports accounted for an estimated 14 percent of production in the six countries. The only pipeline exports from the region are from Iran to Turkey, which commenced in 2001. The volume reached 5.6 Bcm/a in 2006 when it accounted for 1.5 percent of total regional production. In 2007, the first significant trade of gas within the region started when gas flowed from Qatar to the UAE through the Dolphin pipeline.

LNG Exports

The Middle East's history as an LNG producer dates back to 1977, when the first two production trains at the Abu Dhabi Gas Liquefaction Company (ADGAS) plant on Das Island came into operation. The plant was originally designed to supply 2.06 Mtpa of LNG to the Tokyo Electric Power Company (TEPCO) in Japan. The plant was built to use gas produced in association with offshore oil production, which had previously been flared. It was a further seventeen years before the ADGAS plant was expanded and nearly twenty years before the Middle East's second liquefaction plant, in Qatar, came into operation. Map 8.1 shows current and future LNG pipelines and infrastructure.

Much of the Middle East's abundant reserves of natural gas can be defined as stranded in the sense that the potential production, in aggregate, far exceeds local demand. The distance to the main natural gas markets in north-east Asia, Europe and North America made the development of LNG or pipeline export projects uneconomic at prevailing market prices until the late 1990s. In addition, events such as the overthrow of the Shah of Iran in the late 1970s, the Iran–Iraq war in the 1980s and the first Gulf War in 1991 were a reminder to

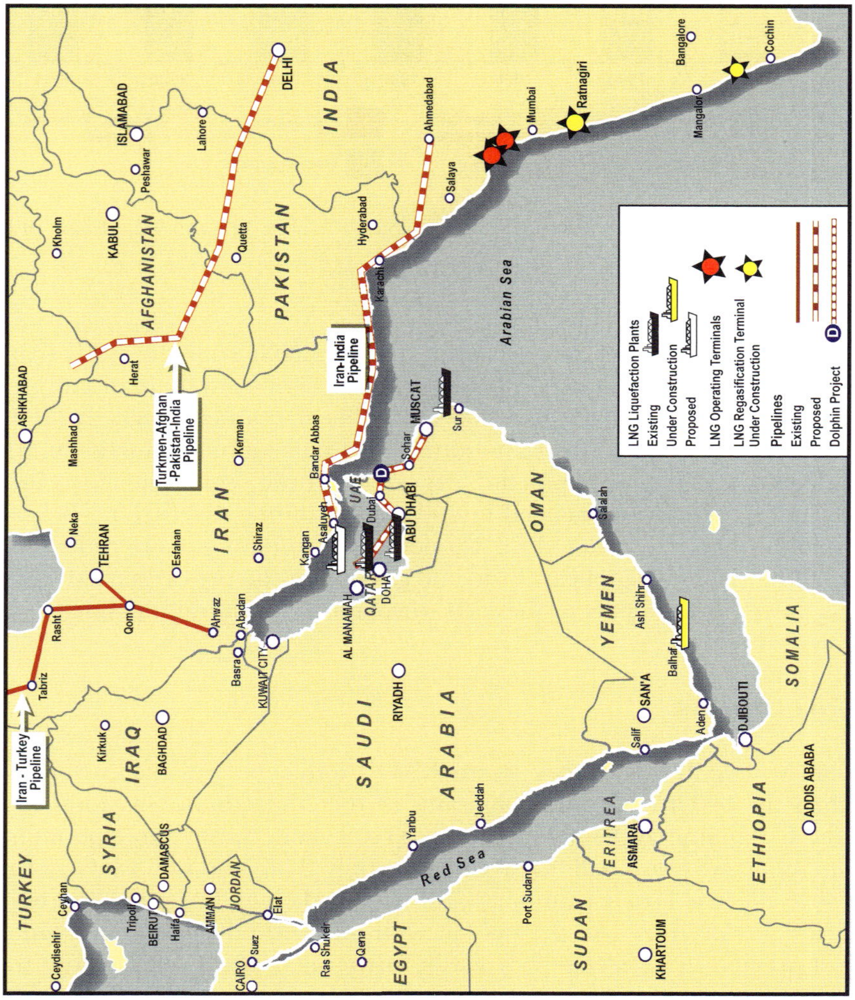

Map 8.1: LNG Pipelines and Infrastructure, Existing, under Construction and Proposed

potential buyers seeking secure gas supplies of the region's instability. Attempts to develop LNG exports from Iran in the 1970s and from Qatar in the 1980s made no progress against the challenges of costs and buyer and investor concerns over the geo-political risks.

The main growth of LNG globally between 1969 and the late 1990s was focused on supplying markets in north-east Asia. Initially Japan was the only importer in the region but it was joined by Korea in 1986 and Taiwan in 1990. These countries do not have significant indigenous reserves and are too far from major gas fields for pipeline imports to be technically or economically viable.[1] This left LNG as the only option to meet the growing demand for a fuel to support policies of diversifying away from dependence on imported oil and of reducing air pollution in major conurbations. With no competition from local or imported pipeline gas, LNG is a price setter in these markets and, from the 1970s to around 2000, the buyers there were generally prepared to pay higher prices than prevailed in, for example, Europe and the USA to secure the supplies they required.

Asia's LNG needs in the 1970s and 1980s were largely met from new projects in South East Asia, with Australia's North West Shelf project adding to regional supply from 1989. Buyers in Japan, Korea and Taiwan had no need to look outside the Pacific Basin for LNG and, with the exception of the small Abu Dhabi project, they were not prepared to, nor did they have to, turn to the more politically unstable Middle Eastern region for their LNG supplies.

The markets of Europe and the USA did not offer a feasible alternative market for potential Middle Eastern suppliers until after 2000. The USA had briefly turned to LNG imports from Algeria in the 1970s to meet what was then seen as a widening gap between domestic production and growing demand. US buyers negotiated contracts for 28 Mtpa (38 Bcm/a) of LNG from Sonatrach in Algeria and four terminals were built to receive the LNG. However, the deregulation of US well-head prices and the impact of the second oil price shock in 1980 closed the gap between natural gas supply and demand and LNG imports were no longer needed. As the US market readjusted to a new price environment, gas prices averaged under $2/MMBtu, too low to make importing LNG from as far away as the Middle East a viable option.

Europe was also well-supplied with gas in the 1970s and 1980s through indigenous production, pipeline imports from Russia and LNG imports from Algeria. LNG supplied to Europe from the Middle East would have had to compete on price with pipeline gas, which made new developments directed at Europe uneconomic.

The prospects for Middle Eastern LNG began to look more promising by the 1990s as the requirements of buyers in Asia continued to grow, outpacing the potential supply from the Pacific Basin. The first project to benefit was ADGAS in Abu Dhabi, which more than doubled its contracted volumes with TEPCO, underpinning the construction of a third production train commissioned in 1994. Qatar formed its first liquefaction company (Qatargas) in 1984 and immediately started active marketing in Asia and in Europe but it was not until 1989 that a buyer, Chubu Electric in Japan, showed serious interest in committing to purchase the LNG. This led to the development of the 6 Mtpa Qatargas project, which came into operation in December 1996. Qatar's second liquefaction project (RasGas) was established in 1993 and started production in 1999. The debottlenecking[2] of the Qatargas and RasGas plants and the construction of new trains at the RasGas plant took Qatar's production capacity to 30.7 Mtpa (41.4 Bcm/a) by the end of 2006, the year that Qatar overtook Indonesia to become the world's largest LNG producer.

Oman became the Middle East's third LNG producer when the two-train Oman LNG plant at Qalhat came into operation. A third train was added in at the end of 2005 taking installed capacity to around 11 Mtpa (14.9 Bcm/a).

Total LNG export capacity in the Middle East at the end of 2006 was 47.4 Mtpa (64 Bcm/a). Actual exports in 2006 were lower at 38.5 Mtpa (52 Bcm/a) since the new trains in Qatar and Oman were in build-up mode during the year (Table 8.3).

Figure 8.3 shows the build-up of Middle Eastern LNG exports over the period 1990 to 2006. In the decade since production commenced in Qatar, the growth rate has averaged 20 percent per annum (Figure 8.4).

Table 8.3: LNG Export Capacity in the Middle East

	Capacity		*Production in 2006*	
	Mtpa	*Bcm/a*	*Mtpa*	*Bcm/a*
Abu Dhabi	5.7	7.7	5.4	7.3
Qatar				
Qatargas	10.0	13.5		
RasGas	20.7	27.9		
Total Qatar	30.7	41.4	24.5	33.1
Oman	11.0	14.9	8.6	11.6
Total	47.4	64.0	38.5	52.0

Source: Author's Own Estimates

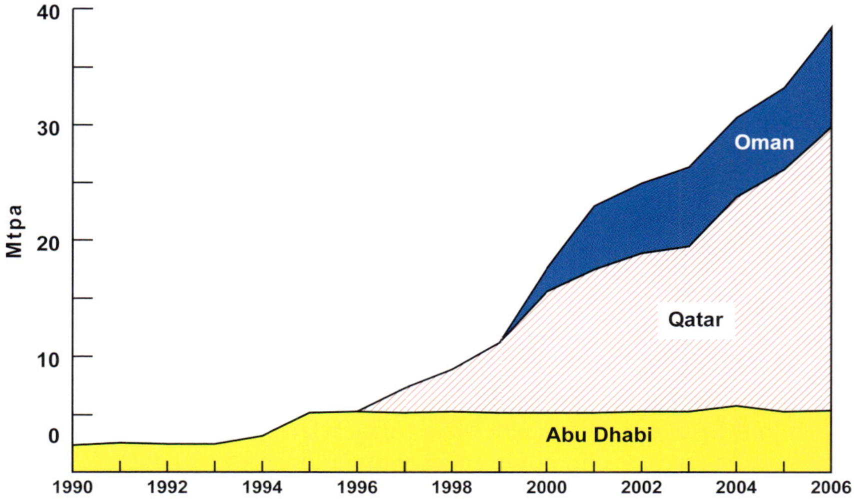

Figure 8.3: Middle Eastern LNG Exports 1990–2006

Source: Author's Estimates and Cedigaz

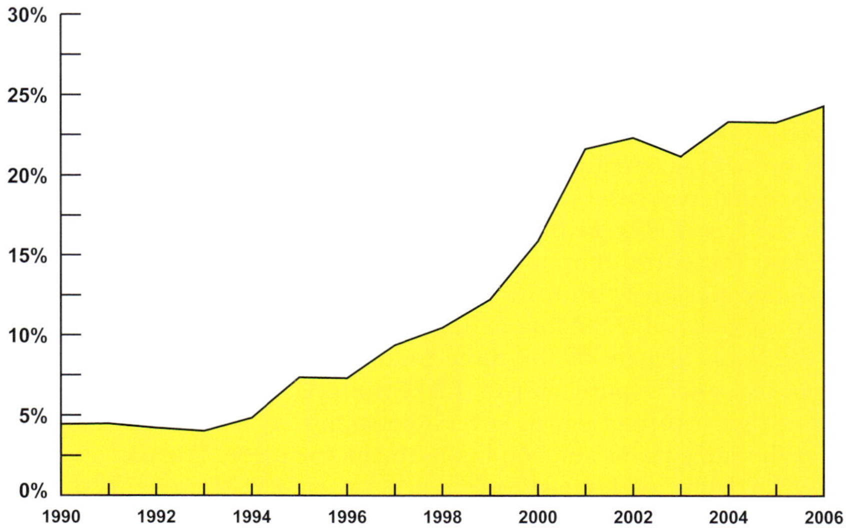

Figure 8.4: Middle East Share of Global LNG Exports 1990–2006

Source: Author's Estimates and Cedigaz

The Middle East's share of global LNG exports has increased from 4 percent in 1990 to close to 25 percent in 2006 (Figure 8.4) and is set to grow further by 2012, as capacity currently under construction in Qatar and in Yemen is commissioned. The increase in the Middle East's share of the global LNG market has come mainly at the expense of the Pacific Basin, which has seen its share decline from 67 percent in 1990 to 41 percent in 2006.

Abu Dhabi

The ADGAS LNG project was established in 1972 as a joint venture between upstream producers BP and Total and the Japanese companies Mitsui and Bridgestone Liquefied Gas, which later became Mitsui Liquefied Gas and its share holding was eventually merged with Mitsui's. The Abu Dhabi government, through the Abu Dhabi National Oil Company (ADNOC), had a minor role in the early stages of development but its share has grown to 70 percent. The project was initially developed as a two-train facility, with a design capacity of 2.06 Mtpa, to supply LNG on a CIF[3] basis to the Tokyo Electric Power Company in Japan.

The original plans for the project were based on the plant being supplied with natural gas produced in association with oil from offshore fields. Production commenced in 1977 but the build-up of output was affected by problems with one of the storage tanks and new tanks had to be constructed. In addition, the supply of associated gas to the plant proved to be unreliable and additional non-associated gas supply had to be brought on stream. Production eventually reached, and exceeded, design capacity by the mid-1980s, peaking at 2.56 Mt in 1991.

In the late 1980s TEPCO approached ADGAS with a proposal to purchase the entire output from an expansion of the plant. ADGAS responded positively and, in 1994, train 3 was commissioned with a design capacity of 2.3 Mtpa. The train performed above capacity taking the total capacity of the three-train plant to 5.7 Mtpa. TEPCO is the sole long-term purchaser of LNG from ADGAS through a unique liquefied gas contract which covers the supply of both LNG and LPG under the same terms and conditions, including price. The base contract is for 5.4 Mtpa of liquefied gas made up of 4.7 Mtpa of LNG and 0.7 Mtpa of LPG. TEPCO has the option of increasing the annual quantities of LNG to 4.9 Mtpa and has agreed to lift a minimum of 4.3 Mtpa. With the plant able to produce at around 5.7 Mtpa, this leaves up to 0.8 Mtpa of surplus capacity. TEPCO has taken delivery of some of the extra volumes and ADGAS has also sold cargoes over the last decade on a short- and medium-term basis to the USA, Europe, India, Taiwan and Korea.

Abu Dhabi's proven reserves of over 6 Tcm and its R/P ratio of 130 years suggest that it is well positioned to expand its LNG capacity. However, it appears to have taken a policy decision to give priority to domestic gas use over the addition of further trains at the ADGAS plant site on Das Island, which would probably require the reclamation of land since space on the island is limited, or the building of a new plant on the mainland.

In a paper presented at the GasTech Conference in December 2006, Fatema Al-Neaimi, Senior Market Analyst in ADGAS's Marketing Division,[4] questioned the future of LNG in the Middle East as a whole, and in Abu Dhabi in particular, as demographics and economics increase local gas demand. The UAE, of which Abu Dhabi is the largest of the seven emirates, is seeing population growth of 7 percent per annum and power demand growth of 5–7 percent per annum. The use of gas for power generation, petrochemicals and industry is expanding rapidly, constraining the amount of gas available for export and, indeed, requiring the UAE to commit to pipeline gas imports from Qatar through the Dolphin project (see below). Imports of pipeline gas from Qatar commenced in mid 2007. It is difficult to understand why a country with such extensive gas reserves has decided not to increase its exports and, moreover, has contracted to receive 18 Bcm/a of imported gas. It is further evidence that the quoted proven reserves figures and the reserves to production are not necessarily a reliable indicator of potential production levels.

The outlook for LNG in Abu Dhabi is for production to continue at up to 5.7 Mtpa until 2019, when the current contract with TEPCO expires. By that time, the two original liquefaction trains will be over forty years old but, provided they continue to be well maintained, there is no reason why they should not be able to remain in operation for another twenty years or more. The reserves offshore Abu Dhabi certainly appear to be sufficient to support such an extension.

Qatar

LNG Exports

The North Field, which was discovered by Shell in 1971, has, 35 years later, transformed the emirate in a way that few could have envisaged. It took several years for the full extent of the reserves in the field to be established but by the early 1980s the proven reserves were estimated to be at least 8.5 Tcm, well in excess of any possible local demand. LNG export was the only realistic option for Qatar to take full advantage of the opportunity that the discovery of the field had created.

The government instigated a tender process in the early 1980s to select companies to partner Qatar Petroleum (then known as the Qatar General Petroleum Corporation) in the development of an LNG project to be supplied with gas from the North Field. BP and Total won the tender and became 7.5 percent partners in the Qatargas project. Qatar Petroleum retained a 70 percent share and 15 percent was reserved for Japanese partners. The first stage of the development of the North Field was to be for domestic use and, as part of the agreement with the government, BP and Total were given an advisory role in the planning and development of the first offshore facilities: a sub-sea pipeline to shore near Ras Laffan in the north of the country and an onshore section to Umm Said to the south of Doha, which at the time was the main industrial centre in Qatar.

Marubeni joined Qatargas, with a 7.5 percent share, soon after BP and Total, and Mitsui took the remaining 7.5 percent share reserved for Japanese partners in 1988. Mitsui's relationship with the Chubu Electric Power Company in Japan was a major factor in the latter company expressing an interest in purchasing the entire 4 Mtpa output from Qatargas's first two trains. A Letter of Intent was signed between Qatargas and Chubu Electric in 1991 and a full Sales and Purchase Agreement (SPA) in 1993. In 1995, Qatargas signed an SPA with a consortium of seven Japanese gas and power utilities for the 2 Mtpa output from the third train. With this agreement, Qatargas's entire planned output from three trains was committed to Japanese buyers. The first cargo of Qatari LNG was loaded in late December 1996 and arrived at Chubu's new Kawagoe LNG receiving terminal and power plant complex near Nagoya on 10 January 1997, launching Qatar on a journey which, by 2006, had already seen it become the world's largest LNG exporter.

The journey was not entirely smooth as the project had to overcome firstly the withdrawal of BP as a partner in 1992 and then the take over of power by the present ruler, Sheikh Hamad bin Khalifa al-Thani, in a palace coup against his father in 1995. Both events could have been very disruptive to Qatar's emergence as an LNG exporter but, with hindsight, both can now be seen as having had a positive influence. The new Emir has been much more reform minded than his conservative father and he has given strong support to the rapid expansion of Qatari LNG exports.

BP's withdrawal from the project came at a time when the company had to reduce its capital commitments to strengthen its balance sheet. The expected returns on the Qatargas project were at the low end of projects in BP's global portfolio of upstream and downstream assets and it was unable to justify further investment in the project. The withdrawal was timed before the SPA with Chubu Electric was signed since BP wanted to minimise the risk of damage to the project's prospects, which it believed would increase after the signature of the SPA and the start of construction of the plant.

Qatar's search for a replacement partner was quickly concluded with Mobil offering to replace BP. At that time Mobil was looking for a new LNG opportunity since the Arun project in Indonesia, which had provided a significant part of the company's worldwide income throughout the 1980s, was facing a long-term decline as production from the Arun field, the plant's main supply source, fell (see Chapter 7). Mobil's entry into the Qatargas project was an important catalyst for the development of a second LNG project in Qatar, known as the RasGas project. Mobil was looking for a project in which it would have a significant stake to replace the effective 30 percent share it had in the Arun project. The 7.5 percent initially on offer in the Qatargas project did not fulfil that requirement. The Government of Qatar was prepared to increase Mobil's share in Qatargas to 10 percent but could go no further without displacing one or more of the other partners or reducing Qatar Petroleum's share to an unacceptable level. The compromise was the establishment of the RasGas project in which Mobil's share was initially 30 percent.

RasGas targeted the Korean market, which by the mid-1990s had grown to the stage where it could support the development of a green-field LNG development. Previously, Korea had obtained its LNG from surplus capacity and additional trains at projects built to supply Japan. In 1995, Korea Gas committed to the 2.4 Mtpa output from the first train. Two years later it doubled the volume to 4.8 Mtpa and, as part of the deal, RasGas agreed to drop the floor price which had been included in the original contract for 2.4 Mtpa. The RasGas project started-up in May 1999 with its first delivery being a spot cargo sold to CMS Energy in the USA rather than to its long-term buyer, Korea Gas, an early indication of the changes that were beginning to take place in LNG markets at that time.

By the time RasGas came into operation, Exxon had merged with Mobil and the shareholder was ExxonMobil. Furthermore, Itochu (4 percent) and LNG Japan (3 percent) had become shareholders and a 5 percent share had been allocated to a Korean consortium as part of the deal for the purchase of the LNG. The resulting shareholdings in RasGas are: Qatar Petroleum 63 percent, ExxonMobil 25 percent, Itochu (4 percent), LNG Japan (3 percent) and KORAS (5 percent).

By 2000, Qatar had five liquefaction trains in operation with a design capacity of 10.8 Mtpa. The trains operated better than expected, and with further additions to capacity achieved through debottlenecking, the capacity of the first five trains reached 16.6 Mtpa by late 2005. Qatar's rapid entry into the LNG export market proved to be a platform for an expansion that has been unprecedented in its scope.

In April 2001, less than two years after the start-up of the RasGas plant, Qatar Petroleum (70 percent) and ExxonMobil (30 percent) announced their commitment to the first of three 4.7 Mtpa trains. The second phase of development has been designated RasGas II and became fully operational in December 2006 when the third of the new trains loaded its first cargo. In 2001, the LNG buyers in Japan and Korea were not in a position to commit to additional LNG imports from Qatar and RasGas II had to turn to alternative markets to sell its output. The first of these was the emerging Indian market. A long-term contract for 5 Mtpa was signed with Petronet, which was looking for LNG supply for its terminal at Dahej in the north-west of the country. After several price renegotiations, a fixed price of $2.53/MMBtu for LNG fob[5] Ras Laffan was agreed for five years. Over the period from five to ten years after start-up (i.e. the beginning of 2009 to the end of 2013) the contract price will revert gradually to indexation with the oil price, as is the case for LNG prices elsewhere in Asia.

India's success in achieving a relatively low, fixed price for the first five years of supply from RasGas ensured that LNG was competitive with domestic natural gas and with alternative fuels, helping to stimulate the demand for LNG. It was achieved at a time when the amount of potential LNG supply from green-field and expansion projects in the Middle East and the Pacific Basin exceeded the global demand for LNG. The competition amongst sellers created a buyers' market, putting downward pressure on prices. Petronet in India was one of the buyers able to take advantage of the situation. By early 2005, some of the potential projects had been delayed or abandoned and others had secured buyers leaving only limited volumes of potential LNG supply on offer to buyers. As competition between buyers for available supply intensified, a sellers' market re-emerged and prices started to move upwards.

A final investment decision for the second 4.7 Mtpa train at the RasGas plant was made in October 2002, with the output committed to the Italian market. The third train followed in July 2004, without any firm sales contracts being in place when the decision was made. However, the LNG has now been committed to India (the contract with Petronet has been increased to 7.5 Mtpa) and Belgium, where the buyer is the main gas marketing company, Distrigas.

RasGas II was the first stage of Qatar's move away from Japan and Korea as the foundation buyers for new LNG trains. The process continued with the commitment to six so-called 'mega-trains' of 7.8 Mtpa capacity announced in stages between mid-2002 and March 2005. The development of 46.8 Mtpa of capacity represents a major addition to global liquefaction capacity, much more than could have been committed under the traditional long-term contracts with buyers in Japan and Korea that had underpinned the initial Qatargas and RasGas developments.

The only way in which buyers could be found for such large volumes of LNG was to target the liquid markets of the USA and the UK. In these markets, buyers for the LNG can always be found if it is price competitive, after regasification, with the domestic gas production and pipeline imports. Natural gas prices are determined by supply and demand and can be very volatile, as experience has shown. For example, in January 2006, UK gas prices were the highest in the world, averaging \$16.40/MMBtu as cold weather and concerns about supply pushed prices upwards. Eighteen months later, in July 2007, the price had fallen to less than a quarter of the January 2006 level as the low demand of the summer months and an increase in pipeline imports led to a surplus of supply. For producers of LNG or pipeline gas targeting a volatile market like the UK or the USA it is essential to be at the low end of the cost curve to avoid losing money or market share if prices fall.

The main opportunity to reduce costs in the LNG business comes from economies of scale, i.e. increasing the size of facilities to minimise unit costs. This is one of the main drivers behind Qatar's decision to opt for the largest possible liquefaction trains – 7.8 Mtpa in the case of the mega-trains – and large LNG ships. In mid-2007, Qatar had on order 45 ships of 210,000m^3 to 266,000m^3 capacity which will each be able to carry between 35 and 72 percent more than the largest ship in operation in mid-2007.

The first two mega-trains, Qatargas trains 4 and 5 (Qatargas II), were announced in 2002 as targeting the UK market and the next four trains, Qatargas trains 6 and 7 (Qatargas II and IV) and RasGas trains 6 and 7 (RasGas III), were first announced as targeting the US market. Subsequent deals have seen some of the volumes committed initially to the US and UK markets being transferred to other markets

Table 8.4: Qatargas Liquefaction Trains

Project	*Trains*	*Capacity in Mtpa*	*Start-up*	*Shareholders*	*Main Markets*
Qatargas I (T1-3)	3	10	1996	Qatar Petroleum 65%, Total 10%, ExxonMobil 10%, Marubeni 7.5%, Mitsui 7.5%	Japan, Spain
Qatargas II (T4)	1	7.8	mid-08	Qatar Petroleum 70%, ExxonMobil 30%	UK
Qatargas II (T5)	1	7.8	1H09	Qatar Petroleum 70%, Total 16.7%, ExxonMobil 18.3%	UK, France, USA, Mexico
Qatargas III (T6)	1	7.8	1H10	Qatar Petroleum 68.5%, ConocoPhillips 30%, Mitsui 1.5%	USA
Qatargas IV (T7)	1	7.8	2H10/ 1H11	Qatar Petroleum 70%, Shell 30%	USA,
Total Qatargas	7	41.2			

Source: Author's estimates

Table 8.5: RasGas Liquefaction Trains

Project	*Trains*	*Capacity in Mtpa*	*Start-up*	*Shareholders*	*Main Markets*
RasGas I (T1-2)	2	6.6	1999	Qatar Petroleum 65%, ExxonMobil 25%, Itochu 4%, LNGJapan 3%, KORAS 5% Mitsui 7.5%	Korea
RasGas II (T3)	1	4.7	2004	Qatar Petroleum 70%, ExxonMobil 30%	India
RasGas II (T4)	1	4.7	2005	Qatar Petroleum 70%, ExxonMobil 30%	Italy
RasGas II (T5)	1	4.7	2006	Qatar Petroleum 70%, ExxonMobil 30%	Belgium, India, Korea
RasGas III (T6-7)	1	15.6	2008/09	Qatar Petroleum 70%, ExxonMobil 30%	USA, Taiwan
Total RasGas	7	36.3			

Source: Author's estimates

and Qatar is signalling to the markets in Asia and elsewhere in Europe that further transfers can be made if buyers are able to offer higher prices or better terms.

In mid-2007, Qatar had eight liquefaction trains in operation and six mega-trains under construction. Start-up of the first of the mega-trains was scheduled for mid-2008, with the remainder following over the next three years. By 2011 Qatar will, on current plans, have 77.5 Mtpa of LNG production capacity in operation making it by far the world's largest producer. Tables 8.4 and 8.5 summarise the status of the liquefaction trains at each of the facilities, their ownership and the main markets for the output from each train.

The entire production from the Qatargas and the RasGas plants is effectively committed to buyers on a medium- and long-term basis as is shown in Tables 8.6 and 8.7. The long-term contracts are for 20 or 25 years duration.

Figure 8.5 shows the regional markets for Qatar's LNG production over the first ten years of production. Although Asia has dominated

Table 8.6: Volumes Contracted on a Medium and Long-term Basis from Qatargas

	Buyer	*Volume in Mtpa*	*Duration*	*Market*
Qatargas I	Chubu Electric	4.00	1997–2021	Japan
	Kansai Electric	0.29	1999–2021	Japan
	Tokyo Electric	0.20	1999–2021	Japan
	Tohoku Electric	0.52	1999–2021	Japan
	Tokyo Gas	0.35	1999–2021	Japan
	Osaka Gas	0.35	1999–2021	Japan
	Toho Gas	0.17	1999–2021	Japan
	Gas Natural	1.40	2002–2012	Spain
	Gas Natural	1.40	2005–2025	Spain
Qatargas I/II	Chubu Electric	1.20	2007-2011	Japan
Qatargas II	ExxonMobil	7.80	2008–2028	UK
	ExxonMobil	2.60	2009–2029	UK
	Total	1.85	2009–2029	France
	Total	1.85	2009–2029	USA/Mexico
	Total	1.50	2009–2029	UK
Qatargas III	ConocoPhillips	7.80	2010–2030	USA
Qatargas IV	Shell	7.80	2011–2031	USA/Europe
	Total Contracted	41.08		

Source: Author's estimates

Table 8.7: Volumes Contracted on a Long-term Basis from RasGas

	Buyer	*Volume in Mtpa*	*Duration*	*Market*
RasGas I	Korea Gas	4.8	1999–2024	Korea
	Endesa	0.7	2005–2025	Spain
RasGas II	Petronet, India	7.5	2004–2024	India
	Edison	4.7	2009–2029	Italy
	Distrigas	2.1	2007–2027	Belgium
RasGas III	CPC	3.0	2008–2028	Taiwan
RasGas III	ExxonMobil	12.6	2008–2028	USA
RasGas II/III	Korea Gas	2.1	2007–2027	Korea
	Total RasGas	37.5		

Source: Author's estimates

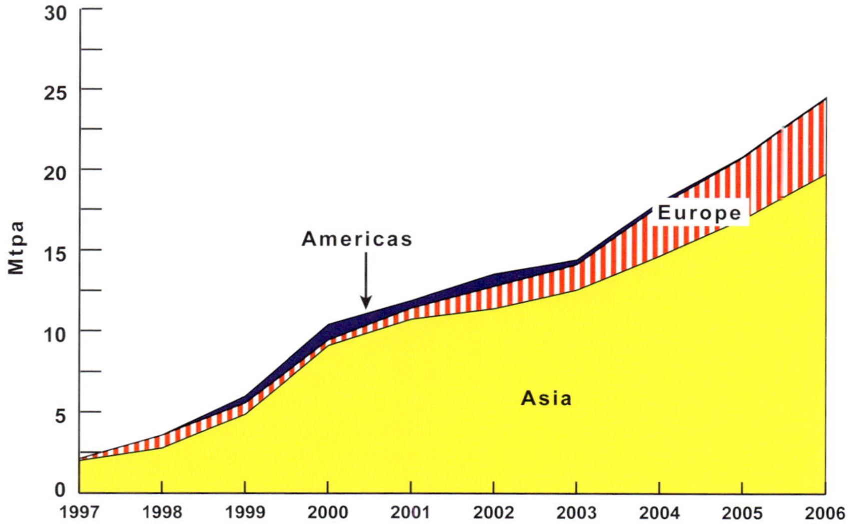

Figure 8.5: LNG Exports from Qatar 1997–2006

Source: Author's Estimates and Cedigaz

Qatar's portfolio of LNG sales, from the first year of operation of the Qatargas project LNG was being sold on a short-term basis to Europe. In 1999, LNG cargoes from Qatar were delivered to the USA for the first time. The Americas (Mexico joined the USA in 2006 as an importer of Qatari LNG) have remained a short-term market for

Qatari LNG with only one cargo being imported in each of the years 2005 and 2006. However, European imports have grown as medium- and long-term contracts with Spanish buyers have come into effect. In 2006, 19 percent of Qatar's production was delivered to Europe, where Spain was the main importer but some cargoes were also shipped to the UK, Italy, France and Belgium.

As Table 8.8 shows, Qatar's sales profile is set to change dramatically by 2012 when all the capacity under construction will be in operation. Asia's share of the volumes will fall to 30 percent and Europe's share will be at a similar level. Qatar's main market will be the Americas, and in particular the USA, which could account for nearly 39 percent of its output.

The data in Table 8.8 have been compiled on the assumption that the LNG will be delivered to the market for which it was contracted. The actual outcome in 2012 could be very different. Over 50 percent of Qatar's production in that year (41.25 Mt out of 77.5 Mt) is contracted into the flexible markets of the UK and the USA, where diverting

Table 8.8: Qatar LNG Sales Profile

	in Mt 2006	*Percent Share*	*in Mt 2012*	*Percent Share*
Japan	7.48		6.00	
Korea	6.74		6.90	
Taiwan	0.38		3.00	
India	5.19		7.50	
Total Asia	19.79	80.6	23.4	30.2
Spain	3.88		3.50	
UK	0.05		11.90	
France	0.43		1.85	
Belgium	0.31		2.10	
Italy	0.03		4.70	
Total Europe	4.7	19.1	24.05	31.0
USA			29.35	
Mexico	0.06		0.70	
Total Americas	0.06	0.2	30.05	38.8
Total	24.55		77.5	

Source: Author's estimates

cargoes to alternative markets can be relatively easily arranged. Contractual commitments downstream of the receiving terminal can be met through the purchase of gas at trading hubs allowing the LNG to be delivered to other markets.

Qatar's marketing strategy over the last few years has created a position where it has options to trade its LNG and enter into contracts with new buyers on a short-, medium- or long-term basis. The process of negotiating with buyers who might be prepared to offer improved terms for diverted cargoes started well before the new trains came into operation. The first of the diversion deals was concluded in November 2006 when Korea Gas agreed to a long-term contract for 2.1 Mtpa at prices which are understood to be above crude oil parity at an oil price of $60/Bbl, a significant premium over the prices in Korea's other long-term purchase contracts.

Buyers elsewhere in Asia are reported to be reluctant to commit to long-term purchases at such high prices but their position could eventually change, especially if alternative supplies are not available. The volume of LNG Qatar will be producing from 2012 and the flexible nature of some of its long-term commitments means that it can not be ignored by Asian buyers as a potential supply source if they are to meet the forecast demand growth in their downstream markets. As a result, the outcome in terms of the share of the regional markets in Qatar's export portfolio after 2012 could be very different from that indicated in Table 8.8, which is based on the long-term contracts in place in 2007.

Other Outlets for Qatar's North Field Gas

The reserves in the North Field are so large that, in addition to LNG, which will remain the main use for the gas, a number of alternative uses are being developed including pipeline exports, Gas to Liquids (GTLs) and domestic consumption.

Domestic Gas Use

The first phase of the development of the North Field came in 1991 when production commenced from the so-called Alpha block supplying gas to the domestic market. Output is around 8.5 Bcm/a, which, after liquids are extracted, yields around 7.4 Bcm/a for local consumption. The next stage in the development of gas supply for the domestic market is the first phase of the Al Khaleej project, a joint venture between ExxonMobil and Qatar Petroleum, which supplies 7.6 Bcm/a of gas for the Oryx GTL plant (see below), a power plant at Ras Laffan and other local uses. The second phase of the Al Khaleej project is scheduled to come on stream in 2009. It will produce a further 12.9 Bcm/a of dry gas for local use including a range of petrochemical plants which are scheduled to commence operations around that time.

The Barzan gas project, which was announced in 2007, will be a further stage of the development of gas production for the domestic market. It is designed to produce 14.2 Bcm/a of gas from early 2012. The project is a joint venture between Qatar Petroleum (90 percent) and ExxonMobil (10 percent). Its announcement coincided with ExxonMobil's decision to cancel its GTL project, on which the company had been working for over a decade. The Barzan project will take gas supply for the local market to over 40 Bcm/a.

Pipeline Exports

The first export pipeline system in the Gulf region has been developed by Dolphin Energy Limited (DEL), a joint venture between Mubadala Development Corporation (51 percent), which is wholly owned by the Government of Abu Dhabi, Total (24.5 percent) and Occidental (24.5 percent). It links Qatar with the UAE and Oman and came into operation in early 2007, initially with supplies from the Al Khaleej project followed by Dolphin-produced gas in July 2007. Dolphin was originally envisaged as a more extensive project linking the members of the Gulf Co-operation Council (Saudi Arabia, Kuwait, Bahrain, Qatar, the UAE and Oman) with a possible spur line to Pakistan. However, opposition from Saudi Arabia to the pipeline crossing its territory meant that Bahrain and Kuwait had to be dropped from the plans. The link to Pakistan was not progressed, leaving a project which connects Qatar to the UAE and Oman.

The capacity of the Dolphin project will be 20.7 Bcm/a when fully operational in 2008. The gas is being produced from the North Field and piped to Ras Laffan where it is processed in a $1.6bn treatment plant. Condensates, LPGs and ethane are separated from the produced gas and sold separately. A 370 km, 48 inch, sub-sea pipeline transports the treated natural gas to Taweelah in Abu Dhabi where it links into an onshore pipeline to distribute it around the UAE and to Oman.

The first stage of the Dolphin project was the pipeline linking Oman and Fujairah which came into operation in January 2004 to transport gas from Oman to Fujairah and Ras al Kaimah. The direction of flow through the line will be reversed in 2008 when the export of Dolphin project gas to Oman commences.

The pipeline from Qatar to the UAE came into operation in March 2007. However, delays in the construction of the onshore gas treatment plant at Ras Laffan resulted in the initial supply of 4.1 Bcm/a coming from gas diverted from the Oryx GTL plant, which faced problems during commissioning. Gas produced by the Dolphin project began to flow to the UAE in July 2007.

Plans for the project to be expanded to 33 Bcm/a and for the possible addition of a second pipeline to meet rapidly increasing demand in the UAE and Oman are on hold because of Qatar's moratorium on the further development of production from the North Field. Qatar is unlikely to favour further exports of gas by pipeline to its neighbours unless prices are much higher than those agreed for the first phase of the project. The price is reported to be $0.87/MMBtu into the pipeline at Ras Laffan, which results in a price of $1.30/MMBtu delivered to the UAE.[6] The price escalates at around 1.5 percent per annum. The Dolphin sales yield considerably less value to Qatar than its LNG exports, which in mid-2007 were selling for an average of around $8/MMBtu.

Gas to Liquids (GTL)[7]

In the early 2000s the Government of Qatar had ambitious plans for the development of GTLs with the stated intention of becoming the 'GTL capital of the world'. Preliminary agreements were signed with ExxonMobil, Shell, Sasol, ConocoPhillips, Sasol-Chevron and Marathon for feasibility studies of GTL plants. However, GTLs became the main casualty of the decision in 2005 to call a moratorium on further developments of production from the North Field. The ConocoPhillips, Marathon and Sasol-Chevron projects were dropped when the moratorium was announced and ExxonMobil's GTL project was cancelled in February, leaving only Sasol's Oryx plant and Shell's Pearl project as survivors.

Cost escalation was probably a major factor in the decision to call a halt to further GTL developments rather than to the expansion of LNG capacity. GTL plants require much higher capital investment per unit of useful energy produced than an equivalent sized LNG plant.

The Oryx plant, a joint venture between Qatar Petroleum and the South African company, Sasol, came into operation in 2006. It has a design capacity of 34,000 barrels per day (b/d) of high grade oil products but problems with the catalyst resulted in a slow build-up of output and in 2007 it was operating at less than one-third of capacity, releasing gas for the early stages of the Dolphin project (see above).

Shell's Pearl project is on a much larger scale than Oryx, with an eventual capacity of 140,000 b/d of oil products. The first phase (70,000 b/d) is scheduled to be commissioned in late 2009, with the second phase (70,000 b/d) a few months later. Some reports put the estimated cost of the plant as high as $18 bn.[8] With work on further GTL projects halted, and uncertainty over Qatar's natural gas strategy at the end of the moratorium on further development of the North Field, it is possible that Oryx and Pearl will be the limit of Qatar's GTL production.

Qatar's Moratorium on Further Developments of North Field Gas

In 2005, Qatar surprised the natural gas world by announcing a moratorium on further development of gas production from the North Field to allow an assessment to be made of the field's performance at a production rate of over 200 Bcm/a, which will be reached soon after 2010.[9] When the announcement was made, the target for the completion of the assessment was 2007 but the date has slipped until 2009 at the earliest and the outcome may not be known until 2012. The result is considerable uncertainty over Qatar's future role as a gas producer and exporter beyond the 77.5 Mtpa of LNG output from projects in operation or under construction in mid-2007 and the 20.7 Bcm/a of Dolphin project pipeline exports.

The moratorium was called because the Government of Qatar wanted to be sure that the North Field reservoir was being managed in a way that maximised the eventual production of gas and associated liquids. Projects in operation and under construction in mid-2007 will produce around 232 Bcm/a of wet gas (that is gas before the ethane, LPGs and condensates have been removed) by 2012, as shown in Table 8.9. At this rate of production, the 25 Tcm of proven reserves in the North Field will last for over 100 years. However, as discussed above, there is a lack of independent studies of the reserves in the North Field and, as a result, there is uncertainty over the accuracy of the 25 Tcm figure. The North Field is Qatar's main hydrocarbon resource and accounts for nearly 99 percent of its proven gas reserves. The government does not want to find that the field has been over-exploited and the availability of production for future generations has been compromised. Even if the 25 Tcm reserves estimate proves to be accurate, Qatar wants to avoid being in a position where over-producing in the early years of operation has reduced the longer-term output. Calling a moratorium on further expansion is therefore a prudent decision by a government concerned to ensure the prosperity of the nation in the future.

The moratorium does, however, create uncertainty for countries that need LNG to meet future gas demand growth. The key questions for

Table 8.9: Expected Gas Production from Qatar's North Field

Project	*Start-up*	*Capacity Bcm/a**	*Utilisation*
North Field Alpha	1991	8.5	Domestic Market
Qatargas I (T1-3)	1996	17.0	LNG Export
RasGas I (T1-2)	1999	11.7	LNG Export
RasGas II (T3-5)	2004	24.5	LNG Export
Al Khaleej I	2005	8.7	Domestic Market and GTL (Oryx)
Dolphin	2007	23.8	Pipeline Exports
Qatargas II (T4-5)	2008	29.0	LNG Export
RasGas III (T6-7)	2008	29.0	LNG Export
Al Khaleej II	2009	15.5	Domestic Market
Qatargas III & IV	2009	29.0	LNG Export
Pearl GTL	2009	17.6	GTL
Barzan	2012	17.6	Domestic Market
Total		231.6	

* Capacity is based on wet gas production. Send-out to the domestic market and LNG and GTL production is lower after taking into account condensate and liquids separated from the gas and the use of gas in the liquefaction and GTL plants

Source: Author's estimates

buyers looking to secure long-term LNG supplies are, when will Qatar make a decision on future LNG developments and when it does, to what extent will it agree to further expansion of its LNG capacity? These are questions to which there were no clear answers in 2007 and it may be some years before the answers are known. It is entirely possible that the government will decide that LNG production of 77.5 Mtpa will meet the longer-term needs of the country and hence no further increase in liquefaction capacity is needed. The other extreme would be a commitment to a rapid expansion of LNG production on a similar scale to the period 2005–2011. The latter strategy would make a major contribution to the increase in global LNG output after 2012 and would ease the concerns of buyers about the longer-term availability of LNG. In the author's view the probability of this outcome is low.

If the outcome of the analysis of the North Field performance is a commitment to further expansion of production, the probability is that it will be at a more measured rate than in the past. Part of the increase from the 77.5 Mtpa capacity of the operating trains and those under construction may come from the six 7.8 Mtpa trains producing at a rate over and above their design capacity. Most LNG trains in operation have produced at least 10 percent over their design capacity, which, if achieved with the mega-trains would add 4.6 Mtpa to capacity. The new trains employ a liquefaction process which is based on technology that has been used in many of the world's operating LNG plants, but with enhancements to increase the output. In addition, compression for the liquefaction cycle is provided by the largest gas turbines ever used in an LNG plant. It is likely that there has been conservatism in their design, which could lead to the trains performing in excess of 10 percent over design capacity.

A possible scenario for the period after 2012 is that Qatari LNG production will exceed the 77.5 Mtpa in operation and under construction in 2007, through over-production by these trains and the building of some new capacity. In this scenario, Qatar's eventual LNG capacity could reach around 100 Mtpa. However, given the uncertainty, it would be a mistake for Asian LNG buyers to rely on the availability of significant additional volumes of Qatari LNG after 2012.

Oman

Oman became the Middle East's third LNG producer when the first of the two trains at the Oman LNG plant came into operation in April 2000. The facilities are at Qalhat, near Sur, on the Indian Ocean coast, south-west of the capital, Muscat. Gas supply is from onshore fields discovered in the late 1980s. Oman LNG was established to develop the project with shareholdings as follows: Oman government 51 percent, Shell 30 percent, Total 5.54 percent, Mitsubishi 2.77 percent, Mitsui 2.77 percent, Partex 2 percent, Itochu 0.92 percent and Korea-Oman LNG (KOLNG) 5 percent. KOLNG is a consortium of Korean companies, led by Korea Gas, whose involvement in the project was part of the agreement under which Korea Gas committed to purchase 4.06 Mt of LNG from Oman LNG for twenty years, a deal which underpinned the development of the project.

The design capacity of the two-train plant is 6.6 Mtpa but, as is the case for many LNG plants, it is able to perform above that level, with maximum output estimated to be around 7.3 Mtpa. In addition to its sales to Korea Gas, Oman LNG secured a contract with Osaka Gas in Japan for 0.67 Mtpa and with Enron's Dabhol project in India for 1.65 Mtpa. However, the collapse of the Dabhol project, shortly after Oman LNG started production, left it with large volumes of un-contracted production, which it has had success in selling under short- and medium-term contracts to buyers in North America, Europe and Asia.

Oman LNG's two trains were constructed at a time when activity in the LNG contracting industry was at a low point and it secured some very competitive bids for construction from contractors. As a result, even after the loss of one of its buyers, the investment was still economic and the returns were enhanced by the project finding alternative markets for much of the un-contracted output. The success of the first stage of the development of Omani LNG production provided a platform for the addition of a third train.

The Government of Oman wanted to increase its share of the new train and decided to establish a separate company, Qalhat LNG, to own the facilities. The shareholdings in Qalhat LNG are as follows: the Government of Oman 46.84 percent, Oman LNG 36.8 percent, Union Fenosa Gas 7.36 percent, Mitsubishi 3 percent, Itochu 3 percent and Osaka Gas 3 percent. The government's direct and indirect shareholding (through Oman LNG) gives it a 65.61 percent ownership of Qalhat LNG. The shareholdings of the Spanish company, Union Fenosa Gas, and of the Japanese companies, Mitsubishi, Itochu and Osaka Gas, were part of the deals under which they committed to purchase LNG from Oman.

Train 3 has a design capacity of 3.7 Mtpa and loaded its first cargo of LNG in December 2005. It increased Oman's total LNG capacity to around 11 Mtpa.

Table 8.10 shows the volumes of LNG contracted on a long-term basis from Oman's three liquefaction trains. As with Qatar, the marketing strategy has changed over time. The long-term contracts for the first stage of development of Oman LNG are with buyers in Asia, whereas nearly 50 percent of the volumes contracted on a long-term basis from train 3 are with Union Fenosa Gas, primarily for the Spanish market.

Table 8.10: Long-term Sales Contracts for Omani LNG Production

Seller	*Buyer*	*Volume in Mtpa*	*Market*
Oman LNG (T1-2)	Korea Gas	4.06	Korea
	Osaka Gas	0.67	Japan
	Itochu	0.70	Japan
Qalhat LNG (T3)	Union Fenosa Gas	1.65	Spain
	Mitsubishi	0.80	Japan/USA
	Osaka Gas	0.80	Japan
	Total	8.68	

Source: Author's estimates

Mitsubishi's contract is based on up to 50 percent of the volumes being delivered to the USA and the remainder to Japan.

In mid-2007, Oman had only committed 79 percent of its LNG capacity on a long-term basis, positioning itself to market the remaining volumes on a short- or medium- term basis into the highest priced markets. In the first seven years of operation, Oman delivered LNG to Taiwan, India, China, Portugal, Belgium, France, the USA and Puerto Rico. The majority of its LNG has been delivered to its long-term buyers in Japan, Korea and Spain, including a significant number of additional cargoes on a short-term basis.

The three liquefaction trains have performed well, exceeding design capacity. However, Oman LNG and Qalhat LNG have been unable to take full advantage of the extra capacity since the government, which owns the gas reserves, has given priority to domestic gas use rather than additional LNG production. The government's policy is to supply the gas needed to meet Oman and Qalhat LNG's long-term contractual commitments, but it will only release extra gas if domestic demand has first been met. LNG output in 2006 was 8.6 Mt, nearly 2 Mt below the maximum volume it could have produced during the build-up of output from Qalhat LNG. In the first half of 2007, production picked up and ran at an annual rate of 9.6 Mt, still below full design capacity.

The main issue for the future of Omani LNG production and its ability to meet Asian demand is the availability of gas supply for the LNG trains. The start of deliveries to Oman by the Dolphin project in 2008 may release some gas for LNG production since, in addition to the delivery of 2.1 Bcm/a to Oman, Dolphin supply from Qatar will replace the gas currently being exported from Oman to Fujairah and Ras al Kaimah.

It is also possible that additional gas reserves will become available in Oman as a result of exploration and production activity directed principally at adding to the proven gas reserves. In April 2006, BG entered into an Exploration and Production Sharing Contract for an onshore block, which contains the Abu Butabul gas and condensate discovery and has the potential for further gas discoveries. In January 2007, BP signed a production sharing contract giving it access to two onshore fields, Khazzan and Makarem, which have not been developed because of the complex geology. However, it is estimated they could yield 0.57 to 0.85 Tcm of natural gas.[10]

Oman is also looking to its neighbour to the north, Iran, for possible gas supply. In May 2007, Oman and Iran signed a Memorandum of Understanding to co-operate in the development of oil and gas resources. The agreement covers the supply of 10.3 Bcm/a of gas from Iran to Oman for liquefaction and the joint development of Henjam/West Bukha gas condensate field, which is in the Straits of Hormuz and straddles the maritime border between the countries. Reserves are estimated at 0.05 Tcm of gas and 400 million barrels of condensate. Despite the progress in co-operation between the two countries, there is a long way to go before any gas flows from Iran to Oman. It would be the first time natural gas supply from one country has been liquefied and exported from a second country.

There are long-term prospects for a significant expansion in the availability of gas supply in Oman, which could support the production of the full 11 Mtpa from the three liquefaction trains presently in operation. If the current exploration activities are very successful, they could lead to a further expansion of the LNG facilities at Qalhat. However, there is considerable uncertainty over when, and if, new gas supplies will be found and developed and, hence, only a low probability can be assigned to new long-term LNG supplies from Oman

Yemen

Yemen will become the Middle East's fourth LNG producer when exports start from the Yemen LNG plant, under construction at Balhaf, on the Indian Ocean coast, in early 2009. The plant is a two-train facility with a design capacity of 6.7 Mtpa. The shareholders are Total, the operator, 39.62 percent, Hunt Oil 17.22 percent, Yemen Gas 16.73 percent, SK (Korea) 9.55 percent, Korea Gas 6.00 percent, Hyundai (Korea) 5.88 percent and GASSP (Yemen's General Authority for Social Security and Pensions) 5.00 percent.

The gas supply will come from the Marib Jawf region of central Yemen. Discoveries of oil and gas were first made in the mid-1980s and the gas, which is mainly associated with oil, was flared when oil production commenced in 1986. In the early 1990s, gas processing facilities were installed and gas re-injected into the reservoir. By the mid-2000s, around 28 Bcm/a of associated gas was being dealt with in this way. A spur line will take natural gas from the fields to the capital Sana'a for local consumption when supply to the LNG plant commences.

The government has granted Yemen LNG exclusive rights to the gas reserves in the Marib area fields. Proven reserves are estimated at 0.26 Tcm with an additional 0.02 Tcm of probable reserves. 0.03 Tcm is allocated to domestic consumption. The remaining reserves can support the two-train plant for twenty years of production but further discoveries of gas will be needed before additional liquefaction trains can be considered.

Hunt Oil and ExxonMobil were the partners in Marib Block 18, with the former as operator, until 2005 when the Yemeni parliament rejected the government's proposal to extend the agreement for five years. The Yemeni company, Safer Exploration and Production Operations Company (SEPOC), took over ownership and operatorship of the block, a decision which, in 2007, was the subject of arbitration in Paris. Uncertainty over the upstream arrangements delayed the financing of the project. Commercial banks and export credit agencies required assurance that technical and operational support will be available to ensure the delivery of gas to the LNG plant before they were prepared to commit funds. The plant requires 11.4 Bcm/a of gas supply, which is well within the installed production and processing capacity in the Marib concession.

Yemen LNG is the first green-field LNG project in the Middle East which has been launched with the majority of its output committed to Europe and the USA rather than Asia. The project spent many years in the late 1990s and early 2000s trying to find buyers for the output. It finally secured commitments to the entire output from three buyers in 2005, allowing a final investment decision to be taken. The buyers and their committed volumes are shown in Table 8.11. The priority market for Total and Suez LNG's volumes is the USA, and both companies are understood to have diversion rights to Europe. Korea Gas's off-take from the project is evenly spread through the year but it has the right to divert 35 percent of the LNG in the summer months (0.35 Mt) to alternative markets, subject to the approval of Yemen LNG. These diversion rights will help Korea Gas manage the seasonality of demand in Korea, which sees imports in the winter months some two and a half to three times their level in the summer.

Although the contracted volumes amount to 6.5 Mtpa, Total is

Table 8.11: Contracted Volumes from Yemen LNG

Buyer	*Market*	*Volume In Mtpa*	*Basis*	*Duration*
Suez LNG	USA/Europe	2.5	FOB	2009–2028
Total	USA/Europe	2.0	DES	2009–2028
Korea Gas	Korea	2.0	FOB	2009–2028
	Total	6.5		

Source: Author's estimates

purchasing the LNG on a DES basis,[11] which means that Yemen LNG has to load an additional 0.15 Mtpa to cover boil-off during the voyage to market.[12] Yemen LNG has, therefore, sold its full design capacity and any surplus LNG that might become available will be the result of the plant performing above design.

It is possible that Yemeni LNG cargoes will be redirected from the US or European markets to Asia, if buyers in Asian markets are prepared to offer prices and terms that make such diversions worthwhile. Yemen LNG has chartered sufficient shipping capacity to deliver all the volumes purchased by Total to the US market and Suez LNG has arranged its shipping capacity on the same basis. Diversions to markets in Asia, where round voyage times are less than for the USA, can therefore be accommodated without the need for additional shipping capacity. Indeed, diversions could release capacity to transport surplus production, if the plant performs above design capacity.

Iran

The Prospects for LNG Exports

The extent to which Iran will produce LNG is one of the major uncertainties in the outlook for Middle Eastern LNG exports after 2012. Iran is the only country in the region with active plans for new liquefaction trains. However, geopolitics, Iran's strategic intentions for LNG exports, escalating costs and lack of clear leadership and decision-making join together to make the range of possible outcomes very wide.

Iran's ambitions to be an LNG exporter date back to the pre-revolutionary times of the 1970s, when a number of feasibility studies were carried out. Plans for LNG were abandoned in the aftermath of the revolution and it was not until the early 2000s that they came back on the agenda as Iran considered how to exploit its enormous gas reserves. Iran's potential as an LNG producer has attracted the attention of investors and buyers looking for new sources of supply.

An added incentive for the Iranian government is Qatar's rise to become the world's largest LNG producer, less than ten years after commissioning its first LNG plant. Qatar's LNG facilities are supplied from the North Field, around one-third of which extends into Iranian waters, where it is known as South Pars. Iran wants to ensure that it produces its fair share of the reserves in the joint field. It is concerned that Qatari production could reduce reservoir pressures on the Qatari side of the field, causing gas and condensate to migrate from the

Iranian side. The risks would be reduced if the field was unitised in the way that fields straddling international borders are in other parts of the world, but there have been no moves for this to happen. Iran's plans for the South Pars field include the development of large-scale LNG production but domestic consumption and the injection of the gas into declining oil fields to enhance production also feature strongly. This differentiates it from Qatar where the much smaller population results in lower domestic demand and where there is not the same need for gas re-injection.

The South Pars field contains just below 50 percent of Iran's gas reserves with the other reserves spread across a number of fields, some of which contain very large accumulations of gas, sufficient to support a world-scale LNG plant on their own. Iran's dependence on the South Pars field is therefore less than Qatar's dependence on the North Field. However, South Pars is an enormous resource base and Iran wants to ensure that it takes full benefit from its exploitation. Qatar started earlier than Iran in producing gas from the field and by the end of 2006 its cumulative production was around four times that of Iran.

The prospects for LNG exports from Iran have attracted interest both from potential project participants and from buyers. In mid-2007, at least six LNG projects were at various stages of the planning process but delays have been the order of the day.

The three most advanced projects in mid-2007 were based on the reserves in the South Pars field:

Pars LNG is planned as a two-train 10 Mtpa plant by the National Iranian Oil Company (NIOC), Total and Petronas. The shares are split on a 50/50 basis between NIOC and the foreign partners. Originally the plan was for Total to have 30 percent and Petronas 20 percent but in early 2005 Petronas announced that it was pulling out of the project, citing poor economics as a reason. By the end of 2005, Petronas indicated that it had reconsidered its decision and would remain in the project.

The date for a final investment decision for Pars LNG has slipped several times since 2004, when there was talk of production starting in 2009. Three years later, the final investment decision (FID) had still not been taken and even a 2012 start-up was looking very doubtful. Total's Chief Executive, Christophe De Margerie was reported as stating in mid-2007 that he could not say when a decision would be made and that 'the cost is so high that it is close to damaging the project'.[13] He added that the cost has more than doubled for the planned 10 Mtpa plant and cited staff shortages, increasing costs of raw materials and a shortage of contractors that can work in Iran. He denied that Total was abandoning the project but did say that geopolitical issues would have to be taken into account before FID can be made. Total and Petronas have agreed to purchase output from train 1 and China, India and Thailand are potential markets for train 2.

Persian LNG is being promoted as a two-train 16 Mtpa project by NIOC (50 percent), Shell (25 percent) and Repsol (25 percent). As for Pars LNG, the FID has been slipping and with it the potential start-up of production. In July 2007, Shell's Chief Executive, Jeroen van der Veer indicated during the presentation of the company's half year results that the FID for Persian LNG was still twelve months away.[14] He pointed to the cost and economics of the project being an issue that has to be addressed and added that when the decision is made political considerations will be taken into account. If a final investment decision is made in the second half of 2008, then start-up would not be before 2013 and there is a high probability of further slippage in the timetable.

The approach to marketing is similar to Pars LNG with the foreign shareholders, Shell and Repsol, committing to purchase the output from the first train and output from the second train being actively marketed in Asia. The project has established an office in Tokyo for the purpose.

Iran LNG is a 100 percent NIOC project, which in 2004–05 appeared to be the front runner of the South Pars based projects, since it commissioned work on Front End Engineering Design (FEED) before the other two projects. A number of companies including BG, ENI and Statoil have, at various times, been linked with the purchase of LNG from the project. In April 2007, it was reported that OMV of Austria had signed a Heads of Agreement with NIOC covering participation in the development of Phase 12 of the South Pars field, the liquefaction plant and the off-take of LNG. Iranian sources said that OMV would have a 10 percent share in the liquefaction plant and purchase 2.2 Mtpa but OMV said the details of the agreement had yet to be finalised.[15]

In addition to the South Pars based projects, at least three other Iranian LNG projects have been announced:

Qeshm LNG is a project promoted by LNG Limited, a small Australian company, which says it can shorten construction times and reduce costs to the extent that small-scale liquefaction trains are economically feasible, even in an environment where the costs of conventionally designed liquefaction trains have escalated rapidly. After carrying out preliminary feasibility studies at a number of sites in the Middle East, Indonesia, Australia and Africa, it is promoting the development of a three-train, 3.45 Mtpa liquefaction plant on Qeshm Island in the Straits of Hormuz. LNG Limited has an agreement with NIOC for the supply of gas for the plant. It says it expects to achieve a 24-month construction schedule, which is extremely optimistic even in a favourable location without adding the complications that are inevitable in Iran.

In late 2006, China National Offshore Oil Corporation (CNOOC) signed an agreement with NIOC for the development of the ***North Pars*** field. The $16 bn deal was said to comprise $5 bn for the development of upstream production and $11 bn for downstream infrastructure including a liquefaction plant. The field will be developed in four phases, the first of which will be for domestic consumption and the other three for export. The announcement of the deal did not give details of the planned size of the liquefaction plant. CNOOC will develop the field in return for 50 percent of the LNG output.[16] The North Pars field is 85 km north of the South Pars field and is estimated to have reserves of around 1.4 Tcm.

In early January 2007, NIOC signed a preliminary agreement with the private Malaysian company, SKS Ventures, for the $16 bn development of the onshore ***Golshan and Ferdos*** gas fields to supply a 20 Mtpa LNG plant. In the announcement of the proposed project, the reserves in the Golshan field were put at 1.4 Tcm and in the Ferdos field at 0.28 Tcm.[17] If these estimates are confirmed, they would be sufficient to support the proposed LNG plant. Both this and the North Pars project are no more than speculative, despite the signature of agreements for their development.

If all the projects described above were developed, Iran would approach Qatar's 77.5 Mtpa of LNG production capacity. This outcome has a very low probability. It is entirely possible that Iran will produce no LNG, as other calls on its gas reserves take precedence over LNG. However, it is likely that it will eventually produce some LNG with at least one of the projects based on gas in the South Pars field being developed; but the time-frame may be nearer 2015 than the 2012 target talked about in 2007.

Despite sanctions making it impossible to sell LNG to the USA, potentially the fastest growing LNG market, Iran has had no difficulty in finding buyers interested in committing to purchase the output from its planned projects on a long-term basis. The foreign partners in Pars LNG and Persian LNG have said they will commit to purchase the output from the first of the two liquefaction trains planned for each plant. Furthermore, a string of Heads of Agreement, Memoranda of Understanding and even Sales and Purchase Agreements have been signed with potential buyers in Asia. China and India have been at the forefront of such agreements along with PTT in Thailand. In Europe, OMV of Austria has also signed a Heads of Agreement committing to purchase 2.2 Mtpa. The Swiss energy group EGL signed a contract for 5.5 Bcm/year to be delivered via the Trans-Adriatic Pipeline starting around 2010–11.[18] Iran's problem has been in moving these projects towards FID in an environment where there is strong opposition to gas exports from some powerful groups in the country and where the geopolitics of its nuclear programme is a source of concern for potential buyers and investors.

US sanctions will restrict Iran's access to liquefaction technology and equipment and to experienced LNG contractors. However, they are unlikely to stop the construction from going ahead since the required technology, equipment and expertise are available from outside the USA. The main effect of the sanctions will probably be to increase costs and extend construction times. A more important factor may be access to finance in an environment where loans to Iran had largely dried up by mid-2007 because of the deteriorating economic situation and uncertainty over the country's energy policy. If the United Nations should decide to impose sanctions on Iran, due to violations relating to the country's nuclear programme, this would be a substantially more serious obstacle to both financing and general trade prospects.

The Development of the South Pars Field

Iran's approach to the development of the South Pars field has been to divide it into twenty-four phases. Each phase is allocated a block in the field from which to produce gas and associated liquids. The production from each phase is typically 10.3 Bcm/a, although some phases have been allocated larger volumes. Since Iranian law does not allow foreign companies to own hydrocarbon reserves, the operations are under a 'buy-back' arrangement with the foreign companies acting as contractors to NIOC in developing natural gas and liquids production. They recover their capital investment and costs and earn an agreed rate of return on their investment through revenues generated by the production. Once the costs have been recovered and the agreed rate of return achieved, operatorship is transferred to NIOC. The typical period for foreign companies to operate the assets, recover their costs and to earn the agreed rate of return, is eight to ten years from first production. This creates problems for LNG buyers and financiers, who are generally looking for a stable project structure throughout the twenty-year or longer life of the typical Sales and Purchase Agreement.

Table 8.12 shows the development phases of the South Pars field. Phases 1 to 5 were in operation in mid-2007 with phases 6 to 10 scheduled to start production by the end of 2008. This will take total output to 110 Bcm/a, of which 72 Bcm/a will be for the domestic market and 38 Bcm/a for re-injection to enhance oil production. The gas for re-injection will be delivered by a 512 km, 56 inch gas pipeline to the Agha-Jari oil field.

Table 8.12: Development Phases of Iran's South Pars Field

Phase	*Start-up*	*Gas Use*	*Volume in Bcm/a*
Phase 1	2003	Domestic Market	10.3
Phases 2 & 3	2002	Domestic Market	20.7
Phases 4 & 5	2004	Domestic Market	20.7
Phases 6, 7 and 8	2007-08	Gas Reinjection	37.2
Phases 9 & 10	2008	Domestic Market	20.7
Phase 11	2012+	Pars LNG	19.6
Phase 12	2012+	Domestic Market & Iran LNG	31.0
Phases 13 & 14	2012+	Persian LNG	31.0
Phases 15 & 16	2010/11	Domestic Market	20.7
Phases 17 & 18	2012+	Domestic Market	20.7
Phases 19, 20 & 21	2013+	To be decided	31.0
Phases 22, 23 and 24	?	To be decided	31.0
		Total	294.7

Source: 'Iran & Qatar: The Beginning of a Race to Produce Shared Gas Reserves', FACTS Global Energy Gas Advisory no. 39, July 2007

Phases 11 to 14, amounting to a total of 81.6 Bcm/a, have been allocated for LNG production, although phase 12 will also supply the domestic market. Phases 15 to 18 will produce an additional 41.4 Bcm/a for the domestic market. The final six phases, amounting to 62 Bcm/a of production have yet to be allocated, although under an agreement with Turkey, phases 22 to 24 could be used to support exports to Turkey and central Europe. Assuming that these phases are developed, and the LNG projects go ahead, production from the South Pars field will be 294.7 Bcm/a. This is some 62 Bcm/a more than Qatar's planned production from its much larger share of the total reserves in the North Field and South Pars. However, the reserves-to-production ratio for South Pars would still be around 45 years.

Domestic Gas Demand

The importance of domestic gas demand in the use of Iran's gas reserves can be seen from the allocation to domestic supply of over 50 percent of output from the first eighteen phases of the South Pars field to be awarded. Iranian gas demand reached 105.1 Bcm/a in 2006, after increasing at an average annual rate of 10.5 percent p.a. over the previous decade. Iran has the highest domestic gas consumption in the region, partly due to its population of around 70 million, which is double the total population of the other Gulf States. Official Iranian statistics show that two-thirds of its population live in the central and northern parts of the country, which experience low temperatures in the winter creating a strong requirement for gas for heating. Low prices are another factor behind the rapid demand growth. In 2005, residential customers paid an average of $0.31/MMBtu for natural gas, about mid-way between the $0.11/MMBtu paid by power plants and the $0.78/MMBtu paid by commercial customers. The low price of gas encourages demand and reduces the amount of gas available for export. However, it would be politically inexpedient for the government to raise prices to international levels.

The expansion of the industrial sector, and, in particular, petrochemicals where some 27 million tonnes of capacity is expected to be brought on stream between 2007 and 2010, is creating a huge demand for gas. A smaller, but not insignificant, impact on gas consumption will come from compressed natural gas (CNG), which the government is actively promoting to reduce the consumption of motor gasoline. Domestic demand is forecast to grow at a rate of around 7.5 percent per annum, taking it to 290 Bcm/a by 2020 or nearly three times the level in 2006. At this level, it will be close to the total planned production from the South Pars field under the first 24 phases of development.

Gas Re-Injection in the Oil Fields

Iran produced 4.3 million barrels per day (mb/d) of oil in 2006 making it the world's fourth largest oil producer after Saudi Arabia, Russia and the USA. Maintaining oil production, let alone increasing it to 5.4 mb/d by 2009, as called for in the fourth National Plan, is a major challenge since many of the oil fields are old and in decline. One way of reversing the decline is to inject natural gas into the reservoirs, which requires large volumes of gas.

Around 31 Bcm/a of gas was injected into oil reservoirs in 2006 and NIOC's plans call for that figure to increase to 100 to 115 Bcm/a by 2010. Some estimates suggest the requirement could be even larger just to maintain production at around the current level. Opponents of large-scale gas exports point to the fact that increasing oil production when oil prices are $60/Bbl is a potentially more valuable use of gas than exporting it as LNG (or by pipeline) where capital investment costs are high and margins generally lower.[19]

Pipeline Exports

Despite having the world's second largest reserves, Iran has been a net importer of natural gas since the late 1990s. Pipeline gas is imported from Turkmenistan to meet demand in the north of the country around Tehran. It is a lower cost option than piping the gas from Iran's own reserves in the south of the country. According to the *BP Statistical Review of World Energy*, Iranian imports in 2006 amounted to 5.8 Bcm.

Iran has also been exporting gas by pipeline to Turkey since 2001. The build-up to the plateau level of 10 Bcm/a has been slow, partly because Turkish forecasts of gas demand growth proved to be optimistic and it over-contracted for pipeline gas from Russia and Iran and for LNG from Nigeria and Algeria. Disputes between Iran and Turkey over quality and price also affected the build-up. In 2006, Iran exported 5.7 Bcm to Turkey; but in January 2007 it stopped gas supplies to Turkey to meet domestic supply, which had increased sharply because of severe weather. Although the stoppage was for only a few days, and Russia was able to increase its deliveries to Turkey to fill the gap left by Iran, it calls into question Iran's reliability as a long-term gas exporter.

A spur from the main Turkish pipeline takes a small amount of gas (around 0.3 Bcm/a) into the Azerbaijan enclave of Nakhichevan, which is matched by the export of the same volume of gas directly from Azerbaijan to Iran. A second spur line takes gas to Armenia under a gas for electricity swap, which came into effect at the beginning of 2007. Under the arrangement Iran supplies from 1.1 to 2.3 Bcm/a to Armenia and receives electricity in payment.

Iran has discussed a number of projects to export gas by pipeline across the Gulf to Oman, Kuwait and the UAE (Dubai, Sharjah and Abu Dhabi). Pipelines to India and Pakistan and to Europe have also been proposed but none, other than the link to Turkey, has as yet progressed to a stage where construction can start.

The most advanced of the projects in mid-2007 was the Iran–Pakistan–Indian (IPI) pipeline, also known as the Peace Pipeline, since it is seen as contributing to improved relations between India and Pakistan. The first proposals for a pipeline linking Iran and India were made in 1989, soon after the discovery of the South Pars field, confirming that Qatar's North Field extended into Iranian waters. In 1993, Iran and India entered into a Memorandum of Understanding to study the feasibility of a pipeline. The studies showed that an onshore route through Pakistan was the lowest cost option. However, Pakistan opposed a pipeline across its territory or through its territorial waters. A deepwater pipeline, avoiding Pakistan, was not economically feasible and the project was put on hold for several years. In the early 2000s, the new Government of Pakistan, under General Musharraf, decided that a pipeline from Iran to India offered the country access to much needed gas supplies and revenues from transit rights.

Initially, negotiations took place on a bilateral basis – Iran and India and Iran and Pakistan – but by 2006 relations between India and Pakistan had improved to the stage where trilateral negotiations could take place. The plan is for a 2300 km pipeline from Assulayeh in Iran (the landfall for pipelines from the South Pars field) to the Indian border; around half the line will be across Iranian territory and half across Pakistan. It will have a capacity of 22 Bcm/a, with 50 percent of the volumes delivered to Pakistan and 50 percent to India. The cost estimate is $7.5 bn and construction will take about four years, meaning that 2011 is the earliest feasible start date, a target which is optimistic.

Price has been a major issue for the parties. Iran has proposed a price that would give it a similar netback as it would earn from LNG sales to Japan or Korea while India has been looking for a cost plus approach, which it estimated would result in a price of between $4.50 to $5/MMBtu delivered to the Indian border. In July 2007, the parties announced that agreement had been reached on a price at the Iranian border linked to the Japanese custom cleared oil price (JCC). The resulting price is reported to be $4.93/MMBtu when the price of oil is $60/Bbl. However, the question of transmission costs and transit fees through Pakistan remained to be resolved.[20]

The progress that is being made in the trilateral negotiations suggests that the probability of the project being developed is increasing but there are still a number of major hurdles in the path of a final commitment to construction. In addition to the issue of transit fees through Pakistan, there is the question of whether Iran will be willing and able to commit the reserves and the 21.7 Bcm/a of gas production required by the project in the face of other demands on its gas resources.

Competition for the gas resources could come from the planned Nabucco pipeline taking Iranian gas from the Iran–Turkey border into Austria where it would link into the wider European gas grid. In July 2007, Iran and Turkey signed a Memorandum of Understanding for the supply of gas from Iran and from Turkmenistan via Iran to Europe.[21] The memorandum also provides for the Turkish upstream company, TPAO, to participate in the development of phases 22–24 of the South Pars field. The understanding is that the 20.6 Bcm/a will be supplied to Turkey and Europe through the Nabucco pipeline, although it is said that the upstream deal is not linked directly with the latter project. The Memorandum of Understanding was signed three days before the announcement that agreement had been reached with India and Pakistan on the price of gas for the IPI pipeline and may, at least in part have been used by Iran to exert pressure on the latter negotiations.

Will Iran Export Gas?

As the sections above have shown, Iran is not short of plans for the export of its massive gas resources either as LNG or by pipeline. Markets in Europe and Asia are being targeted for LNG exports while Europe, Iran's neighbours in the Gulf region and South Asia are being targeted for pipeline exports. Numerous Memoranda of Understanding have been signed with buyers and prospective partners but turning these preliminary agreements into real projects has proved to be a major challenge for Iran. The only export projects in operation are the pipelines to Turkey and Armenia, and the Turkish project has been beset by problems from both sides. Throughput in 2006, five years after start-up, was well below the full contract volumes of 10 Bcm/a. Iran's decision in early 2007 to renege on its contractual commitments by interrupting supply to Turkey, when it was short of gas to meet domestic needs raises questions over how it would perform if some of the larger-scale projects at the planning stage were to be developed and exports were at a much higher level.

Geopolitics can not be ignored in assessing the future of Iranian gas exports. US sanctions will restrict the country's access to technology and expertise and, while this will not necessarily stop the development of export projects, it will almost inevitably increase costs and slow progress. Moreover, raising finance for the development of major export projects will be a challenge for Iran. US sanctions will have an impact on financing but a more serious issue for financiers may be concern over the performance of the Iranian economy and the lack of a clear energy policy.

Asian countries seeking additional gas supplies can not ignore Iran's 28.1 Tcm of proven gas reserves However, Iran has signed agreements with potential importers in Europe and Asia for more gas than it can possibly make available for export after meeting domestic demand and gas re-injection to enhance oil production. Asian buyers will face competition from Europe in securing access to any reserves that are allocated for export. The role that Iran can play in meeting future Asian gas demand is, therefore, very uncertain and most Asian buyers will want to avoid becoming over-reliant on Iranian LNG or pipeline gas supplies.

Saudi Arabia

As Table 8.1 shows, Saudi Arabia has proven gas reserves of almost 7.1 Tcm, the fourth largest in the world. Production in 2006 was a relatively modest 73.7 Bcm/a and at that level the reserves will last for nearly 100 years. Despite the obvious potential for gas exports, the government has opted to retain the gas for domestic use. From time to time there have been some proposals from within Saudi Arabia, and from external organisations, that LNG exports should be considered but they have lacked official support and have not progressed very far.

The potential for further gas discoveries in Saudi Arabia is very large. Jensen Associates has used US Geological Survey data to estimate that the undiscovered reserves could be of the order of 18 Tcm.[22] This would take total reserves to around 25 Tcm and, at that level, they would be on a par with the proven reserves in Qatar and Iran. The Rub al Khali Basin, the so-called 'Empty Quarter', could hold a significant portion of the undiscovered reserves. Exploration in that region has only just commenced but if successful, Saudi Arabia's policy of not exporting gas could be reconsidered. However, the exploration licences for the Rub al Khali Basin were awarded in 2003 and 2004 and it will be many years before a clear indication of the reserves position is known. Consequently, the prospects for Saudi Arabia exporting LNG to Asia can be discounted until well into the next decade.

The Role of Middle Eastern LNG in Meeting Asian Gas Demand

The Middle East has become an important source of LNG supply to the markets in Asia. Over the ten-year period 1996 to 2006, its share of Asian LNG imports increased from 7.4 percent to 32.6 percent. In the process, the Middle East accounted for close to two-thirds of Asia's incremental LNG supplies. The Middle East's four operating LNG plants were all originally commissioned to supply Asian markets – ADGAS and Qatargas for Japan and RasGas and Oman LNG for Korea. As Table 8.13 shows, in 2006 Middle Eastern LNG accounted for 24 percent of Japanese LNG imports, 47.5 percent of Korean LNG imports and 88.3 percent of Indian LNG imports. Its role in supplying Taiwan with LNG is smaller, only a 7.5 percent share in 2006, but it is set to rise when Qatar's 3 Mtpa contract with the Chinese Petroleum Corporation (CPC) comes into effect in 2008. China is a newcomer to LNG imports, having commissioned its first receiving terminal in May 2006. It imported its first Middle Eastern LNG cargo, a spot purchase from Oman, in April 2007.

Table 8.13: The Share of Middle Eastern LNG in Asian Markets

Market	*Volume Imported from Middle East in Mtpa*		*Percent Share of LNG from Middle East*	
	1996	*2006*	*1996*	*2006*
Japan	4.3	15.2	9.4	24.4
Korea	-	12.2	-	47.5
Taiwan	-	0.6	-	7.5
India	-	5.4	-	88.3
China	-	-	-	-
Total	4.3	33.4	7.4	32.6

Source: Author's estimates

Middle Eastern LNG production is set to increase dramatically between 2008 and 2012 as Qatar's six mega-trains add 46.8 Mtpa to operating capacity and Yemen LNG brings its 6.7 Mtpa capacity plant into operation. Figure 8.6 shows the author's projections of Middle Eastern exports over the period 2007 to 2020 from the projects in operation or under construction in 2007. Output is expected to increase from 38.5 Mt in 2006 to just over 100 Mt in 2013, by which time all the trains under construction in 2007 should be operating at, or above, design capacity.

The probability of production over the period 2007 to 2102 being close to that shown in Figure 8.6 is high, since the forecasts are based on LNG facilities in operation and under construction in 2007. The main downside risks to this outcome are the failure of newly commissioned trains to perform as expected, a major technical problem with

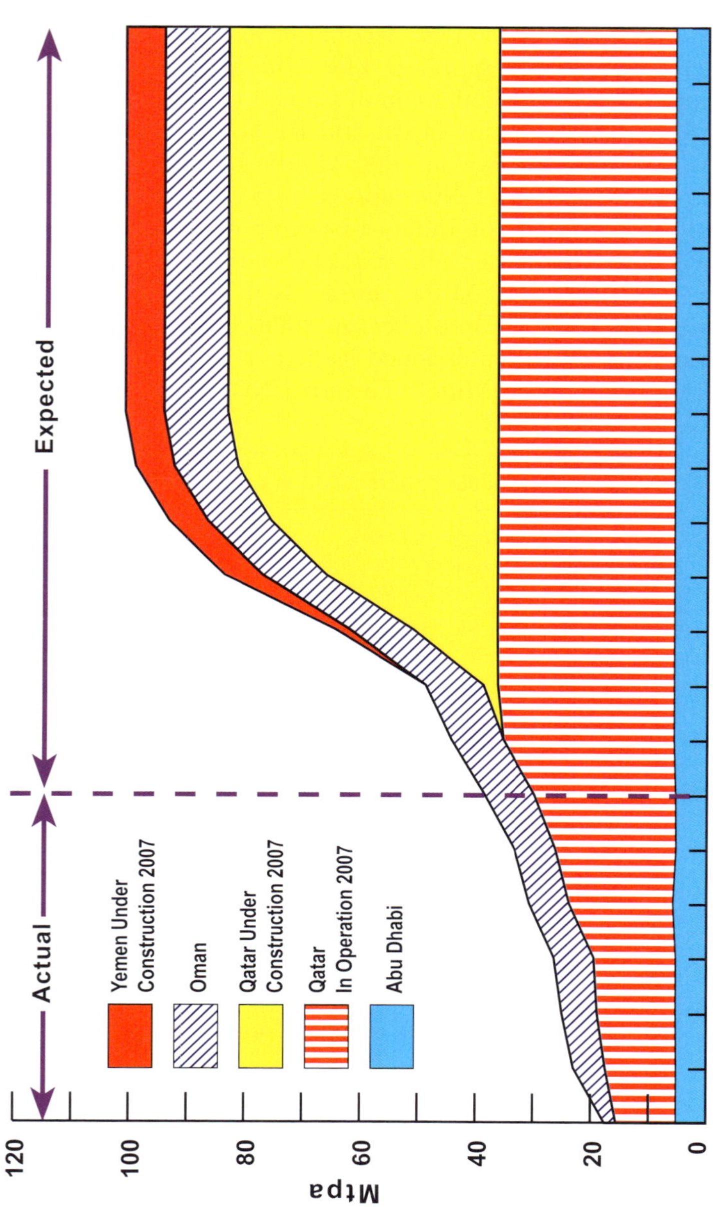

Figure 8.6: Middle Eastern LNG Exports 2000–2020 from Projects in Operation and Under Construction in 2007

Source: author's estimates

an operating plant or an interruption to production for political or gas supply reasons. The construction time needed for new liquefaction capacity means that even an early final investment decision on, for example, planned capacity in Iran, could not have a significant impact on output before 2013. The main upside to the above outcome before 2013 comes from newly commissioned trains operating at above design capacity. The probability of this happening is high based on the performance of trains in operation in the Middle East and elsewhere. A 10 percent improvement in performance, which would be in line with the performance of trains commissioned between 1997 and 2007, could add around 7.3 Mtpa of additional production by 2013.

The upside for Middle Eastern LNG production between 2013 and 2016 arises mainly from the possible commissioning of new trains in Iran. After 2016, there could be additional production from Iran and Qatar but estimates of the volume and the timing can only be speculative. Figure 8.7 shows a possible profile for Middle Eastern LNG assuming that the Pars and Persian LNG projects in Iran are commissioned in 2013 and 2014 respectively and that a further 25 Mtpa of capacity in Iran and/or Qatar is commissioned between 2016 and 2020.

As Figure 8.7 shows these assumptions would result in Middle Eastern LNG exports growing steadily from 2013 to 200, when it could reach close to 160 Mtpa. It must be emphasised that the outcome after 2013 is uncertain and it is much too soon for Asian buyers to be able to plan on the basis that this additional LNG will be available. Any expansion of capacity after 2013 will be a bonus for buyers looking to secure supply to meet growing natural gas demand in their markets.

Asian buyers largely missed out on the second phase of expansion of Middle Eastern gas production, which is scheduled to come on stream between 2008 and 2011. They will not want to miss out on a third phase, if it happens. Table 8.14 shows how Asia's share of Middle Eastern production could change based on contracts in place in mid-2007. Over 70 percent of the 47 Mtpa of capacity in operation in 2007 was committed to Asian buyers but that share falls to 16.6 percent for the capacity under construction. Overall, assuming there is no change to the contractual arrangements, only 42 percent of the Middle Eastern LNG production in 2013 will be delivered to Asian markets.

Over 50 percent of the capacity under construction is, at least nominally directed to markets in the Americas, mainly the USA. The flexibility of the US market means that the LNG can be diverted to other markets on a short-, medium- or long-term basis provided any costs incurred by the buyer, for example through commitments to receiving terminal capacity, are covered. Similarly, most of the 25

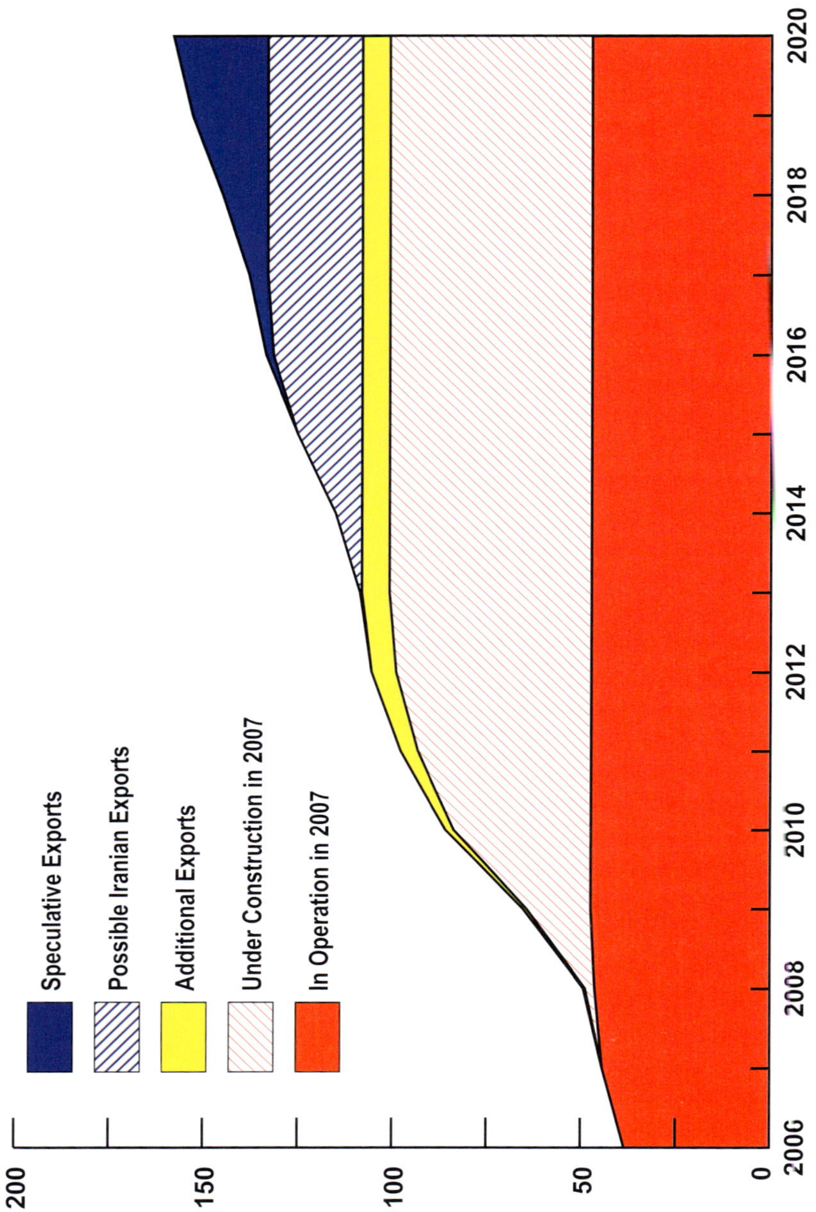

Figure 8.7: Possible Middle Eastern LNG Exports 2006–2020

Source: author's estimates

Table 8.14: Shares of Middle Eastern LNG Output

Region	*LNG Trains in Operation in 2007*	*LNG Trains Under Construction in 2007*	*Possible Shares in 2013*
Asia	71.3%	16.6%	42.0%
Europe	26.9%	25.8%	25.3%
Americas	1.8%	57.5%	32.7%

Source: Author's estimates

percent of Middle Eastern capacity committed to Europe is targeted at markets where diversions of cargoes are also possible.

The contracts with buyers in the flexible markets of the USA and the UK provide opportunities for Asian buyers to secure LNG supply from the second phase of the development of Middle Eastern LNG, having missed out when commitment to the development of the capacity was originally made. However, the sellers will only be willing to divert cargoes if the buyers offer prices that yield a higher netback to the liquefaction plant than sales to the original destination. Korea Gas was the first buyer to offer such prices on a long-term basis through the contract with RasGas agreed in November 2006. Buyers in Japan and elsewhere in Asia have been prepared to pay a premium for cargoes on a short- and, in some cases, a medium-term basis but have resisted committing to long-term supplies at what they see as unacceptably high prices. The next few years will determine whether Asian buyers are able to meet the demands of their customers, without having to compete with buyers in Europe and the Americas for Middle Eastern LNG.

Conclusions

The increasing supply of LNG from the Middle East over the period from 1997 and 2006 made an important contribution to meeting natural gas demand in Asian markets, at a time when the established suppliers in South East Asia were unable to respond. The commitment of buyers in Japan and Korea, who were seen as reliable off-takers and have strong credit ratings, enabled Middle Eastern producers to finance a rapid expansion in liquefaction capacity. Between 1996 and 2006, Middle Eastern LNG production increased by close to 800 percent and in the process met two-thirds of Asia's incremental LNG imports.

The Middle East has turned to the more flexible markets of Europe and the USA for the second phase of LNG development. Over 80 percent of the capacity under construction in the Middle East in mid-2007 is contracted to non-Asian markets. However, there is the potential for some of these volumes to be transferred back to Asia, if the buyers are prepared to compete on price. The second phase of development will take Middle Eastern LNG production to over 100 Mtpa by 2013.

There is considerable uncertainty over whether there will be a third phase in the expansion of Middle Eastern LNG and pipeline exports and if so how much more capacity will be added. Governments in the Middle East are increasingly debating whether to increase exports or give priority to the domestic use of gas. Saudi Arabia has opted not to export gas, preferring to use it to substitute for oil in industrial and power generation uses, releasing oil for export. Abu Dhabi has indicated that it will not expand its LNG exports and it has become a gas importer with the start-up of the Dolphin project, which brings gas by pipeline from Qatar. Oman and Yemen's reserves are too small to support further exports beyond the capacity in operation or under construction.

Qatar and Iran have by far the largest proven reserves in the region and have the potential to lead the next phase of export expansion – Qatar through LNG and Iran through either LNG or pipeline gas or both. Qatar has imposed a moratorium on further development of the North Field to allow it to assess its performance when annual production rates reach over 200 Bcm/a around 2010 or 2011. Iran is ambitious to join Qatar as a major gas exporter. However, its ability to meet growing domestic demand, to provide gas for re-injection to boost oil production and to export large volumes as LNG or by pipeline is uncertain. If it finds it is unable to meet all the competing demands, exports could be the first option to be cut.

The Middle East will continue to play a role in meeting Asian gas demand. However, growing domestic demand and government policies to conserve their reserves will result in the pace of growth slowing and creating an environment where Asian buyers will face competition from Europe and the Americas for available supplies.

References

BP Statistical Review of World Energy 2007, Annual Publication by BP plc., London, UK, June 2007.

2006 Natural Gas Year in Review, Annual Publication by Cedigaz, Paris, France, April 2007.

Natural Gas Market Review 2007: Security in a Globalising World, International Energy Agency, Paris, France, April 2007.

Notes

1 For prospects of Russian pipeline gas to Japan and Korea see Chapter 6.

2 Debottlenecking is the process of increasing the capacity of an LNG plant by removing constraints on production identified during operations. Examples of the modifications that are made include piping changes, increasing the power of compressors and increasing the capacity of gas treatment units. The addition to capacity through debottlenecking varies from plant to plant but has been as much as 25%. The unit cost is generally well below that of a new train or plant.

3 Cargo, Insurance, Freight (cif) is a contractual arrangement under which the seller transports the LNG to the buyer's receiving terminal. The cost of transporting the LNG and insuring it during transit is included in the price. Ownership of the LNG passes to the buyer either on loading or at some point on the voyage to the receiving terminal.

4 'Natural Gas Developments in the Arabian Gulf', presented by Fatema Al-Neaimi at the Gastech 2006 Conference in Abu Dhabi, 4–7 December 2006.

5 Free on Board (fob) is an arrangement under which the buyer arranges the transportation and purchases the LNG as it is loaded on to the ship at the liquefaction plant.

6 See 'Dolphin Will Sell Gas to Customers at $1.30–$1.40/mn BTU, Says Sayegh', *Middle East Economic Survey*, vol. XLIX, no. 19 (8 May 2006).

7 Gas to Liquids is a process that converts natural gas into liquid oil products, typically naphtha and diesel oil. GTL products compete in oil markets with refinery output and generally command a premium because of their clean burning characteristics. Converting natural gas to oil products is more complex than the LNG process and the plant costs are typically much higher. However, the product can be transported in conventional oil carriers and, as a result, shipping costs are much lower than for LNG. The output from GTL and LNG plants is a liquid but whereas in the LNG process the physical properties of the gas have been changed, and the LNG can be converted back to a gas by applying heat, in the GTL process the chemical properties are changed to those of an oil product.

8 See 'Sharp GTL Jolt from QP, Exxon', *World Gas Intelligence*, vol. XVIII, no. 8, 21 February 2007, p. 1.

9 There appears to be no definitive official statement about the moratorium. It was apparently first announced by Qatar's Energy Minister, Abdullah bin Hamad Al-Attiya, at the Gas Exporting Countries Forum in Trinidad and Tobago in April 2005. 'Qatari Delays', *World Gas Intelligence*, 4 May 2007.

10 See 'BP Signs Oman Deal Which Could Yield Some 20 to 30 Trillion Cubic

Feet of Gas', BP press release dated 22 January 2007 (www.bp.com).

11 DES – Delivered Ex-Ship is an arrangement in which the seller is responsible for shipping the LNG to the buyer's receiving terminal. Ownership of the LNG transfers to the buyer as the LNG leaves the ship. This differentiates it from a cif sale, under which the ownership transfers at the loading point or on the voyage to the receiving terminal.

12 Boil-off gas: around 0.1–0.15% of the LNG carried by an LNG ship boils-off each day. This gas is burnt to raise steam for the turbines that power the ship, typically providing around 50% of the fuel required by the ship, with the remaining requirements coming from fuel oil. One ship in operation in September 2007 has on-board facilities to re-liquefy boil-off gas and 45 ships being brought into service from late 2007 to transport Qatari LNG production will have similar facilities.

13 See 'Total Scraps LNG Timetable as Costs Top $4bn' , Heren LNG Markets HLM 3.4.15 April 2007.

14 See 'Shell Plans FID on Iranian Liquefaction Plant in 12 Months: CEO', *Platts LNG Daily*, vol. 4, no. 145, 30 July 2007.

15 See 'OMV Seeking Iranian LNG or Pipe Gas?', *World Gas Intelligence*, vol. XVIII, no. 17, 25 April 2007.

16 See 'NIOC and CNOOC Sign $16bn Deal to Develop North Pars Field and LNG Plant', *Gas Matters Today Asia*, Thursday 21 December 2006.

17 See 'Iran Finds New Malaysian Partner', *World Gas Intelligence*, vol. XVIII, no, 2, 10 January 2007.

18 The Trans-Adriatic Pipeline will be an extension of the Turkey–Greece Interconnector and will run from Greece across the Adriatic Sea to Italy. 'Swiss Firm signs 25 year Gas Deal with National Iran Gas Export Company', *Gas Matters Today*, 5 June 2007.

19 International Energy Agency, *World Energy Outlook 2005*, OECD/IEA: 2005, Figure 11.11, p.365 shows that at an oil price of $28/bbl equivalent, it is far more profitable for Iran to reinject gas than export it as LNG.

20 See 'Price Formula Agreed for IPI Pipeline but Terminal Fees Still Undecided', *Gas Matters Today*, Tuesday 17 July 2007.

21 See 'Turkey's Answer to Russia: Iranian Gas', *World Gas Intelligence*, vol. XVIII, no. 29, 18 July 2007.

22 See 'Outlook for Global Trade in Liquefied Natural Gas Projections to 2020' prepared for the California Energy Commission by Jensen Associates and published in August 2007.

CHAPTER 9

CONCLUSIONS

Jonathan Stern

Although this book is a second edition, very few paragraphs have survived intact from its predecessor. In reality, this is a new book which reflects the dramatic changes that have occurred in the energy sectors and gas industries of the region during the 2000s. The next important point to make is that, to a much more significant extent than the first edition, this book is about gas and energy, rather than international political, developments. This is not because international politics have become less important or active, but more a reflection of the huge increase in gas activity throughout the region which has compressed the space available to deal with other issues.

The first section of the conclusions looks at some common themes which have emerged from the preceding chapters. The second part summarises the findings of the individual chapters on suppliers and markets and the final section provides some conclusions to the questions that were posed in the introduction to the book. Whilst drawing heavily on the text from the preceding chapters, this chapter contains the views of the author for which he alone takes responsibility.

Thematic Issues

Energy, Gas and LNG Markets in the 2000s

The economic, energy, natural gas and LNG outlook has changed fundamentally since the first edition of this book. In the early 2000s, Asia had only recently emerged from the 1997 crisis which affected virtually the entire region. Over the past decade, Asian economies have been among the leaders of a sustained period of global economic growth, with rates of 6–10 percent per annum. Partly related to this surge of global and regional economic growth, the energy, natural gas and LNG outlook in 2007 was entirely different from that of the early 2000s. In the early 2000s, crude oil prices were in the range of

$20–30/bbl with a consensus that they would remain at this level for the foreseeable future. Natural gas and LNG prices tended to follow those of oil, with only North American gas markets showing early indications of the price increases that were soon to overtake global energy markets. Long-term LNG contracts had been concluded between Middle East and Australian suppliers and Chinese and Indian buyers which guaranteed relatively low prices over extended time periods in what was generally considered to be a 'buyers' market'. LNG project costs were on a steadily decreasing curve to the point where, in the early 2000s, it was generally believed that LNG could be delivered from most liquefaction to most regasification plants at costs around $3/MMBtu.

In barely more than five years, virtually all elements of this outlook changed dramatically. In 2007, crude oil prices fluctuated in a range of $60–100/bbl and conventional wisdom suggested that they were unlikely to fall significantly below $50/bbl in the period up to 2020. Natural gas and LNG prices followed oil, and LNG rapidly became a 'sellers' market' as a result of limited availability of new supply at a time when demand was increasing significantly. In 2007, Japanese and Korean long-term LNG import prices were in the range of $7–10/MMBtu with the Japan Crude Cocktail (the marker price for Asian LNG) at $10–12/MMBtu, and spot cargoes had been traded well above long-term prices.[1] In the first half of the 2000s, a very large amount of new liquefaction capacity commenced construction backed by long-term contracts which will start by 2012. After that date, it was not clear from where the next major expansion of gas supply for Asian markets would originate and the number of candidates was relatively limited.

Costs for natural gas pipeline and LNG projects had risen dramatically as a result of huge increases in (principally) steel and nickel prices and cost inflation due to a shortage of contractors and skilled labour able to deliver large and complex projects to demanding deadlines.[2] These increases in costs had become a substantial contributing factor to the delays involved in bringing new LNG projects on stream as well as affecting economic viability. Serious competition for energy, natural gas and LNG resources between international and national companies had developed, the latter very often representing the interests of governments of importing countries. It is for this reason that many of the chapters in this book are dominated by the view that the gas supply/demand balance in Asia for much of the 2010s will be tight, and that market power will lie with the sellers.

Short-Term LNG Supply and Continuing Globalisation of LNG and Gas Markets[3]

This book is not about the future of global gas and LNG markets, but so important is the impact of that future on Asian gas markets, that the subject is central to the conclusions of the book. By the mid to late 2000s, it had become conventional wisdom that, international gas trade, and certainly LNG trade, was developing into a global market.

Since the first edition of *Natural Gas in Asia* was published, long-distance movements of LNG, linking Asia with Atlantic gas markets, have grown significantly. Spot and short-term LNG trades have emerged in both Asia and the Atlantic with a growing trend towards global trading and price arbitrage.[4] Global spot and short-term LNG trade increased from 1.8 Bcm (1.3Mt) in 1997 to nearly 33.8 Bcm (25Mt) in 2006. In that year, Asian LNG suppliers accounted for more than half of LNG spot/short-term cargoes traded (Figure 9.1). This represented a threefold volume increase on previous years when short-term Asian trade had never risen above 30 percent of the global total.

While cargoes outside long-term contracts only amounted to around 14 percent of total LNG trade in 2006, the development of a degree of price arbitrage – i.e. LNG cargoes being shipped to buyers willing to pay higher prices than the originally intended buyer – had opened up the possibility of a commercial paradigm change for natural gas trade. With substantial new liquefaction capacity under construction in the Middle East in the late 2000s, most of which was committed to the Atlantic Basin (i.e. Europe and the United States), there was the possibility that, given sufficiently large price differentials, substantial numbers of (particularly) Middle East cargoes could be diverted to Asia–Pacific buyers. Because the Middle East is roughly equidistant between the Atlantic and Pacific Basins it is the global 'hub' for LNG trade and price arbitrage. On a much smaller scale, Sakhalin LNG can be arbitraged between Asian buyers and the west coast of North America, an option which will become more relevant if and when more receiving terminals are built on that coast.

In 2006, around 56 LNG cargoes, or 4.7 Bcm (5–6 percent of Asian LNG supply), was supplied from the Atlantic Basin into Asian markets. While this is a small percentage of total Asian LNG supply, these cargoes represent a growing shift on the part of more dynamic LNG players towards shorter-term contracts. The origins of spot and short-term LNG supplies in Asia can be found in the Korean need for seasonal peak supplies. But Japanese nuclear power problems have

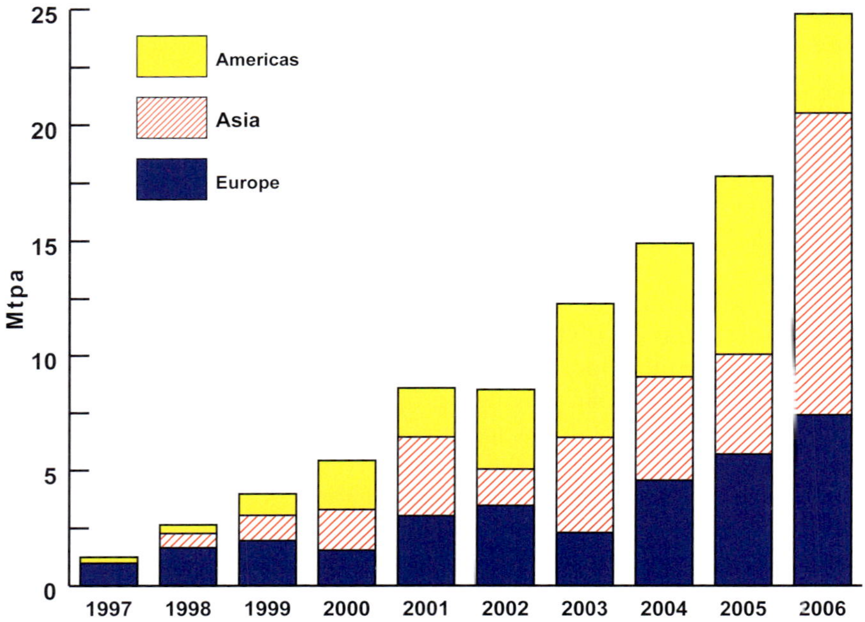

Figure 9.1: Global Spot and Short-term LNG Trade, 1997–2006 (Mt**)

* 1Mt of LNG = 1.35 Bcm of gas

Source: Flower LNG

created a larger and 'longer term' market for such supplies, to the point where short-term trade has become a permanent, and probably growing, feature of the business.

To some extent, even in the late 2000s, short-term trades in Asia were still the tip of a 'long-term contract iceberg', with long-term 'trust' relationships still a major part of the business. While all parties want continuity of LNG trade – buyers to meet gas and power demand, and sellers to meet financing commitments and the requirements of national treasuries – the nature of the business has fundamentally changed from its origins in the 1970s and 80s.

During the late 1990s and increasingly in the 2000s, it became clear that a combination of:

- advances in technology and the growing flexibility of the different components of the LNG chain,
- the need for buyers to access flexible supplies at short notice,

- changing business models and the entry into LNG markets of commercial players oriented towards short-term trading and profitability,
- the growing influence of price arbitrage between Henry Hub (USA), NBP (UK) and JCC[5]

had irreversibly changed the traditional rigid international LNG business. While this does not necessarily mean a *rapid* transition to a global LNG market and the development of spot pricing on a worldwide basis, it is becoming increasingly difficult to argue against that eventual outcome.

While Asia may be able to continue its traditional JCC/'S curve' long-term contract pricing for some years, it is difficult to imagine that by 2020 – and perhaps substantially before – Asian LNG trade will not have moved to spot and futures pricing based on a Japanese price (or perhaps a mix of Japanese and Korean prices).[6] The traditional resistance of Asian gas players to competitive pricing is being undermined:

- on the buyers' side to have access to larger quantities of short-term gas. In the case of Japanese power companies this was due to problems with nuclear plants; in the case of Kogas due to seasonal requirements and the deficit of long-term contract gas;
- on the sellers' side by the profitability from a standard cargo of LNG worth around $20 million at a price of $6/MMBtu, which means that at 2007 prices, spot cargoes were being sold for around $40 million; representing very significant earnings for both LNG exporting governments and project investors.

With the protection afforded by the 'S Curve' formula no longer available for new contracts in 2007, buyers were facing demands for crude oil parity pricing. Whether such demands will succeed is uncertain, but this would significantly affect the competitiveness of gas in many Asian markets and, should the JCC formula continue to result in much higher prices than Atlantic Basin markets (allowing for transportation differentials), this might give rise to revolutionary thinking about pricing. Such thinking may be accelerated as contracts with 'S curve' provisions begin to expire over the next decade and will probably not be replaced.

The Asian LNG market can be thought of as evolving in four phases:[7]

1. 1970–85 Japan only;
2. 1986–2003 Japan+South Korea+Taiwan;
3. 2004–present Japan+South Korea+Taiwan+China+India+Mexico (west coast);

4. A future of 'global competition'.

The market is currently somewhere between phases 3 and 4. The increasingly 'global' nature of the LNG market is becoming evident in terms of longer-term supply and demand conditions in its two major regions. Pacific Basin LNG participants increasingly need to take into account what is happening in the Atlantic Basin, not simply in relation to short-term pricing and destination of cargoes, but more fundamentally how longer-term trends on the other side of the world will affect their market. Failure of the United States to make progress in accessing Alaskan gas supplies, and continued political problems between Europe and its pipeline suppliers, especially Russia, will accelerate demand for LNG in the Atlantic Basin during the 2010s. With significant uncertainty about the ability and willingness of traditional Atlantic Basin LNG suppliers to increase their exports, US and European demand for LNG could create more intense competition for global supplies.

An additional uncertainty is the extent to which Asian markets will be impacted by growth of LNG imports on the west coast of North America (as opposed to the east and Gulf coasts which are part of the Atlantic Basin). Should a number of regasification terminals be built on the west coast, close to the United States, these could pull LNG away from Asia and give those importers additional exposure to Henry Hub prices.[8] Because most Asian gas markets have no immediate alternative to LNG imports, a fact that LNG exporters are well aware of, importers are placed at a significant disadvantage in a competitive market and may create greater urgency to develop access to pipeline gas.

Bilateral and Multilateral Political Relations

One of the themes of the first edition was the problem of international relations in the region as a whole, and in sub-regions, which have created barriers to pipeline trade. These have largely been problems in traditional bilateral relationships and it can be argued that the lack of successful multilateral organisations in Asia – akin to the European Union or NAFTA – has allowed these traditional relationships to continue, without the possibility of regional mediation. But the 2000s may have already witnessed some progress in this respect.

South Asia provides classic examples of such problems with the traditional India–Pakistan rivalry and the more mysterious (but similarly problematic) India–Bangladesh relationship, having created significant obstacles to international pipeline trade. In the late 2000s, bilateral India–Pakistan relations had improved to the point where pipelines from either Iran or Turkmenistan were under serious consideration. However, political problems between Iran and the international community, and the ongoing military operations in Afghanistan – which would be a transit country for Turkmen gas – continued to complicate these options. Additionally in late 2007, political instability in Pakistan seemed likely to prevent any rapid decision making in that country. In the late 2000s, with Indian domestic gas production and LNG imports set to increase substantially, urgency to develop pipeline gas trade had significantly diminished, and the most likely scenario was that Pakistan would develop pipeline imports which, in time, might be extended to India. All of this was much more positive than any potential trade with Bangladesh, the prospects for which seemed to have been eliminated as much by internal Bangladesh gas requirements, and the likelihood that pipeline gas from Myanmar would go to China (see below) rather than India, as by the lack of improvement in the political relationship.

The development of gas and power interconnections between the Association of South East Asian Nation (ASEAN) countries provided the sole example of multilateral energy cooperation in Asia prior to the 2000s. Further development of these interconnections has slowed partly because the 'easy' ones have been achieved, but also because of the increase in domestic energy demand in exporting countries. A specific problem for ASEAN may be caused by widespread publicity given to violent government repression of political protestors in Myanmar during 2007, which may have made an expansion of energy trade more complicated. The Myanmar government's subsequent decision to direct new pipeline gas exports to China – which defended the government's actions – rather than by pipeline to India or by ship to Korea, can be seen as a political, as much as an economic, statement.

However, the major political dynamic of gas in Asia concerns four countries: Russia, China, Japan and South Korea (including unresolved problems on the Korean Peninsula). A broad generalisation about the early to mid 2000s would be that, at least as regards the prospect of gas pipeline development, relationships between these countries have deteriorated. This applied particularly to the Japan–China relationship in the Koizumi era, exemplified by the inability of the two countries to negotiate an amicable division of a small offshore gas field, ownership of which was claimed by both countries. The most important relationship is Russia–China and the least which needs to be said is that political relations between the countries have not been strong enough to promote gas pipelines. While it is difficult to separate political from commercial issues, it is too simple to see the lack of progress either in terms of disagreement over gas prices, or as part of a larger commercial negotiation strategy for both Gazprom and CNPC (important as those issues undoubtedly are). There are fundamental issues of lack of trust between the countries stretching back through the twentieth century and beyond. From an energy and pipeline perspective, this lack of trust was not helped by what the Chinese side perceived as unreliable contractual behaviour in respect of the East Siberian oil supplies when the Kremlin refused to allow the Angarsk–Daqing pipeline to go ahead (partly) because Yukos was the Russian partner.

Chinese companies are proactively developing pipeline gas supplies from non-Russian sources, specifically Central Asian countries and Myanmar which, in part, demonstrate a Chinese political preference for dealing with energy suppliers from a position of political strength; a position which Beijing clearly feels it does not have with Moscow. Continued delays to large-scale pipeline gas development could create economic problems for both Russia and China with strong political overtones. The problem for China is that large-scale pipeline gas may be essential for urban air and water quality improvements in cities far from the east coast. For Russia, the problem may be that the economic renaissance which is planned for the eastern part of the country over the next two decades, partly around gasification and gas-based industries, will be too expensive and likely to be endlessly delayed unless driven by export projects. Such delays may create additional political problems for Moscow given the expectations which have been raised in these regions.

While in 2007 both sides were waiting for concessions by the other, the costs of such inaction are likely to escalate while the mutual benefits of reaching agreement seemed convincing. Continued failure to reach agreement on pipeline gas will gradually push Russia towards an LNG 'solution' for East Siberian and Far East gas. The LNG solution will probably be more costly for both Russia and China and would represent a continued failure of bilateral political relations. But it will suit Japan, where the continued failure to normalise post-Second World War political relations and outstanding territorial disputes continues to play some part in the reluctance of utilities to take Russian pipeline imports seriously, despite the commercial opportunities which pipeline gas could have delivered. Having rejected the pipeline supply option, factoring more Russian gas into LNG portfolios would improve diversity of supply for many Japanese utilities with a supplier who still supports their traditional business model of long-term contracts. An LNG solution would not suit South Korea, which alone has struggled to accelerate pipeline imports of Russian gas, but has been thwarted by the failure of the Russia–China relationship and the lack of a political solution on the Korean Peninsula. The immediate frustrations from failing to realise the pipeline gas option in East Asia will therefore be felt most keenly in Seoul, but the longer-term energy consequences will be felt throughout the region.

However, alongside this relatively pessimistic scenario, there is some optimism that the development of regional multilateral organisations may have a role to play in future energy and gas cooperation. The Shanghai Cooperation Organisation (SCO) was created in 2001 with China, Russia, Kazakhstan, Tajikistan, Kyrgyzstan and Uzbekistan as full members; Mongolia, India, Pakistan and Iran have subsequently joined as observers, and the SCO has rapidly become an organisation of significant regional importance.

Despite much western media speculation on the possibility that the SCO could become a 'regional energy club', the Organisation's energy aspirations expressed in the 2007 Bishkek Declaration were somewhat bland and general:[9]

> The heads of the SCO member states note the important role of energy sector as a basis for steady economic growth and security, and attach special significance to strengthening interaction in this direction. Reliable and mutually beneficial partnership in various fields of energy sector will help strengthen the security and stability across the SCO region and the wider world alike. Comparison of energy strategies in the SCO framework is a pressing task. Considering the current resources, demand, capabilities and the potential, the SCO member states will continue to promote dialogue on energy issues and practical cooperation among energy-producing, transit and consumer states.

While it is clear that energy is a part of the SCO agenda, it is equally clear that security and terrorism, rather than economic and commercial relations, are the main business of the Organisation. It is too soon to suggest that the Organisation can play any kind of mediating role between its most powerful members, notably Russia and China, to resolve differences and encourage progress on regional energy supply and transportation issues. But it is not farfetched to suggest that it could have a role to play in multilateral energy relations, which could be extremely positive for gas development.

Liberalisation, Competition and Security of Supply

The revolutionary fervour of utility market liberalisation originating from (principally) North America and the UK in the 1980s and 90s was coolly received by governments and utilities throughout most of the OECD, but particularly in Asian countries dominated by the twin gas security preoccupations of:

- how to ensure sufficient future gas supplies and, given the overwhelming dependence of most countries on imports,
- how to ensure continuity of gas flows from a diverse group of suppliers.

However, none of the Asian countries studied here – with the possible exception of China for which, as an emerging gas market, liberalisation was less relevant – could completely ignore the promise of lower prices from greater efficiency and competition. The previous edition suggested that liberalisation could have a significant impact on both the Japanese and Korean gas markets particularly in relation to the latter's privatisation and restructuring programme. In the event, the impact has been much less than expected but still far from negligible, especially in Japan.

The opening of electricity and gas markets to competition in Japan did take place but stopped short of residential markets which are not expected to fully liberalise. As the policy pendulum swung back towards security of supply, it became clear that the structure of both the gas and electricity industries would remain essentially intact, with the traditional regional dominant players retaining their positions. But liberalisation has had a significant effect, particularly in the gas market, not so much in relation to customer switching, but in terms of new entrants capturing demand expansion, alliances between new and established companies, and the changing business models of all players.

Electric utilities were already in a very strong position with regard to LNG imports, and are increasing their influence in downstream gas markets. While city gas companies still dominate their markets, competition from regional electric utilities, and alliances between the latter and city gas utilities, LPG companies, and oil majors, have become important features of the gas market. This competition has had an impact on prices, as new entrants have been forced to make attractive offers in both power and gas markets in order to capture customers.

As far as LNG imports are concerned, the traditional purchasing consortia of power companies, gas companies and trading houses have largely disappeared and been replaced by corporate alliances that meet the individual needs of the partners. This happened because companies needed to establish their own identities in the wake of liberalisation. It has allowed a greater focus on the specific needs of buyers: flexibility to cope with future dips in nuclear power availability in the case of power companies, price competitiveness for city gas companies.

In Korea, the privatisation of Kogas and associated liberalisation and competition reforms lost momentum in the 2000s. Market conditions in the late 2000s and early 2010s will make it extremely difficult to return to this agenda, particularly because of the accusation that the supply crisis of the late 2000s has largely been caused by government refusal to allow Kogas to sign new long-term contracts (because of its anticipated privatisation). While some competition has been introduced by allowing competitors to build LNG terminals and purchase new supplies, by 2007 that trend was on hold.

The role of the new Indian Regulatory Authority will be critical in determining the pace at which gas markets develop in the coming years. While the Authority has powers in relation to setting transportation tariffs, the regulatory regime is at an extremely early stage of development and few conclusions can be drawn as to how it might evolve.

Overall, governments in Asia have not given much priority to liberalisation and competition and – as in many other countries – it seems that the agenda of the 1990s has been overtaken both by traditional security of supply concerns (which never disappeared in Asia to the extent they did elsewhere). Liberalisation may improve efficiency and produce prices lower than would have been the case in monopoly markets, but it restricts the ability of the state to develop and implement national energy policy. In conditions of tight supplies and high prices, governments are concerned about energy security defined as obtaining an adequate supply of energy for their own country. This view of the global energy market as a zero-sum game where each government is concerned to secure national supplies, undermines prospects for inter-state energy collaboration in Asia, and means that competition and liberalisation are considered at best less important and at worst unworkable.

Local, Regional and Global Environmental Issues

As in so many other regions, environmental pressures will grow in importance in Asia over the next two decades and this will assist the development of gas markets. The countries included in this book are divided in their environmental focus, with climate change and carbon reduction looming large in Japan, but local and regional air pollution concerns dominant in China and India.

Urban pollution is a particularly serious problem in China with over 200 cities considered to have moderate to severe air pollution. These persistent environmental problems could spur a concerted effort to accelerate the import of natural gas, but expanding natural gas consumption beyond the 200 Bcm expected by 2020 would only have a noticeable environmental impact if the bulk of the gas were directed to power generation, where displacement of coal would be the highest.

Even if China could achieve an additional 100 Bcm of gas imports by 2020, and the entire volume were dedicated to power generation, this additional capacity could replace only about 15 percent of the growth in coal-based capacity expected by 2020.

The Indian situation is similar to that of China but not so severe. Nevertheless, the conversion of public transport to CNG in many Indian cities is a clear indication of environmental pressures. Those reading this book in the hope that gas development in China and India might be able to displace large volumes of coal in the energy balances of these countries, will be disappointed. Gas will have a role in reducing emissions from coal, but it is likely to be marginal given the size of the carbon reduction problem. In both countries, gas is predominantly aimed at improving urban air quality, just as it was in Japan and the majority of OECD countries when the fuel was introduced many decades ago.

By contrast, in Japan there is considerable concern about meeting carbon reduction targets and gas is likely to play a significant role in trying to achieve the country's commitments. However, to the extent that gas-fired generation has to substitute for nuclear stations, this will increase Japanese carbon emissions. Should China and India decide to make carbon limitation commitments in a global post-Kyoto climate regime, this would enhance the value of gas in their energy balances, but at the time of writing there was no sign of such developments.

Finally on environmental issues, there is the question of the continued acceptability of LNG – a technology with an excellent safety record – and how this might be affected by any major accident, and the perceived vulnerability of facilities to terrorist attack.

The Markets and Their Suppliers

The Markets

China

Consumption of all forms of energy has been growing rapidly in the country since 2000, with natural gas having the fastest growth rate of 13.8 percent per year. Gas's share of primary energy demand rose by 50 percent in the first half of the 2000s, but from such a low base that in 2005, it comprised only 3 percent of the total. Over 90 percent of China's thermal electricity generation is coal-based. High and steady growth in electricity consumption has been a major driver in the surging demand for coal, slowing the rise in the share of natural gas. China's goal of raising the proportion of cleaner natural gas in its energy mix to 10 percent by 2020 is best understood in terms of government priorities for further expansion of the natural gas market. Even for China to achieve the projected natural gas consumption of 200 Bcm, requires resolution of constraints on resource availability, a growing domestic economy, and a stable international and national political climate under which long-term decisions and investments can be made with some confidence. Much needs to change before these conditions are met, and the consequence of not meeting the 10 percent target for gas may be an even greater reliance on coal for power generation.

Looking out to 2020, a key uncertainty in determining the degree to which a successful natural gas expansion programme could expand its role in the total primary energy mix is the outlook for coal demand. By 2005, the consumption of coal in China's households had decreased by over 100 million tonnes from its peak in the late 1980s, replaced largely by natural gas, coal gas, and LPG. In northern cities such as Beijing, with major urban pollution problems, district heating plants were largely converted to natural gas. Natural gas also penetrated the transportation sector, with fleets of buses and taxis rolled out in Beijing, Xi'an, Qingdao, Guangzhou, and other cities.

This type of market development shifted the focus from high-volume/low-price consumers towards low-volume/high-price users requiring major investments in distribution infrastructure and equipment conversion. This increased seasonal swings with Beijing unable to meet peak demand until storage could be built. Only natural gas from LNG terminals was directed in significant volumes to power generation, but the slowing pace of LNG construction has hampered growth in gas power generation from this source as well.

Under the scenarios laid out in this book China fails to reach its planning target of 10 percent penetration of the primary energy mix, and would need to import an additional 70–100 Bcm of natural gas by 2020 to reach a 10 percent level of natural gas use, requiring the construction of two to three additional baseload natural gas import pipelines which can only come from Russia. With the likely development of a major pipeline from Central Asia and a smaller pipeline from Myanmar, China is maximising its non-Russian supply options. But continued failure to develop a large-scale Russian pipeline relationship will mean that China will be overwhelmingly dependent on LNG for its future gas imports.

Natural gas could play a larger role than the 200 Bcm of consumption projected in 2020. But accelerating the consumption of natural gas highlights the limited options China has to respond to global warming in the context of its energy structure. Local and regional (as well as global) environmental problems suggest that diversification away from coal would be beneficial. Electricity policies, however, are concerned with keeping prices as low and stable as possible. Widespread gas uptake in the power sector remains hindered by pricing policies that put gas at a disadvantage to coal in much of the country, by electricity tariffs that give priority to lowest-priced fuels, and by the widespread lack of enforcement of policies to clean up emissions from power plants.

China's natural gas industry has developed remarkably in the last decade and has laid the foundation for continued strong growth over the next decade. The fact that the share of gas in total energy demand will only increase from 3 percent in 2006 to 7–8 percent in 2020 should not obscure the fact that – with a projected growth of 150 Bcm over the same period – the country will not only become the largest gas market in Asia, but the third largest in the world after the USA and Russia.

Japan

Since the early 2000s, all the major assumptions needed to project future gas demand have become much clearer and in 2007 the outlook is more positive. Previously it was assumed that increased demand could be achieved only if gas prices could be substantially reduced by the rapid introduction of competition. Ironically, however, as the prospects for radical restructuring of the utility sector receded and oil prices increased, gas became increasingly competitive against oil products, thereby substantially raising gas demand. The only less positive aspect of gas development compared with the early 2000s, is that pipeline imports from Sakhalin will not proceed without a reversal in the position of (particularly the electricity) utility companies.

There is no national pipeline network in Japan and no plans to build such a network from one end of the country to another. Nevertheless interconnections between regional pipeline networks are growing, and coastal LNG shipping in small-scale tankers is reaching locations which did not previously have access to gas. Partly for this reason, there is confidence that gas demand will grow both in power and non-power sectors, partly due to fuel-switching and partly to carbon reduction measures. In the electricity sector, much will depend on the progress of nuclear power and the load factor that nuclear stations can achieve. By 2020, the difference between nuclear stations running at 60 percent and 90 percent load factors will be around 30 Bcm/year of LNG demand. If electric utilities believe there is a chance that nuclear plants will continue to experience the problems encountered in the 2000s, they will need to provide for additional LNG-fired power capacity and supply. In the city gas sector, a substantial increase in demand can be achieved if future price formulae ensure that regasified LNG remains competitive with oil products.

South Korea

Korean gas demand will grow at lower rates than during the 1990s, in the wake of the completion of the 2500 km nationwide pipeline network, but the Korean government's 2007 demand projections of 50–55 Bcm in 2020, accounting for 15–16 percent of primary energy consumption, look realistic. The major immediate challenge for the Korean gas market is the gap between contracted supplies and anticipated demand, which is expected to reach 8–8.5 Mt in 2011 rising to a peak of 11–12 Mt in 2015. While Korea remains keen to import pipeline gas from Russia – and specifically from the Kovykta gas field – before 2015, Russia's Eastern Gas Programme ruled out this option. It will be difficult for Russian pipeline exports to Korea to start prior to 2017, and these are most likely to be from Sakhalin 3 or Chayanda (Sakha) gas with a pipeline passing through North Korean territory, assuming a political solution on the Peninsula. Failure to achieve such a solution would require supply via an offshore pipeline with substantially increased costs. Should pipeline options fail to make progress this will accelerate Korean initiatives to acquire equity gas via involvement in exploration and production which could potentially lead to a merger between KNOC and Kogas.

India

The future of the Indian gas sector is inexorably linked to changes in other parts of the economy and most particularly the pace of reforms in the power sector. For political reasons, power and fertiliser subsidies will not be phased out over the time horizon of this study which means that the majority of gas customers will be protected from market pricing. But as the distribution network extends across the country, increasing numbers of domestic, commercial and industrial consumers will convert to gas. The availability of gas to create and build its own markets may, even in the case of city gas drive, rather than follow, the reform process.

Price reform may also be driven by worsening electricity supply, particularly a peak shortage which in some states exceeds 25 percent. Industrial and commercial users facing increasingly uncertain supplies may prefer power based on probably more expensive, but secure, domestic or even imported gas, as supplies become available in the coming years. The major uncertainty is the size of the market which will be willing and able to use this gas on an ongoing basis. For the period up to 2020, gas demand in India will therefore be supply and price-driven. A change in these dynamics awaits government ability and willingness to move beyond restrictions on pricing and allocation. For these reasons, it is particularly difficult to make an estimate of Indian demand in 2020 and the likely share of gas in primary energy demand. The figures in Table 9.1 should therefore be viewed in that context.

The success of India's exploration policy means that supply growth may come as much from future domestic gas development as imports. Although the gas sector will continue to suffer from severe price-led supply constraints, it is at a major turning point. By 2009, Reliance Energy's east coast production will double domestic supplies; further discoveries would mean that imported gas will face severe price competition from domestic gas. For this reason, although Indian companies continue to show interest in potential new LNG developments, particularly in Sakhalin, the urgency to import additional LNG and pipeline gas has receded. Projects not underway in 2007 may be delayed until the market moves to the next level of growth, but that will be difficult if it needs to be based on imported gas and LNG at 2007 prices. Nevertheless, with two operating LNG terminals and two more under construction, plus active diplomacy on international gas pipeline imports, India is closer to being integrated into international gas markets than at any time in the past.

The Suppliers

Russia and Central Asia

One of the main conclusions of the first edition was that Russian LNG exports to Asia would commence by 2010, but volumes would be unlikely to exceed 10 Bcm/year. This has turned out to be too pessimistic; LNG exports starting in 2008 will exceed 14 Bcm by 2010. The other conclusion from the first edition that large-scale pipeline exports of Russian gas to Asia on the scale of 20–50 Bcm/year seemed – as it has for the past forty years – at least a decade away, remains robust.

The major change since 2002 is in Central Asia, where a pipeline taking gas from (principally) Turkmenistan and Kazakhstan to China seems likely to be completed by the early 2010s. While commercial negotiations between Central Asian countries and China had not been finalised in late 2007, there was great confidence that this would be achieved, in comparison to the Russian negotiations which appeared to have reached a commercial impasse.

Gazprom's legal monopoly on exports and the introduction of the 'single export channel' mean that importers cannot negotiate for gas from a specific field but only for a volume of Russian gas, after which Gazprom then determines the most appropriate source. Despite this principle, the publication of the 2007 Eastern Gas Programme made clear that Sakhalin and Yakutia (Sakha) were firmly designated to be the sources of pipeline gas exports to both China and Korea. The late 1990s and early 2000s mainly focused on the prospect of pipeline exports from the Kovykta (Irkutsk) field to both China and Korea. Kovykta, still seen by the majority of observers – including this author – as by far the most commercially attractive export option, will be used for local demand and possibly be piped west to connect with the main network carrying gas to western Russia and beyond. This means that pipeline exports to Korea, about which both the Russians and Koreans are enthusiastic, must either be achieved via North Korea or an offshore pipeline from Vladivostok, both of which present difficult political and commercial problems.

This outcome is a serious disappointment given the optimism which was expressed about the Russia–China gas trade in the mid-2000s. There is still time – and an abundance of resources – for Gazprom's projection of 68 Bcm of pipeline gas to be delivered to China, with additional pipeline gas to Korea, by (or even before) 2020. But in order to achieve such an outcome, both sides will need to show much greater urgency and willingness to compromise than was evident by late 2007. With China having selected Central Asian gas as the external source of supply for the second West–East pipeline, any immediate prospects for the Altai Pipeline from Western Siberia have been seriously diminished. Fundamental disagreement on prices – with Gazprom sticking to an oil-related formula and the Chinese insisting on a formula reflecting local energy (coal-related) conditions – meant that as this book was completed, the two sides were reported to be far apart. An additional commercial issue is that Chinese companies clearly want direct access to resources and the construction of gas infrastructure within the exporting country, which Gazprom is unwilling to agree.

It may therefore be that Russian gas exports to Asia will be easier to achieve in liquefied form. The Sakhalin 2 project will begin deliveries in 2008, but the Eastern Gas Programme only envisages a doubling of this capacity by 2030, which seems extremely conservative in relation to available resources. It would be possible for much larger-scale LNG exports based on already identified Sakhalin resources, to be delivering to Asia by 2020, and it may be that LNG can resolve the political and commercial barriers to pipeline gas exports. With the exception of Korea, Asian governments and companies have shown insufficient enthusiasm and urgency to move forward, but on the Russian side, the government has played a key role in slowing down development by annulling the Sakhalin-3 licences and failing to introduce the necessary legislation to hold new tenders for Sakhalin 3 and Chayandinskoye.

With the resources which have been identified, and the gas prices seen in international trade since 2003, there is no reason why Japan, China and Korea could not be importing at least 100 Bcm of gas a year from Russia and Central Asia by 2020 (indeed a higher figure would be possible). But in 2007 only 14.5 Bcm of LNG from the Sakhalin 2 project was definite, with an additional 30 Bcm from Central Asia seeming highly likely. There seemed no reason why political and commercial obstacles to substantially increased gas trade could not and should not be resolved, but the evidence of the early to mid 2000s gave no grounds for over-confidence. Hence, there is no credible China or Asia 'card' that Russia can play with European gas customers, by suggesting that current or future supplies will be 'switched' to Asia. It will be many years – and probably decades – before Russian exports to Asia could assume European proportions and even if that happens, the sources of supply for the two markets will be substantially different.

South East Asia and Australia

The major gas event in this region over the past five years was the decision taken by Indonesia to curtail LNG export contracts in favour of domestic demand. This may become a trend throughout the region albeit one which is more likely to affect the expansion of exports rather than current contracts. Malaysia has always followed such a policy through development of industry and power in Peninsular Malaysia based on offshore gas resources. The Western Australian condition that 15 percent of new gas reserves should be diverted to the domestic market, demonstrated the increasing sensitivity of State politicians to local needs. In 2007, governments were focused on domestic gas development, driven by higher oil and gas prices (and therefore reduced need for increasing exports), and a political imperative to support domestic and regional economic and energy demand growth.

The choice of market for gas depends on domestic industrial development, political decision making related to employment creation and gas prices. In 2006, when Indonesia started reducing its LNG export volumes, gas availability was only part of the reason. Other factors included the government's need to maintain political stability by meeting growing domestic energy demand in the diverse islands of the archipelago. By contrast, new countries entering the LNG business, such as Papua New Guinea, are seeking export earnings as a means to develop the domestic economy. However, in the key countries, the remote location of reserves that will be developed for LNG exports means that future domestic demand will most probably be based on new reserves and fed by local pipeline gas.

By 2020, LNG exports from these countries could be as high as 185 Bcm, more than double 2006 levels. The principal uncertainties will be how fast Australian exports can be expanded and whether Indonesian exports can recover, and potentially increase again, in the mid to late 2010s. Sellers will not have a problem finding markets for additional LNG volumes, and will continue to seek long-term contracts for 80–90 percent of project volumes to enable financing and the necessary support for project development.

LNG will only be exported out of the region when high gas prices in the Atlantic Basin or the US West Coast 'pull' cargoes away. But exports are by no means guaranteed to take priority as the political conflict with domestic gas usage continues. Because countries are looking to retain a larger share of their gas to meet domestic demand and employment creation, expansion of ASEAN regional gas interconnections is unlikely. If additional gas should become available then it will probably be used domestically. Where there are exportable surpluses, sellers may prioritise exports to ASEAN countries, but not at prices less than the netback they could achieve from LNG exports.

Post-2003 oil and gas price levels have undoubtedly made possible the development of new gas export projects. When the first edition of this book was written in 2001, export of LNG from Papua New Guinea was not seriously considered. Seven years later, in an environment where Japanese buyers are willing to pay $8–10/MMBtu for LNG, such projects have become commercially realistic. But stability of all kinds – not just political but also economic and fiscal – will be necessary to deliver and maintain gas projects over the next decade. The approach taken in Malaysia, where Petronas has managed the country's energy resources for over thirty years, stands in contrast to Indonesia, where the shift of this responsibility from the national oil company Pertamina to BP Migas, may have been partly responsible for gas availability problems in the 2000s.

Middle East

The Middle East has made an important contribution to the natural gas requirements of Asian markets over the past decade, a period during which the established suppliers in South East Asia struggled to meet increasing demand. From 1996 to 2006, the Middle East accounted for nearly two-thirds of Asia's incremental LNG imports. By 2013, Middle

Eastern LNG exports will exceed 135 Bcm, but over 80 percent of the capacity under construction in the Middle East is contracted to non-Asian markets. Nevertheless, some of these volumes may be transferred to Asia, if those buyers are prepared to offer higher prices.

There is considerable uncertainty over whether there will be a further significant expansion of Middle Eastern LNG and pipeline gas exports post-2013, and if so how much more capacity will be added. Governments in the Middle East are increasingly debating whether to increase exports or give priority to the domestic use of gas. Saudi Arabia has opted not to export gas, preferring to use it domestically and release oil for export, a highly profitable strategy in the post-2003 period. Abu Dhabi has indicated that it will not expand its LNG exports and with the start-up of the Dolphin project bringing pipeline gas from Qatar, it has become a gas importer. The currently known reserves of Oman and Yemen are too small to support export projects in addition to those currently in operation or under construction.

Only Qatar and Iran have the potential to lead the next phase of Middle East export expansion – Qatar through LNG and Iran through either LNG or pipeline gas or both. Qatar has imposed a moratorium on further development of the North Field until around 2011. Despite Iran's ambitions to become a major gas exporter, its ability to meet growing domestic demand (including reinjection to enhance oil production) and to develop substantial exports of LNG and pipeline gas, is uncertain. Equally uncertain is whether sanctions by the USA and the UN, in relation to Iran's nuclear programme, will permit large-scale investments and gas trading relationships with the international energy community.

The Middle East will continue to play a substantial role in meeting Asian gas demand. However, despite the region's tremendous resource base, the growing share of production sold to domestic markets at unrealistically low prices, combined with government policies to conserve reserves, is putting a brake on export development which will affect the development of global, not just Asian, LNG markets.

What Future for Asian Gas?

Finally we return to the main question posed in the introduction as to whether the share of natural gas in Asian countries is likely to approach the OECD average of around 20 percent of primary energy demand over the next two decades. Our general numerical conclusions (Table 9.1) suggest that, with the exception of Japan, this is unlikely, but that in all countries share of gas will increase significantly. Table 9.1 shows that all countries will significantly expand their gas demand. For the four countries in total, our projections show demand more than doubling driven by the quadrupling of Chinese, and doubling of Indian, consumption.

Table 9.1: Gas Demand Scenarios for the Major Asian Markets, 2020 (Bcm)

	2006 Actual		*2020*	
	Demand (Bcm)	*percent PED**	*Demand (Bcm)*	*percent PED**
Japan	84.6	14.6	101–116	18–20
Korea	34.2	13.6	50–55	16–17
China	55.6	2.9	205	7–8
India	39.7	8.5	82	10–12
TOTAL	214.1		438–458	

* primary energy demand

An interesting issue in Table 9.1 is the extent to which these demand levels depend on increased imports. In the Japanese and South Korean cases the answer is near totally, but for China and India, that question requires a good deal of judgement in relation to ability and willingness to pay for increasing volumes of gas at the prices of the post-2003 period. Our demand projection suggests that India looks likely to import a maximum of 20 Bcm by 2020 which may all be in the form of LNG. Chinese imports are projects in the range of 50–100 Bcm with pipeline gas likely to account for not more than 30 Bcm.

With the likelihood that LNG will become a global market well within the time-frame of this study, importing countries may not be able to ensure supplies unless they are prepared to outbid competitors. The problem faced by countries largely dependent on LNG for their imported gas – which seems likely to be the position of all four countries with the possible exception of China – is that only one Asian supplier, namely Australia, is certain to substantially increase LNG exports over the next decade above currently contracted volumes. While it is certainly possible that Russia (Sakhalin), Qatar and Iran could significantly increase exports, this is in doubt for different reasons. Sakhalin development depends on the Russian government to accelerate its licensing process and agreement between the resource holders and Gazprom. Qatar and Iran, because of their Middle East location, could as easily deliver to the Atlantic as Asia, just as supplies from Australia and Russia could (more expensively) be delivered to the west coast of North America rather than Asia. However, irrespective of actual destinations, increasing exports would take the pressure off the supply side of the global LNG market.

By contrast, should substantial additional LNG exports fail to materialise, Asian markets will be constrained unless pipeline supplies can be mobilised. To the extent that large-scale Russian gas pipelines continue to be blocked by the twin impasses of the Moscow/Beijing relationship and the lack of a political solution on the Korean Peninsula, a much needed energy resource for the entire region will be at best delayed for decades, and at worst remain stranded in the eastern part of Russia.

Given the attention that is being given to the growing importance of China and India on future international energy markets, we conclude that China and India will become more important, but not dominant, players in Asian LNG markets over this period.[10] Both countries have serious constraints on the prices they will be willing to pay for LNG supplies. Should Asian LNG prices remain at the levels of 2006–07, it will be very difficult for both countries to *significantly* increase their imports. Intense regional competition for LNG supplies should increase interest in access to pipeline gas, assuming that the latter will be made available at less than LNG parity prices. This assumption is probably not tenable for Middle East pipeline supplies, but for East Siberian gas it will be a very relevant consideration because of the additional cost of piping gas to the coast and building a liquefaction plant, where a direct pipeline link will be significantly less costly. High prices and intense Asian and global competition for LNG could therefore be a reason for a major reassessment of (particularly) Chinese pipeline gas trade with Russia.

Finally, given that the changes over the past six years have been so great that it has been necessary to completely rewrite the second edition of the book, the validity of any projections which look forward twelve years must be questionable. An important lesson from the first edition is that both market conditions and conventional wisdom in relation to natural gas and LNG can change very quickly. We are clear that gas will become a much more important fuel in East Asia both in absolute terms and as a percentage of energy demand. We may be too pessimistic in believing that the undoubted potential which exists to expand Asian gas markets by means of large-scale pipeline trade with Russia will remain unfulfilled, at least up to 2020.

Notes

1 LNG Focus, December/January 2008, p.20.
2 For details see Appendix 1 of this volume.

3 I am grateful to David Ledesma for some of the ideas and text in this section.
4 'Spot' LNG is defined as a single cargo and 'short-term' as LNG sales of less than two years duration.
5 Henry Hub is the reference price for North American gas and the NBP (National Balancing Point) is the reference price for the UK. These are the daily spot and futures prices against which LNG buyers and sellers can buy and sell LNG cargoes and make comparisons with the Asian market.
6 For an explanation of the JCC/S curve see Appendix 2.
7 I am indebted to Michael Bradshaw for this observation.
8 The Costa Azul regasification terminal on the west coast of Mexico, near the border with California, is due to be completed in 2008.
9 *http://www.sectsco.org/html/01753.html*
10 As this study was completed, the International Energy Agency's *World Energy Outlook 2007: China and India Insights*, IEA/OECD: Paris, 2007 had just been published, which showed a significantly increased role for the two countries in international energy markets.

APPENDIX 1

COST INCREASES FOR LNG PROJECTS

David Ledesma

The energy industry has seen an unprecedented increase in raw materials driven by a growth in demand from the energy and other global industries. In all areas of the industry from upstream exploration and production rig rate to pipeline and FEED equipment costs there have been increases in costs of up to 400 percent (Figure A.1). A key area which has affected the whole energy business is the cost and availability of manpower with project management rates increasing by nearly 100 percent and infrastructure construction costs increasing by over 50 percent.

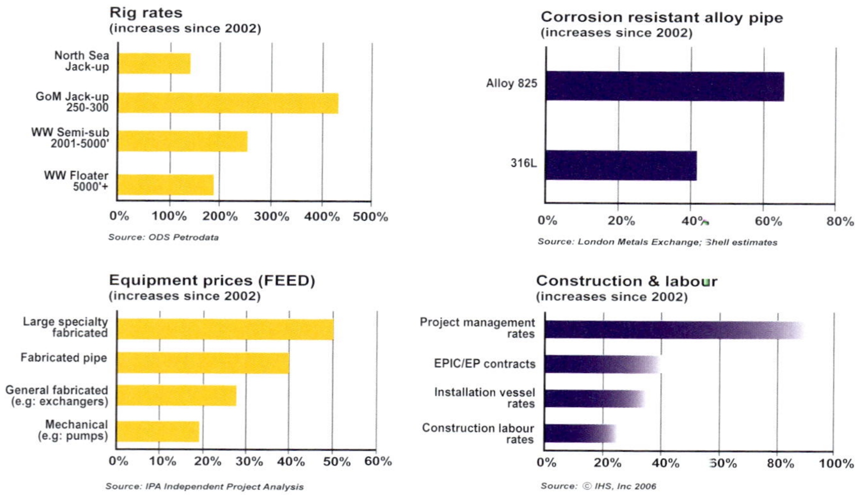

Figure A.1.1: Industry Cost Escalation

The effect of this has been an increase in the cost of supply which, in the main, has been matched by rising LNG and gas prices. The issue is that, as these costs rise further, project sponsors are questioning whether to proceed with a project or to defer it in case market prices fall below economic levels. In addition, there is the simple fact that rigs are not available to carry out exploration activities and it is the reduced gas availability, together with higher LNG and gas export capital costs, that is putting strain on LNG supply. National governments and energy companies argue that in this business environment international energy companies and investors should be willing to accept lower returns – not a story that the companies necessarily want to hear.

The cost of LNG production facilities is a function of various factors including plant location, labour costs, feedgas composition and LNG product specification. The cost of the liquefaction plant is a critical factor for any LNG export project since it usually accounts for around 50 percent of the chain costs. Figure A.2 below shows an indication of the cost of liquefaction capacity since 1965, shown in US dollars per tonne per year of production capacity (i.e. if a plant cost $5bn and has a capacity of 5 Mtpa then the cost/tonne is US$1000.

Figure A.2 shows that whereas during the period from the mid-1960s to the mid-1980s there was a reduction in plant capacity costs due to increased train size and improved technology, costs increased in the period 1985–1990 (a main reason for this was the high cost North-West shelf plant). Since 1990, plant costs have fallen in unit terms due to competition within the LNG engineering industry as well as to technology which permits larger LNG trains (production facilities).

New projects, however, are experiencing higher costs, as noted above. In 2007 it was announced that three LNG projects would proceed, at differing capital costs: the Skikda rebuild (a brownfield development)

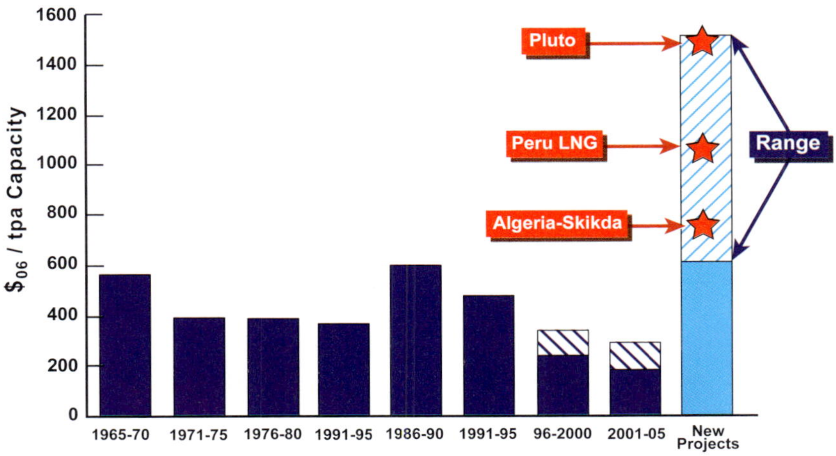

Figure A.1.2: Liquefaction Plant Capital Costs

Source: Flower LNG

reported at ~ $700/Mt installed capacity, Peru LNG reported at ~ $1,000/Mt and Pluto (Australia) reported at ~ $1,400/Mt. These higher costs mean that LNG sellers will need to secure more competitive commercial terms from the markets, specifically higher prices and maybe floor prices, to support the development of this new LNG capacity.

It is expected that these high cost trends will continue for some years.

APPENDIX 2

LNG PRICING IN ASIA – JAPAN CRUDE COCKTAIL (JCC) AND 'S'-CURVES

Andy Flower

The prices in the earliest contacts for the supply of LNG into Japan were fixed for the 15 or 20 year duration of the agreement. However, after the first oil price shock in 1973, LNG prices moved to a linkage with crude oil and by the mid-80s price formulae in most contracts were of the form:

P(LNG) = A * P(Crude Oil) + B

Where:

P(LNG) is the price of LNG in $/MMBtu
P(Crude Oil) is the price of crude oil in $/Bbl
A and B are constants

In most of the contracts with Japanese buyers, the oil price index is the monthly JCC price. JCC is the Japanese Custom Cleared crude oil price, often referred to as the Japanese Crude Cocktail. It is the average price of crude oil imported into Japan and published by the Japanese Ministry of Finance each month. The main exceptions to the use of JCC in Japanese LNG contracts are those with Pertamina for the supply of LNG from the Bontang and Arun plants. These use the average price of Indonesian crude oil (ICP). JCC has also been adopted as the oil price index in long-term contracts with Korea Gas and CPC in Taiwan. More recently it has been used for contracts with Indian and Chinese buyers, although in the case of the contract between Petronet in India and Qatar's RasGas project, JCC indexation will not become effective until 2009.

If the LNG price was priced on a crude oil parity basis, i.e. a Btu of LNG was the same price as a Btu of crude oil the price formula would be:

P(LNG) = (1/5.8) * P(Crude Oil) = 0.172 * P(Crude Oil)

since the average barrel of crude oil has an energy content of around 5.8 MMBtu. In many of the Japanese price formulae negotiated in the late 1980s for delivered ex-ship (DES) sales, the multiplier of crude oil was 0.1485 and the constant in the formula was between 0.6 and 0.9. For free on board (fob) sales the constant was around zero and in some cases the multiplier of crude oil varies marginally from the basic 0.1485. However, after taking in account the buyer's cost of shipping, the fob formulae generally yield a similar cost to the buyers for LNG delivered to Japan as DES sales.

The relationship of the price of LNG delivered to Japan and the crude oil price is shown diagrammatically in Figure A2.1. The LNG price is above crude oil parity at low oil prices but the premium erodes as the oil price increases and is eliminated at a price of between $26 and $30/Bbl, depending on the size of the constant in the formula. This approach to pricing was adopted in the late 1980s, a time when $20/Bbl was seen as the long-run average price of crude oil. Prices above $29/Bbl and below $11/Bbl were seen as exceptional circumstances and the buyers and sellers agreed that they would 'meet and discuss' how to price the LNG in the event that oil prices moved outside the so-called 'applicable range'.

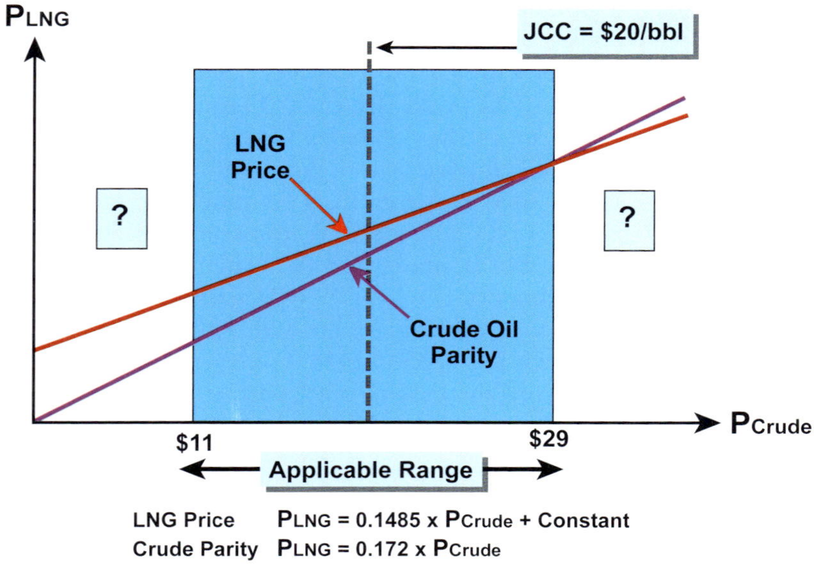

Figure A.2.1: Basic Japanese LNG Price Formulae

Korea and Taiwan entered the LNG market in 1986 and 1990 respectively and they adopted a similar approach to pricing as the Japanese buyers. However, the contracts with Korea and Taiwan did not generally include an applicable range.

The approach in contracts with Japanese buyers of a premium for the seller at low oil prices and that premium declining as the oil price increases, reflects the relationship nature of the LNG business in the 1980s where buyers and sellers were locked together for 20 years or more and had to recognise the need to be mutually supportive. The additional income for the sellers at low oil prices helps ensure that the economic viability of the project is maintained. However, at higher oil prices the sellers no longer require help and the premium erodes and eventually disappears, giving the buyers a more competitive product.

The concept of working together to share the upside and downside effect of oil price movements was taken one step further in the early 1990s, when many Japanese prices came up for renegotiation, through the introduction of the so-called 'S'-curve. As Figure A2.2 shows, the 'S'-curve results from varying the relationship of the LNG with the oil price over the applicable range. The basic formula:

P(LNG) = 0.1485 * JCC + Constant

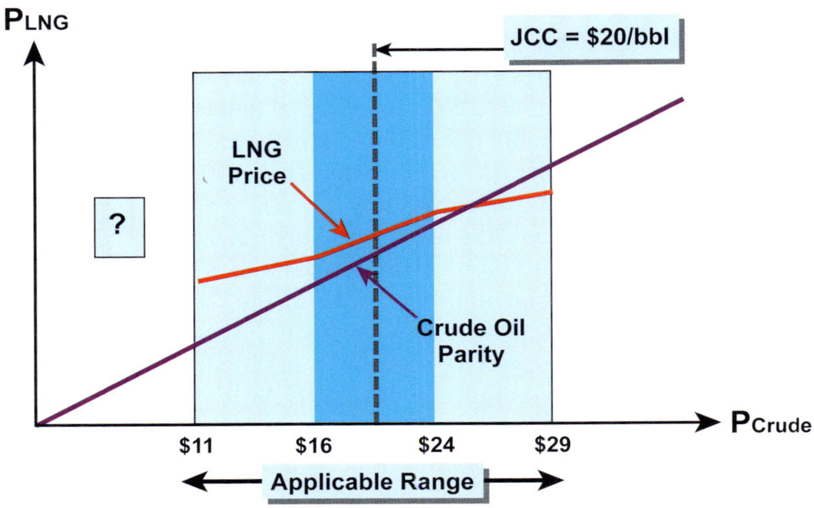

Figure A.2.2: 'S'-Curve Price Formula

applies in the mid-range of oil prices, typically from around $16 to $24/Bbl, while the so-called 'kink pints' vary amongst the contracts. Between $11 and $16/Bbl the relationship approximately halves so that, with every $1/Bbl reduction in the oil price, the LNG price declines by around $0.07/MMBtu rather than $0.1485/MMBtu in the basic formula. Between $24 and $29/Bbl there is a reciprocal flattening of the relationship between the LNG price and the oil price. The result is a significantly enhanced premium for the LNG suppliers at low oil prices matched by a larger discount for the buyers at high oil prices. The concept of $20/Bbl being the long-run average price of crude oil and prices below $11/Bbl and above $29/Bbl being 'exceptional circumstances' was retained in most cases.

The 'S'-curve was incorporated into many of the price formulae for the supply of LNG to Japan by the early 1990s, the main exception being the contracts with Pertamina in Indonesia. However, the 'S' curve was introduced into some, but not all, Indonesian contracts around 2002. Korea and Taiwan maintained the straight line relationship between the LNG price and the oil price until after 2000, when the 'S'-curve was adopted in some contracts.

The linkage to JCC oil prices and the 'S'-curve was still the agreed pricing basis for much of the LNG delivered into Japan, Korea and Taiwan at the end of 2007. Price re-negotiations between LNG suppliers and their Japanese buyers in the late 1990s and early 2000s generally left the basic price formula in place, with the main change being to the level of the constant in the price formula. The upper limit of around $29/Bbl for the applicable price range was generally retained.

JCC exceeded $29/Bbl in November 2003 and remained above that level through to the end of 2007, creating a need for the two sides to decide how to deal with a situation, which had been considered exceptional only a few years beforehand. In some cases the buyers and sellers moved quickly to agree how to deal with oil prices above $29/Bbl, whereas in others the issue was still not resolved some four years later. The approach taken by the buyers and sellers who reached early agreement was to put another 'kink' in the 'S'-curve through a further reduction in the relationship with oil prices to around 0.05. These agreements were reached in buyers' market where the volume of LNG supply available exceeded the demand. As the market tightened from 2005, sellers found themselves in a stronger negotiating position and 'S' curves began to disappear from new deals to be replaced by a return to a straight line relationship between the oil price and the LNG price up to oil prices of $60/Bbl and higher. In the negotiations to deal with prices above the top end of the 'S'-curve, sellers were pressing for the slope to be increased rather than reduced. By late 2006, agreements were being finalised with the LNG price returning to close to oil price parity and in at least one case higher, even at oil prices above $60/Bbl.

Many of these new price agreements have yet to be implemented which means that the majority of LNG being delivered to buyers in Japan, Korea and Taiwan still benefited from the reduced relationship with oil prices at the top end of the 'S'-curves. The result has been an increasing discount between the price of LNG and JCC prices. In the 22 months from January 2006 to October 2007, the average price of LNG imported into Japan was nearly $4/MMBtu less than the average JCC price. The discount for Korean and Taiwanese LNG imports was smaller at around $2.50/MMBtu.

The gap between the cost of LNG and oil imports provided a boost to LNG demand. However, as new price formulae negotiated from late 2006 come into effect and the agreement is reached on how to price LNG above the $29/Bbl threshold in many Japanese contracts, the gap will decline.

INDEX

Abu Dhabi 222, 332, 335-8, 340–1, 351, 357, 367, 376–7, 398
Abu Dhabi Gas Liquefaction Company (ADGAS) 335, 338–40
Afghanistan 85–7, 113, 223, 385
Agha-Jari oil field Iran 365
Algeria 127–8, 213, 337, 367
Al Khaleej project Qatar 350–1
Altai pipeline Siberia 224, 243, 256, 263, 395
Amadeus Basin Australia 305
Angarsk-Daqing pipeline Siberia 386
APEC (Asia Pacific Economic Cooperation) 176, 212–4, 254
ARCO (Atlantic Richfield) 17, 18, 231
Arun field Indonesia 287, 343
Arrow Energy 308
Arrow Australia 82
Association of Southeast Asian Nations (ASEAN) 279–301, 306, 327–8, 385–7
Asia Gas Pipeline Network 282
Asia–Pacific LNG market 54–5, 118, 127, 130, 154, 216, 233–6, 248–54, 275, 305, 328 381
Asia-Pacific Research Centre (APERC) 176–7, 214
Asian Development Bank 72, 76, 84, 85–6, 88, 113
Australia 6, 11, 13, 20–2, 36, 40, 55, 82, 128, 134, 148,187–9, 192, 199, 222, 279–84, 299, 305–9, 312–17, 320–4, 363, 396–9, 404
Australian LNG Ltd 21, 289

BHPBilliton Australia 312–4
Bangladesh 1, 3, 4, 88, 89, 384, 385
Bahrat Power Corporation 112–4
Bahrein 330, 351
Barnaul pipelines Siberia 246, 254
Barrow Island Australia 307, 312
Bassein gas field India 68
Beijing 12, 18, 24–6, 28, 31, 35, 45, 49, 50, 62–4, 223, 256, 271, 275, 386, 391, 400
BaoSteel Shanghai 27
Bintulu LNG complex Malaysia 301–5
Bohai gas field China 12, 34
Bontang 282–3, 287–9, 293–7, 325–6, 405
Brazil 106, 115
BP 38, 50, 66, 81–2, 113, 199, 203, 232, 239–43, 259, 271, 289–93, 310–2, 340–3, 357, 376–8
BP-UK see ARCO
Browse project Australia 307, 313–5, 317, 326
Brunei 128, 148, 187, 279–81, 296, 302
Burmah-Shell 68
Burrup Peninsular Australia 307, 310, 313

Caltex 68, 189, 200, 213, 313
CLP (Hong Kong) 313
Canada 13, 165
carbon reduction 3, 137, 142, 166, 389–92
Cauvery Basin 78, 101, 113
Cedigaz France 267, 284, 297, 316, 326–9, 339, 348, 376
Central Electricity Authority of India 71, 112
Changqing gas field China 18, 26
Chayanda (Sakha) Siberia 210, 251, 272, 393
Chayandinskoye field Siberia 227, 244, 250, 253, 272, 396
Chevron 12, 189, 288–9, 293–5, 310–2, 319 (Sasol-Chevron) 352
China 1–65 86–9, 106, 115, 135, 169, 193–4, 201–71, 274, 283–5, 289–93, 302–22, 357 36–401
China Gas 1, 21, 27, 29, 193–4, 214, 275, 278, 395
China Light & Power 29
China Cooperation Organisation (SCO) 41–2, 64, 387
China National Offshore Oil Corporation (CNOOC) 17–8, 21–4, 33–8, 43–4, 194, 289–90, 303, 311, 363, 378
China National Petroleum Corporation (CNPC) 11, 17–8, 31, 37–8, 43–51

64, 201–5, 210,217, 237–40, 255–6 260, 275, 313–4, 320–2, 328, 385
China State Economic and Trade Commission 38
China United Coal-Bed Methane Corporation (CUCBM) 12–3
Chongqing 14–5, 26, 41, 45, 52
Chongqing-Wuhan pipeline China 26
Chubu 124–5, 132–4, 144–5, 169
Chubu Electric 125, 131, 134, 140, 161–70, 236, 303–4, 311, 338, 342–3, 347
Chubu Gas 131
Chuetsu offshore earthquake China 153, 163, 166
Clean Development Mechanisms (CDM) 13–4
coal-bed methane 10–4, 81, 82, 326
coal gas 23–4, 31–3, 62, 308, 391
coal substitution 55
coal supply 23, 65
compressed natural gas (CNG) 31, 76, 104, 112, 366, 390
condensate fields, 30, 47, 227, 277, 287, 354, 357, 360
ConocoPhillips 293, 298, 301, 310, 315, 347, 352
Cooper Basin Australia 305
Costa Azul project Mexico 401
CUCBM 12, 13

D-6 field India 95–6
Dabhol LNG Terminal India 74, 89, 90, 95, 102, 109, 355
Daesung Group Korea 193–4, 214
Daewoo International 192–3, 202, 319
Dagang oilfield China 31
Dahej LNG project India 69, 89–95, 100–2, 114, 213, 344
Dahej-Uran-Dabhol India pipeline 109
Daqing oilfield China 14, 16, 54, 386
Daulatabad field Turkmenistan 85, 87, 261
Degolyer & McNaughton 85, 267
DPRK 206 207–8, 217–9, 244
Delhi 76, 108, 112–5
diesel fuel 58, 143, 144, 171, 377
Director General of Hydrocarbons (DGH), India 73, 82, 111
Distrigas Belgium 345, 348
Dolphin pipeline Qatar–UAE 335, 341, 351–4, 357, 376–7, 398
Dongguan Fuel 21 China
'Donghae 1' field Korea 189

EBRD (European Bank for Reconstruction and Development) 234, 269
Egypt 127–8
EIA (US Energy Information Administration) 66, 108
East Asia Gas Company 203, 239
East China Sea oilfield 11, 12, 37
Eastern Gas Programme Russia 210, 244–9, 252–3, 263–8, 272, 393–5
Electricity Industry Committee Japan 137, 170
Elk field Australia 321
EMDC (Energy Data and Modeling Centre) 119, 169, 171
Energomachexport, Russia 29
Energy Silk Route 41, 260
ENI 81, 293, 310, 362
environmental issues 3, 7, 13–4, 39, 41, 53, 62–5, 76, 117–8, 125, 143, 156, 168–70, 216, 227, 233–5, 245, 256–8, 265–6, 272–5, 306–8, 314, 317, 323–5, 389–92
ESPO (East Siberia-Pacific Ocean) pipeline 251–3
ExxonMobil 29, 38, 39, 202, 204, 206–9, 237–8, 287, 293–4, 298, 312–5, 321–2, 329, 343–52, 359
Exxon Neftegas Ltd 204, 216, 230, 237

Far East Energy 13
Ferdos gas field Iran 363
fertiliser producers 10, 14, 16, 26, 31, 33–5, 45, 56–7, 69, 71–5, 92–4, 97–8, 105, 110, 112, 305, 393
Foshan Gas, China 21
fossil fuels 1, 61
fuel oil A 143–4, 150, 171
fuel oil C 143, 150–1, 171–2
Fuj refinery China 39
Fujian Investment and Development Company 22
Fujian LNG 22, 289

GAIL (Gas Authority of India Ltd) 68–9, 74, 81–5, 89–95, 100–5, 109, 112–4, 193, 319
gas consumption 17, 24, 69–73, 97, 119, 121–3, 147, 179, 197, 330, 366
Gas Utilities Japan 126

Gaz de France 29, 112
Gazprom 29, 48–51, 135, 201–2, 204–10, 217–8, 224–30, 234–51, 254–9, 261–3, 265–77, 385, 395, 399
global warming 8, 61, 118, 155, 172, 391
Gorgon gas field Australia 36, 55, 148, 188, 312, 313, 317, 326
Golshan gas field Iran 363
Greka Inc 12
Guangdong China 15–23, 29, 34, 44–6, 52, 192, 310
Gujerat State India 68, 70, 76, 89, 92, 100–1, 213
Gujerat State Petroleum Corporation Ltd (GSPC) 78, 80–1, 100–1
gas fields:
  China 68, 85 293, 385
  India 69, 89, 92, 93,
  Korea, 134, 155, 186–9, 192, 203, 214–7, 337, 393
  Russia 202–3, 220–31, 244–5, 249, 251, 256, 274
  Kazakhstan 261–2
  Australia 134, 315–7
  Brunei 318
  Malaysia 280, 287, 363
Gas Linkage Committee (GLC) India 73–4
gas reserves:
  Australia 279, 305–6, 312, 323, 396
  China 4, 11–2, 28, 37, 39, 57, 322
  China Sea 293
  India 78–82, 85
  Indonesia 285–6, 289, 321–8
  Iran 331, 360–3, 365–6, 360–6, 370
  Japan 287–9, 312, 318
  Kazakhstan 262, 267
  Malaysia 319, 323–5
  Papua New Guinea 328
  Philippines 280–5
  Sarawak 30–2, 302
  Thailand 279, 322
  Turkmenistan 85–7, 260–2, 267, 276–8
  Uzbekistan 213
  Vietnam 279, 322
Gorgon field Australia 312
Greka 12
Gujerat State Petroleum Corporation (GSPC) 78, 80, 81, 111
GDPs:
  Australia, 296 305, 321
  Bangladesh 88
  Brunei 279, 318
  China 5, 42, 67
  India 72, 97, 112
  Indonesia 279, 285
  Korea 174–5
  Japan 117, 119, 168
  Malaysia 279, 296
  Myanmar 319
  Pakistan 84
  Papua New Guinea 320
  Russia 225
  Thailand 279
  Vietnam 279

Hazira-Vijaypur-Jadishpur (HVJ) India 88–95 100–1, 111
Henry Hub USA 191, 383–4, 401
Hiroshima 124–6, 132, 141, 145
Hiroshima Gas, 141, 236, 288, 304
Hong Kong 9, 13, 18, 20–1, 29, 312–3
Hong Kong & China Gas 21, 29
Hong Kong Electric 21
Hong Kong Town Gas 13
Houston Inspection International 29
Huabei oilfield China 16, 18
Hunt Oil 214, 358–9
Hyundai Corporation 202, 214, 358

Ichthys LNG Australia 134, 317
India 1–6, 27, 41, 51–2, 66–115, 169, 193, 213–405
Indian Oil Corporation (IOC) 112
Indonesian Petroleum Law 285
IEA (International Energy Agency) 43, 108, 176, 185–6, 2 1–5, 277, 378, 401
IEEJ (Institute of Energy Economics Japan) 144. 162, 173
International Monetary Fund (IMF) 76, 329
Interros Russia 240, 243
Irkutsk Oblast 46–9, 203, 207, 217–27, 239–46, 250, 253–4, 263, 268–74, 395
iron and steel industry 31, 60–1, 72–3, 184, 196
IEP Report 74-5, 97–8,106, 111–5
INPEX Holdings 134, 289, 307, 310, 314
Iran 4, 27, 41–4, 83–5, 188, 213,

330–78, 384–7, 398–9, 399
Iran-Pakistan-India gas pipeline (IPI) 83, 113, 368, 369, 378
Itochu Petroleum Inc Japan 29, 343–6, 355–6
Iwatani International Corporation 131, 171

Japan 1–6, 43–54, 67, 106, 113–73, 280, 286–9, 298–319,327, 335–58, 368–409
Japan Corporation 203, 240
JCC (Japan Crude Cocktail) 84, 113–4, 130, 146–50, 256, 276, 368, 383, 401, 405–9
Japan Gas Association 127, 172
Japan National Petroleum Corp (JNOC) 203, 240
Japan Nuclear Energy National Plan 117, 151–2, 164, 172, 173
Japan Nation Oil Corp (JNOC) 203, 240
JAPEX 136, 140, 161, 171, 304
Java 281, 293
Jungar field China 12

KG (Krishna Godavari) Basin India 74, 78–80, 90, 95, 98–101, 107, 111–4, 120, 192
K-Power 191, 199–200, 289
Khabarovsk pipelines Russia 49, 202, 226, 236, 245–50, 253–4
Kakinada pipelines India 78, 89, 101, 109
Kalimantan-Java pipeline 282–3
Kansai Japan 124, 125, 131–4 140, 144–5, 165, 169
Kansai Electric Company 124, 131, 163–4, 169–72, 288, 304, 310–12, 347
Karachnaganak gas field Kazakhstan 261
Kashagan gas field Kazakhstan 261
kerosene 143–4, 213,
Kikeh gasfield Malaysia 301
Kochi gas terminal India 90, 102, 107, 115, 145
Kazakhstan 11, 40, 41, 46–53, 259–65, 277–8, 387, 394
Koizumi (era) 136, 158, 385
Korea 1–6, 25, 46–55, 165, 169, 174–219, 221–90, 303–4, 310–29, 337, 341–59, 368–71, 375–400, 405–9
Korea Energy Economics Institute (KEEI) 175–8, 180, 187, 191 194–7, 211, 215–6
Korea National Oil Corp (KNOC) 189, 193, 197–8, 202, 207, 211
Korean Gas Company (Kogas) 4, 48, 165, 169, 175–97, 203–9, 244, 257, 271, 276, 304, 318, 383, 388–9, 393
Korea Petrol Development Corporation (PEDCO/KNOC) 207
Kovytka (Irkutsk) 203–5, 216, 239, 243–4, 250, 261, 271–4, 393, 395
Kovytinskoye field 48, 203, 244
Krishna Godavari (KG) Basin India 78–80, 90, 95, 98–101, 107, 111–4, 192
Kutch Basin India 81, 83
Kyoto Protocol 118, 125, 153, 155
Kyrgyzstan 41, 47–8, 387

LNG (liquid gas) terminals
  China 8, 20–3, 34, 36, 44, 52–7, 62, 73, 310, 394
  India 4, 44, 74, 89, 93–5, 103, 109, 115
  Iran 362–5, 373
  Japan 131–5, 138, 155, 181–5, 171
  Korea 181–4, 196, 389, 391
  Malaysia 304
LPG (liquefied petroleum gas) 20–4, 31–3, 62, 107, 138, 160, 180–3, 212, 340, 388, 391
Longgang field China 11, 14
Lowell Oil Company 12

Macao LNG terminal 44
Mahanadi India gas project 78
Malampaya gas field Philippines 280
Malaysia-Singapore pipeline 280
Maret Islands Australia 307, 314
Marib Jawf Yemen 358–9
Marubeni Corporation Japan 29, 193, 214, 319, 342
METI statistics 120–3, 136–9, 147, 162–8, 170–3
methane gas 12, 14, 56, 63, 81–2, 213, 286, 305, 308
Migas (BP Migas) 293–5, 327, 397
mineral processing (non-metal) 31–2
Ministry of Land and Resources China 39

Mitsubishi 29, 173, 194, 229, 235, 257, 260, 289–90, 302, 310, 318, 327, 355–6
Mitsui 29, 139, 229, 235, 289, 310, 340, 342, 346, 355
Mongolia 10–11, 15, 26–8, 41, 50–4, 203, 241, 244, 270, 387
Mongolian Petroleum Administration 240
Myanmar 41, 51–3, 89, 192–3, 279–80, 319–20, 324, 328–9, 385–6, 391

Nadym-Pur-Tazov-Xinjang gas pipeline Russia 49
naphtha 71, 94,95, 143, 144, 377
National Development and Reform Commission China (NDRC) 23, 28, 35, 38–9, 43–6
National Coal Corporation China 13
National Iranian Oil Company 361–4, 378
National Thermal Power Corporation (NTPC) India 73–4, 95, 98, 111–2
Natuna gas field Indonesia 193, 281, 285, 293–4
natural gas consumption 9, 31, 32–3, 59, 62, 118–9, 121–5, 139, 197, 389, 391
Natural Gas Research Society (NGRS) DPRK 205–6, 217
NBP (UK) 383
New Exploration Licensing Policy (NELP) 77–82, 96–9,107,111–3
Nigeria 127–8, 367
Ningbo LNG China 24
Nippon Oil Corporation 131, 135, 140–1, 170, 289, 302, 319–21
Nissho Iwai Japan 29
North-West Shelf project (NWS) Australia 187, 188, 192, 310
nuclear power 5, 9, 19, 22, 48, 59, 117, 119, 121–3, 137,141–2, 148–53, 156, 163–8, 171, 175–7, 215, 381, 388, 392
nuclear weapons crisis 47, 84, 168, 207–8, 210, 218, 364, 398

Oceanway Project California 312
Oil and Natural Gas Commission India (ONGC) 68–70, 73, 77–82, 89–96, 105, 111–4, 193, 229, 232, 319
Oil India Ltd (OIL) 68, 111
Oil Search 321–2, 328–9
Okayama Gas Japan 134, 135, 145, 161
Oman 127–8, 148, 169, 187–8, 194, 212–3, 300–8, 351–8, 367, 371, 376–7, 398
OMV Austria 362, 364 378
Ordos Basin Mongolia 10–12, 26, 28, 37, 54
Orion Sweden 12
Oryx GTL plant Qatar 350, 353
Osaka Gas Company 124, 134, 140, 149–50, 161, 170, 236, 288, 303–4, 311, 315, 318, 347, 355–6
Osman gas field Turkmenistan 260

Pacific Oil 44
Pakistan 1, 3, 41, 81–9, 113, 330, 336, 351, 367–9, 384, 387
Panna-Mukta field India 77, 93–7, 112–4
Papua New Guinea 279–80, 320–1, 327–8, 396–7
Pars: North Pars field 187, 331–54; South Pars field 361–9
Peru 46, 192, 214, 404
PetroChina 26–38, 201–2, 217, 275
Petroleum and Natural Gas Regulatory-Board India (PNGRB) 103, 107, 176
Petroleum Association of Japan 171
Petroleum Law, Indonesia 285
Petronet LNG 2, 69
Petronet Ltd 89, 93–7 100–1, 107, 111, 115, 213, 313, 344–5, 348, 405
Phillips 12
Pilbara project Australia 305, 314, 326
Pertamina Indonesia 286–96, 327–8, 397, 405, 408
Pinghu gas field China 19, 23
Pluto project 148, 306–17, 326, 329, 404
pollution 7, 18, 26, 36, 41, 55, 62, 63, 112, 146, 172, 286, 337, 389–91
POSCO Korea 184, 190–6, 199, 200, 201, 289, 290
Poten LNG 329
PPS (power producers and suppliers) 137–9, 157
pricing policy
Asia 383–4, 392, 405
China 13, 16, 19, 27–36, 53, 57–8 63, 67
India 71, 81, 84, 90–2, 97, 101, 105, 393–4

Japan 5, 276, 406
Russia 49
Turkmenistan 52
Teheran 84
Primorye region Russia 205, 245
Puguang gas field China 2, 11, 14, 44–5

Qaidan Basin China 11–2, 26
Qatar 89, 94, 114, 127–8, 133, 148, 187, 191, 199, 205, 213, 286, 317, 330–65, 370, 373, 376, 398–9
Qatargas 338, 342, 343, 345–8, 354, 371
Qatar Petroleum 342, 344, 346, 350–2
Qeshm LNG (Iran-Australia) 362–3
Qingdao LNG terminal China 44, 53, 62, 391
Qinghai pipeline China 10, 14 15, 26
Qinshui field China 12
Qiondongnan oilfield China 11, 12, 17

RasGas 346, 347, 348, 354, 371, 375, 405
Ras Laffan 342, 350–2
Ras Laffan Company 21, 187, 342, 344
Korea Ras Laffan LNG Ltd 187
Ravva field 77, 94, 97, 112, 114
Reliance Industries Ltd India 78, 81, 89–90, 98, 111
Reliance Gas Transportation Infra-structure Ltd (RGTIL) 81, 111
Rosneft 228–32, 238, 247–51, 259, 269, 273–4, 277, 300
Royal Dutch Shell 29, 36, 229
Rusia Petroleum 50, 203, 227, 239–43, 270–1
Russia 3, 6, 27, 29, 41–54, 87, 148, 174, 187, 201–9, 216–8, 223, 226–67, 276–78, 337, 366–7, 378, 384–7, 391–400
Russia–China gas pipeline 275

Sabah, Borneo 296–8, 301–2
Sakha Republic 46, 202–10, 221–4, 227, 244–9, 253–4, 263, 268, 272–4, 393–5
Sakhalin projects 44, 46, 49, 53–4, 135–6, 148, 155, 159, 187–8, 191, 199, 203–12, 216–7, 220–78, 381–401
Santos 308, 310, 321
Sarawak Borneo 296, 298, 301–2, 323,
Sasoil 352
Saudi Arabia 330–5, 351, 366 370–6, 398
Saudi Aramco 39
security issues 2–3, 17, 40–1,48, 52, 63–4, 84–5, 117, 129, 137, 142, 153, 159, 186–9, 207, 211–8, 265–6, 280, 301, 358, 377, 387–9
Sempra Energy Korea 289–90
Senoro LNG project Sulawesi Indonesia 290, 295–7, 327
Seto Inland Sea 134, 145, 157
Shaan-Jing gas pipeline China 45, 54
Shaanxi-Beijing gas pipeline China 10–15, 18, 26–31
Shanghai Cooperation Organisation (SC0) 41–2, 64, 387
Shanghai Securities News 40, 64
Shenergy 23
Shengli oilfield China 14, 16
Shenzhen Gas China 21, 32
shutdowns (plant) 121, 163, 167, 186
Shwe project Myanmar 51, 319, 329
Sichuan, Central China 8, 10–12, 15, 26, 37, 44–5, 52
Sichuan Basin 14
Sichuan–Chongqing pipeline 26, 52
Sichuan–Shanghai pipeline 45
Sichuan gas reserves 10, 11, 44
Shizuoka Gas 124–6, 134, 161, 304, 311
Showa Shell 140, 170
Siberian pipeline network 250, 255–6
Sibneft Oil Company 230, 250, 268, 274
Sidanco (BP) China 203, 239–40
Sinopec China 11, 17, 33–46, 64, 229, 232
Sodeshi LNG Japan 132, 169
Songliao gas field China 12
South Pars Iran see Pars
Statoil Norway 239, 269, 362
subsidies 4, 5, 77, 98, 105–7, 114, 290, 296, 318, 393
Suez LNG Egypt 359–60
Sulige gas field Mongolia 11, 28
sulphur 5, 41, 71, 150–1,171–2, 298
Sumimoto Corporation 29, 203
Sunrise project Australia 315, 326–9
Surgil gas field Uzbekistan 192, 213
Surgutneftegaz 247, 249
Sweden 38, 46

Taiwan 1, 55, 235, 280, 288, 302–4 337,

341, 346–9, 357, 371, 383, 405–9
TAP Consortium 87
Tanggu gas field Indonesia 289
Tapti gas field India 77, 93–5, 97, 112–4
Tarim Basin China 11–2, 28–30, 37, 47, 265
Teikoku Oil 131, 134, 161
Tenghiz gas field Kazakhstan 261, 277
terrorism/terrorist threat 2, 3, 7, 41, 48, 117, 165 175, 196, 203–8, 224, 387, 390
thermal power generation 69, 71–3, 98, 111, 121, 131, 134, 139–40, 152–4, 207
Tiga project Malaysia 148, 302–5
Timor-Leste Australia 315, 329
TNK BP 49, 50, 230 240–6, 249–51, 268–74
Toho Gas Japan 125–6,134, 140,161, 169, 288, 304, 311, 347
Tohoku Electric Power Japan 131–2, 140, 163–5, 170, 236, 287, 290, 304, 311, 347
Tohoku Natural Gas Japan 140, 145, 161, 170
Tokyo Electric Power Company (TEPCO) 124, 163, 167–71,287, 304, 311, 318, 335, 338–41
Total 39, 288–9, 293–5, 307, 314, 319, 338, 342, 347–51,358–62, 378
TotalFinaElf 29
town gas *see* LPG
Trans-Adriatic pipeline 364, 378
Trans-ASEAN gas pipeline (TAGP) 281
Transgas 29
Transneft 250, 274
transnational pipelines 47, 67, 73, 86, 93, 107
Trinidad and Tobago 127–8, 377
Turkmenistan-China gas pipeline 278
Tyumen Oil Company Russia 226, 240, 268

UAE/ United Arab Emirates 127–8, 330–5, 341, 351–2
United States of America 13, 41, 54–5, 84, 218, 222, 381, 384
United Technologies 29
United Nations 14, 52, 207–8, 398
Union Fenosa Gas Spain 356
UNOCAL 40, 288, 293, 295
Urals Energy 230–2, 246, 269, 273
Uzbekistan 41, 48, 51–3, 213, 259–62, 265, 277–8, 387
urea (fertiliser) 71, 73, 286

Venezuela 46

West-East pipeline, China 4, 7, 8, 13–4, 18, 23, 26–8, 30–3, 44–6, 49–54, 201, 208, 242, 255–6 262–3, 395
West Siberia Pipelines 246–250
Woodside Petroleum 36, 306, 310–5, 317, 323
World Bank 14, 22, 76, 34, 87, 195, 279
World Gas Conference 2003 50, 254–6, 265–6
World Fact Book (CIA) 329
World Gas Intelligence 328–9, 377–8
World Trade Organisation 38–9

Xianjiang China, 14–5, 28, 30, 46–9, 52–3, 255

Ya 13-1 China gas field 17–20
Yadaveran oilfield Iran 44
Yaduna gasfield Iran 280
Yamal Nenets/Peninsula Myanmar 223–5, 240, 244, 254, 267–8
Yetagon gas field Myanmar 281
Yemen 187–8, 192, 214, 330–5, 340, 358–60, 371–2, 398
Yemen LNG Project 187–8, 214, 358–60, 371
Yingehai China 11
Yoloten gas field Turkmenistan 260
Yukos Russia 48, 227, 386
Yunnan pipeline China 15, 52

Zapolyarnoye field Siberia 224, 234, 254
Zhejiang LNG project China 23–7, 45, 53
Zhuhai China LNG Project 22, 44, 53